P9-CEU-114

Integrated Digital Electronics

PRENTICE-HALL SERIES IN ELECTRONIC TECHNOLOGY

Charles M. Thomson and Joseph J. Gershon

Integrated Digital Electronics

WALTER A. TRIEBEL

Training Specialist
Sweda International, Inc.

PRENTICE-HALL, INC.,
Englewood Cliffs, N.J. 07632

Library of Congress Cataloging in Publication Data

TRIEBEL, WALTER A
Integrated digital electronics.

Bibliography: p.
Includes index.
1. Digital electronics. 2. Digital integrated circuits. I. Title.
TK7868.D5T73 621.3815 78-12809
ISBN 0-13-468900-3

Editorial/production supervision and interior design by Donald Rosanelli and Gary Samartino.
Cover design by Ernest Socolov.
Manufacturing buyer: Gordon Osbourne.

Printed in the United States of America

10 9 8 7 6 5 4 3 2 1

PRENTICE-HALL INTERNATIONAL, INC., *London*
PRENTICE-HALL OF AUSTRALIA PTY. LIMITED, *Sydney*
PRENTICE-HALL OF CANADA, LTD., *Toronto*
PRENTICE-HALL OF INDIA PRIVATE LIMITED, *New Delhi*
PRENTICE-HALL OF JAPAN, INC., *Tokyo*
PRENTICE-HALL OF SOUTHEAST ASIA PTE. LTD., *Singapore*
WHITEHALL BOOKS LIMITED, *Wellington, New Zealand*

To my father, ADOLF A. TRIEBEL

Contents

Preface

Integrated Digital Electronics is written to provide a general study of modern digital electronic theory, circuits, and devices. Throughout the book, integrated circuit devices and circuits are used instead of discrete transistor circuitry. The textbook is written for the electronic technology curriculum offered at community colleges and technical institutes. However, because of the integrated circuit approach to digital electronics, the text can be a valuable reference for practicing engineers and technicians.

The text provides a comprehensive study of digital electronics. It starts with the basic topics of number systems, logic gates, integrated gate devices, and combinational logic theory and builds through an extensive study of combinational logic networks, sequential circuitry, and medium-scale integrated circuit devices. The book concludes with a chapter on microcomputer systems and large-scale integrated devices. Numerous examples and exercises are provided in each chapter.

In the text, standard TTL and CMOS integrated circuit devices are introduced for each of the basic types of circuitry covered. TTL and CMOS were selected because they are the most widely used types of integrated circuitry in industry.

The technical level of the material in the book requires minimal prior knowledge of basic electricity and electronics. Little mathematical background is needed.

In courses where time is limited, Chapter 14 (pulse circuits) and Chapter 15 (flip-flops) can be covered earlier in the program. In this way, background is obtained in the areas of fundamentals of digital electronics, combinational logic, and sequential circuitry without completing the text.

I would like to express special appreciation to Edilberto Austria for his many worthwhile comments on my text and to thank my wife Frieda for her assistance in preparation of the manuscript.

WALTER A. TRIEBEL

Introduction to Digital Electronics

1

1.1 INTRODUCTION

In this book, we shall study *digital electronics* from an *integrated circuit* point of view. This chapter begins our study with a survey of the equipment and the type of electronic components used in the digital field.

For this purpose, the list of topics that follow are included in this chapter:

1. Digital computers and other digital equipment
2. Discrete and integrated circuitry
3. Integrated circuits and their packaging
4. Small-, medium-, and large-scale integrated circuits
5. Monolithic technology

1.2 THE DIGITAL COMPUTER

In the past, digital electronics has been closely identified with the study of electronic computers. However, the range of modern equipment in the digital category is more broad. Today, the same type of circuitry used in computers is found in many other kinds of electronic equipment.

Digital Computers

We shall begin our look into digital electronic equipment with the *digital computer*. A computer is an electronic data processing system. Data or information is inputted into the computer in some form, processed within the computer, and outputted or stored for later use. Figure 1.1 (a) shows a modern computer system.

Computers cannot think about how to process data. They must be told exactly how to operate. The process of telling a computer how to work is called *programming*. When operating, the computer follows a set of instructions called its *program*.

For example, a large department store can use a computer to take care of its charge accounts. In this case, data such as the price and department of items purchased by a customer are entered into the computer daily by an operator. This information is stored in the computer under the customer's account number. On the next billing date, a tabular record of each customer's account is outputted by the computer and mailed to the customer.

Electronic computers first became available in the 1940s. These early computers were built with vacuum tube digital electronic circuits. In the 1950s, a second generation of computers was made. During this period, transistor electronic circuitry was used to produce more compact and reliable systems.

When the integrated circuit came into the electronics market during the 1960s, a third generation of computers appeared. With the integrated circuit (or IC, as it is known), industry could manufacture very complex, high-speed, and reliable computers.

Block Diagram of a Computer System

To study the computer, the system is divided into several fundamental units. The block diagram of Fig. 1.1(b) shows a simplified digital computer system. Here, we see that the computer contains four major sections:

1. Input unit
2. Central processing unit
3. Memory unit
4. Output unit

Each of these sections has a special function in terms of overall computer operation.

Of these units, the *central processing unit* is the heart of the computer. It is used to perform all arithmetic operations and logic decisions initiated by the program. The central processing unit is also known as the CPU.

On the other hand, the *input and output* sections are used to enter information into the processing unit of the computer and to output data, respectively. For

Figure 1.1 (a) Large-scale computer installation (International Business Machines Corporation).

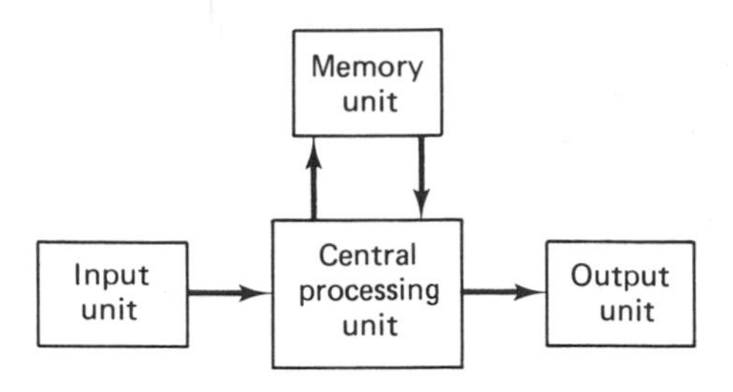

Figure 1.1 (Cont'd) (b) Computer block diagram.

instance, a typewriter keyboard can be used to enter the program and other information. After processing, results can be produced in a printed output form with a high-speed printer.

The *memory* is used to store or hold information such as numbers, names, and addresses. These data can be recalled and processed at a later time. In large computer systems, magnetic tape units are the most widely used type of memory.

Equipment attached to the main processing unit by wires are called *peripheral* devices. Input/output devices such as typewriter keyboards and high-speed printers are examples of peripherals.

Minicomputers

In many applications, the computing capability of a large digital computer is not needed. For this reason, an industry has developed around a smaller-scale system known as a *minicomputer*. A modern minicomputer system is shown in Fig. 1.2.

The block diagram of a computer system shown in Fig. 1.1(b) is also correct for a minicomputer. It also contains an input unit, central processing unit, memory unit, and output unit. However, each unit is made with a smaller working capacity for use in a minicomputer system.

Microcomputers

The *microcomputer* is a recent development in digital electronics. Like the minicomputer, a microcomputer contains the basic building blocks of a large-scale computer: central processing unit, memory, input, and output sections. The central processing unit in a microcomputer is known as a *microprocessor*.

The working capacity of each section in a microcomputer system is reduced in size as compared to the minicomputer. Many of the available microcomputer systems are formed from several electronic devices. However, a complete microcomputer system can be built into a single electronic device.

Figure 1.2 Minicomputer system (Digital Equipment Corporation).

1.3 OTHER DIGITAL EQUIPMENT

In recent years, many other digital electronic products have become available. For instance, the mechanical calculator has become a thing of the past. It has been replaced with a modern electronic calculator such as that shown in Fig. 1.3. Using digital electronics, the calculator is less expensive to make, easier to repair, and much more compact.

Another system that was traditionally mechanical is the cash register. However, in recent years the electronic cash register and point-of-sale terminal are rapidly taking over the marketplace. A modern cash register is shown in Fig. 1.4.

Electronic instrumentation is another field where the use of digital circuitry is becoming popular. Figure 1.5 shows a digital electronic multimeter. This instrument is hand-held and battery-operated unlike the large ac-operated instruments used in the past.

Figure 1.3 Hand-held calculator (Texas Instruments Incorporated).

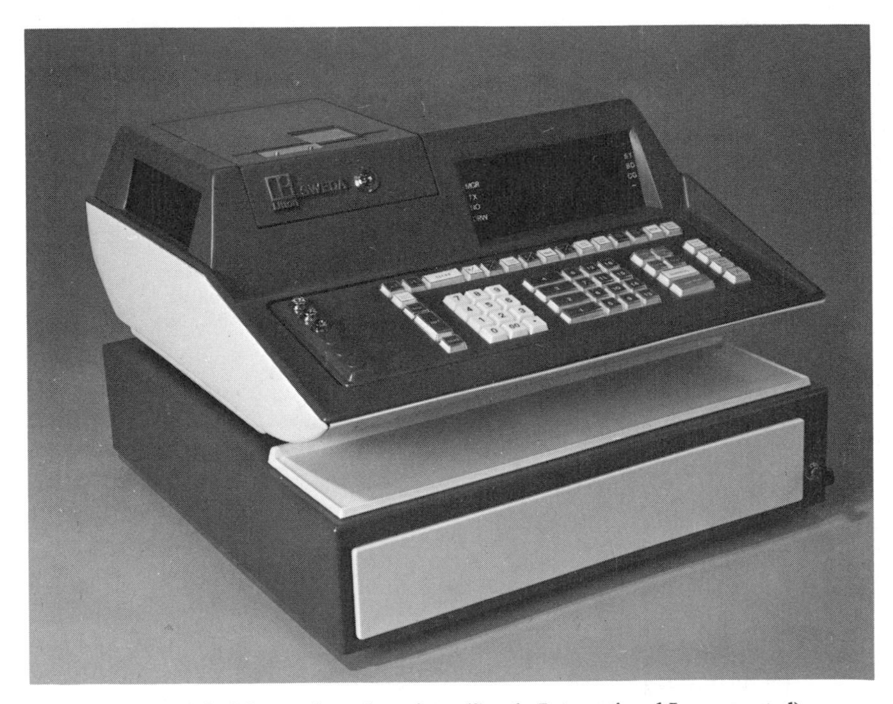

Figure 1.4 Electronic cash register (Sweda International Incorporated).

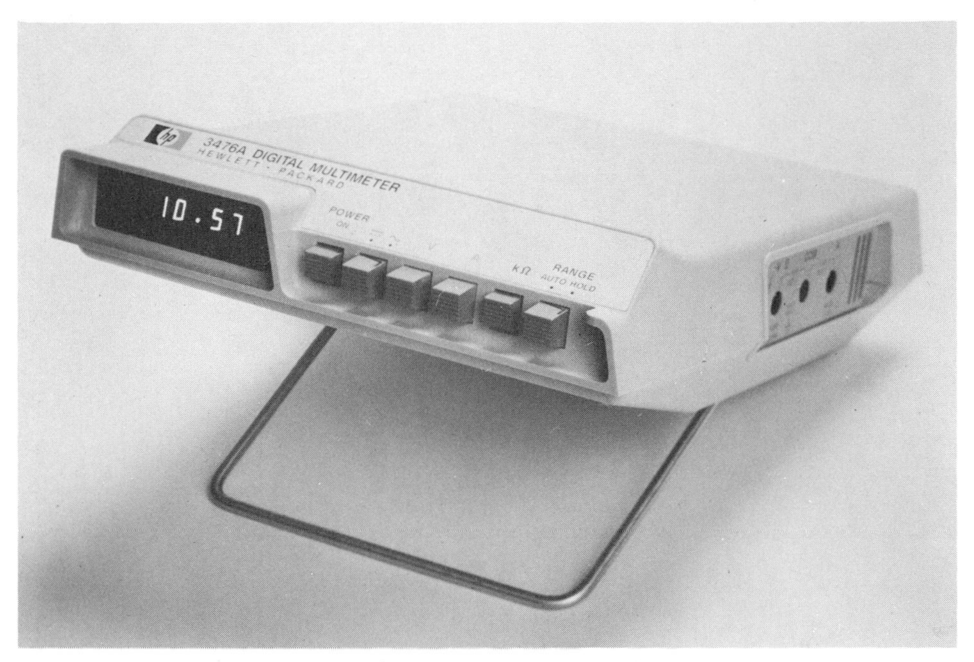

Figure 1.5 Digital multimeter (Hewlett-Packard Company).

1.4 DISCRETE AND INTEGRATED CIRCUITRY

The different parts of an electronic system are constructed using what is known as *electronic circuitry*. A circuit is made by interconnecting electronic devices such as the resistor, capacitor, diode, transistor, and integrated circuit to perform some useful function. The circuitry used in digital equipment can be grouped into two areas. They are called *discrete circuitry* and *integrated circuitry*.

A discrete digital circuit is one made by wiring electronic devices such as resistors, diodes, and transistors together. On the other hand, an integrated circuit is a digital circuit built into a small package. In this case, the complete circuit is purchased instead of built from separate electronic devices.

Integrated-type digital circuits are most widely used in modern electronic equipment. Some integrated circuit devices are shown in Fig. 1.6.

Integrated Circuit Packaging

Integrated circuits, or ICs, as they are better known, are available in several differently shaped packages. Two of the most popular are the *dual-in-line* or DIP package shown in Fig. 1.6(a) and the *flat* package of Fig. 1.6(b). Of these two types, the DIP package device is more widely used. The reason for its wide use is the ease with which it can be mounted.

(a)

(b)

Figure 1.6 (a) DIP package IC (Texas Instruments Incorporated); (b) Flat package IC (Texas Instruments Incorporated).

Some of the more common DIP packages have 14, 16, and 24 pins. The pin layout of a 24-pin IC is shown in Fig. 1.7. Looking at this diagram, we find that the pins are numbered in a counterclockwise direction starting from the left side on the end with the indicator mark.

The casing of an IC is made from either plastic or ceramic materials. Devices with a ceramic package have better temperature characteristics and can be used to make equipment that must operate over a wide range of temperature.

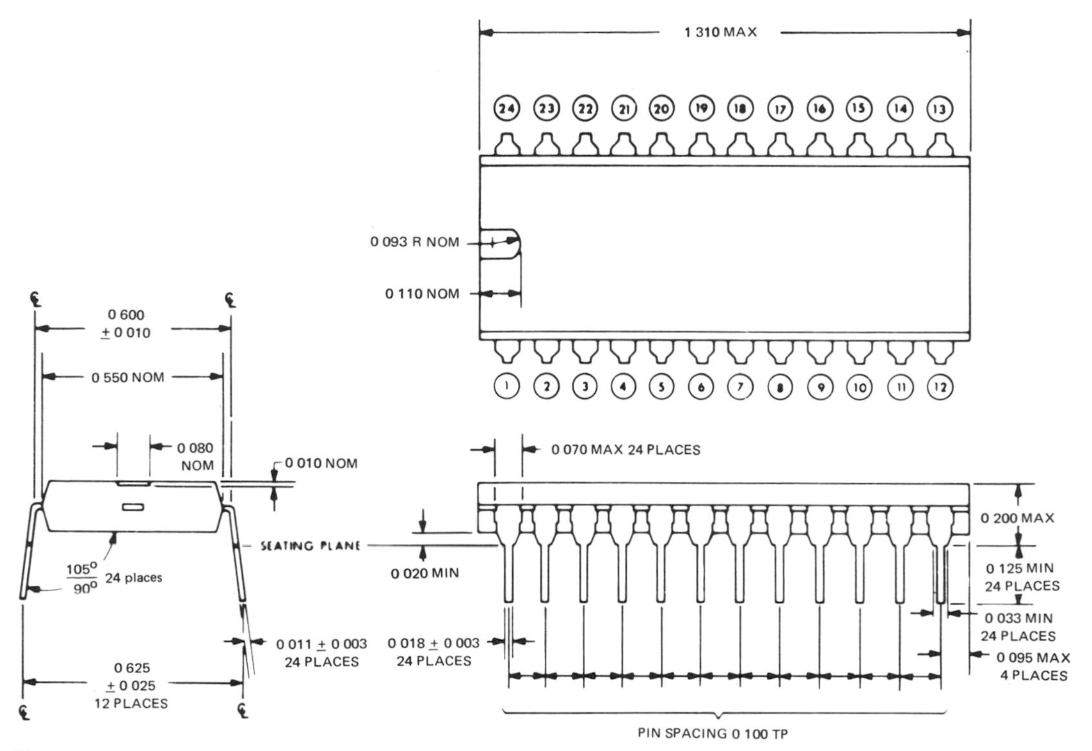

Figure 1.7 DIP pin numbering (Texas Instruments Incorporated).

1.5 SSI, MSI, AND LSI CIRCUITS

The circuitry within an integrated circuit can be constructed from different types of electronic devices. In this section, we shall introduce some of the more popular types of integrated digital circuitry.

The names of types of digital circuitry found in ICs are based on the kind of devices used to construct the circuit. Some of the more common types are listed in Fig. 1.8. Looking over this list, we see that one type of circuitry is called DTL for diode-transistor logic. The circuits within a DTL integrated circuit are built using resistors, diodes, and transistors.

In modern digital equipment, TTL (transistor-transistor logic) and CMOS (complementary metal oxide semiconductor logic) are the most widely used types of integrated devices.

Many different standard digital circuits are available as integrated circuit devices. Here, we shall group devices into three categories based on the complexity

DL	Diode logic
RTL	Resistor-transistor logic
DTL	Diode-transistor logic
TTL	Transistor-transistor logic
CMOS	Complementary metal oxide semiconductor logic
ECL	Emitter coupled logic

Figure 1.8 Types of integrated logic circuitry.

of their circuitry. The three categories we use are called *small-scale* ICs, *medium-scale* ICs, and *large-scale* ICs.

Small-Scale Integrated Circuits

The simplest digital circuits are grouped in the category called small-scale integrated (SSI) circuits. This category contains the basic logic and switching circuits used to build larger digital circuits and systems. The AND gate and OR gate are the names of two widely used circuits available in SSI devices. A detailed list of standard SSI devices is given in Fig. 1.9(a).

Medium-Scale Integrated Circuits

The integrated digital circuits in the next level are grouped into a category known as medium-scale integrated circuits. The MSI category contains circuits

Inverter
AND gate
OR gate
NAND gate
NOR gate
Buffer
Exclusive-OR gate
AND-OR-invert gate
Monostable multivibrator
D-type flip flop
J-K flip flop

(a)

Magnitude comparators
Parity checker/generator
Decoders
Adders
Multiplexers
Data latch
Counters
Shift registers

(b)

Read only memory (ROM)
Random access memory (RAM)
Programmable read only memory (PROM)
Arithmetic logic unit (ALU)
Central processing unit (CPU)
General purpose input/output device (GPI/O)

(c)

Figure 1.9 (a) Standard SSI devices; (b) Standard MSI devices; (c) Standard LSI devices.

with a complexity of from 12 to 100 SSI circuits. For instance, a digital decoder circuit can be made from 8 NOT gates and 10 NAND gates. This gives a total of 18 SSI circuits and puts the decoder circuit into the MSI category. Some of the widely used MSI circuits are listed in Fig. 1.9(b).

Large-Scale Integrated Circuits

A large-scale integrated (LSI) circuit device contains a complete digital subsystem. The complexity of an LSI device exceeds the equivalent of 100 SSI circuits. Three of the most commonly used LSI circuits are the read only memory, random access memory, and central processing unit. A more detailed list of standard LSI devices is provided in Fig. 1.9(c).

1.6 MONOLITHIC TECHNOLOGY

The way in which the circuitry inside an integrated circuit is made is called a *technology*. The circuitry of TTL and CMOS integrated circuit devices is made by the *monolithic* technology. This technology is the most widely used process in the manufacture of integrated circuits.

The monolithic technology uses a multistep process to build circuitry onto a single chip of silicon crystalline material. This piece of silicon is called the *substrate*. Using the monolithic process, the resistors, diodes, transistors, and interconnections needed to construct a circuit are formed. Figure 1.10 shows a picture of the structure of a monolithic device.

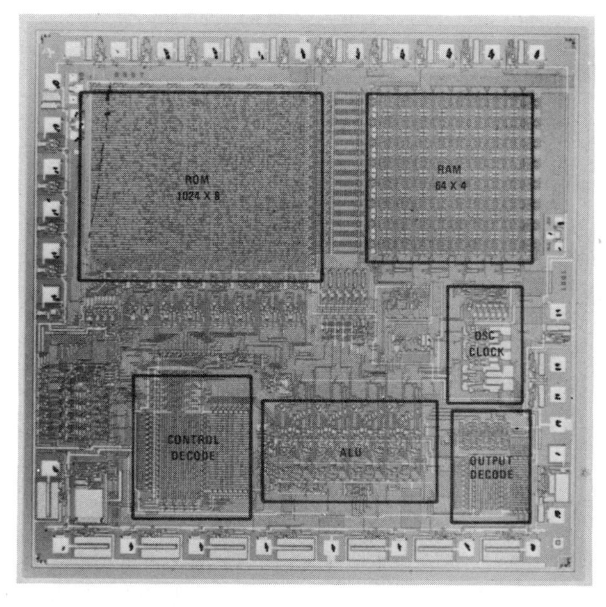

Figure 1.10 Monolithic device structure (Texas Instruments Incorporated).

After this, gold or aluminum wires are used to connect the circuitry on the chip of silicon to the pins of the package and it is enclosed within a plastic or ceramic case.

ASSIGNMENTS

Section 1.2

1. Can a computer think?

2. What device is the building block of third-generation computers?

3. Give the four parts of a computer system.

4. In which block of a computer system would a visual display unit (television screen) be used?

5. What is the smallest scale computer called?

Section 1.3

6. Name five different pieces of digital equipment.

Section 1.4

7. What term is used to refer to digital circuitry made with individual resistors, diodes, and transistors?

8. Give the name of the most widely used IC package.

9. How many pins do some of the standard integrated circuits have?

10. What two materials are used for the case of ICs?

Section 1.5

11. What does the notation DTL stand for?

12. Give the names of the two most widely used types of integrated circuitry for modern equipment.

13. List the names of five circuits available in SSI devices.

14. Name five circuits available in MSI devices.

15. Give three LSI devices.

16. What does the abbreviation ROM mean?

Section 1.6

17. What technology is used to make the circuitry in TTL and CMOS integrated circuits?

18. What is the segment of material on which the IC is built called?

19. What type of material is used to make the circuitry within an IC?

Number Systems and Information Organization

2

2.1 INTRODUCTION

In this chapter and the two that follow, we shall develop the basic theory needed to start our study of digital circuit operation. In this chapter, we shall cover the topics of number systems and information organization. In the chapters that follow, the basic logic gates and TTL integrated logic gate circuitry are introduced. With this knowledge at hand, we can study the operation of logic networks and more complex integrated circuitry.

In digital systems, the operation of circuitry, data, and instructions are expressed using numbers. The types of numbers used are not normally the decimal numbers we are familiar with; instead, binary, octal, and hexadecimal numbers are widely used. For this reason, we shall begin our study with these three number systems and their use in representing digital information. The topics included in this chapter are as follows:

1. A number system
2. Binary numbers
3. Octal numbers
4. Hexadecimal numbers
5. Information organization

2.2 A NUMBER SYSTEM

Here we shall use decimal numbers to develop the general characteristics of a *number system*. A number system is formed by selecting a set of symbols to represent numerical values. When doing this, we can select any group of symbols. The number of symbols used is called the *base* or *radix* of the number system.

For example, let us look at the decimal number system. It is made by selecting the numerical symbols 0 through 9. These symbols are shown in Fig. 2.1(a).

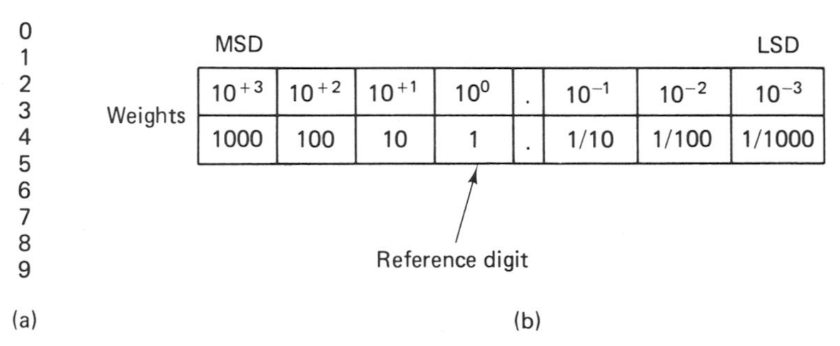

Figure 2.1 (a) Decimal number system symbols; (b) Digit notation and weights.

Here, we find that 10 different symbols are used, so the base or radix of the decimal number system is 10. Each of these symbols indicates a different numerical quantity. In the decimal system, 0 is the smallest quantity and 9 the largest quantity.

Digit Notation

With just the 10 basic symbols of the decimal number system, we cannot form every quantity needed in mathematics and science. For this reason, *digit notation* is used. An example of a decimal number written in digit notation is

$$735.23$$

Here the same basic symbols are used to form a larger multidigit number. This number has symbols entered into five different digit locations.

Weight

When digit notation is in use, the value of a symbol depends on its location in the number. The positional value of a digit is known as its *weight*. For instance, in the number 735.23 the symbol 3 occurs in two locations. Because the weights of the digits in which 3 lies are different, each takes on a different positional value.

In Fig. 2.1(b) we have shown some digit locations of the decimal number system and the corresponding weights. Here the digit just to the left of the decimal point is used as the reference digit and its weight is 10^0 or 1. This location is called the *units* digit.

Looking at the decimal weights, we find they are formed by raising the base of the number system to a power. For decimal numbers, the base is number 10. The exponent or power of the weight can be + or −. The value of this exponent is found by counting the number of digits to the units location. All digits to the left of the units digit are considered to have a weight with a positive exponent of the power of 10. For the digits to the right of the reference digit, the weights have a negative exponent.

As an example, let us look at the second digit to the left of the units digit in Fig. 2.1(b). This location has a weight of 10^{+2}, or multiplying out, we get 100. For this reason, it is called the *hundreds* digit.

For another example, let us take the second digit to the right of the units location. In Fig. 2.1(b), we find this digit has a weight of 10^{-2} or $\frac{1}{100}$. This location is also known as the *one-hundredths* digit.

Having introduced the weight of a digit, let us now look at how it affects the value of a symbol in that location. The value of a symbol in a digit other than the units digit is found by multiplying the symbol by the weight of the location. In our earlier example, 735.23, the symbol 7 is in the hundreds digit. Therefore, it represents the quantity $7 \times 10^{+2}$ or 7(100) equals 700 instead of just 7.

On the other hand, the 3 in the one-hundredths digit stands for 3×10^{-2} and has a value of $\frac{3}{100}$.

Most Significant and Least Significant Digit

Two other terms needed to talk about numbers and number systems are called the *most significant digit* and the *least significant digit*. The leftmost symbol in a number is located in the most significant digit. This location is indicated with the abbreviation MSD. On the other hand, the symbol in the rightmost digit is said to be in the least significant digit location or LSD.

In the number we have been using as an example, the symbol in the MSD location is 7 and its weight is 10^{+2}. Moreover, the LSD has a weight of 10^{-2} and the symbol in this location is 3.

EXAMPLE 2.1

What are the symbols and weights in the MSD and LSD of the number 6045.044?

SOLUTION:

The leftmost digit is the MSD. Here the symbol and weight are 6 and 10^{+3}, respectively. The LSD is the rightmost digit. Its symbol is 4 and weight 10^{-3}.

2.3 THE BINARY NUMBER SYSTEM

Digital electronic devices and circuits operate in one of two states, ON or OFF. For this reason, the *binary number system* instead of the decimal is used to describe their operation.

The base of the binary number system is 2, and just two symbols are used to form all numbers. These symbols are the numbers 0 and 1, as shown in Fig. 2.2(a). In an electronic circuit, binary 0 can represent a switching device such as a transistor that is turned ON. On the other hand, a 1 can be the same transistor turned OFF.

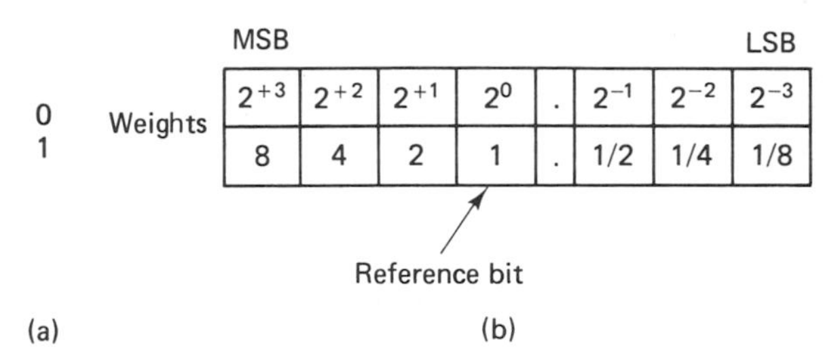

Figure 2.2 (a) Binary number system symbols; (b) Bit notation and weights.

Bits and Weights

To make a large binary number many 0s and 1s are grouped together. The location of a symbol in a binary number is called a *bit* instead of a binary digit. The term bit is a contraction of the words "binary digit."

As an example, let us take the binary number

$$1101.001_2$$

Looking at this number, we find it has 7 bits. The number 2 written to the right and slightly below tells it is a base 2 or binary number.

As with decimal numbers, each bit location in a binary number has a weight. These weights are expressed as the binary base 2 raised to a positive or negative exponent. In Fig. 2.2(b), the weights corresponding to some binary bits are shown. From this diagram, we see the 2^0 bit is the reference bit and its weight is 1.

Bits to the left of the 2^0 bit have weights with positive exponents and those to the right negative exponents. For example, the weight of the third digit left of the point is 2^{+2} or 4. If a 1 occurs in this bit location like in the number 1101.001, it stands for a decimal value of 4.

Most Significant and Least Significant Bits

In a binary number the leftmost bit is called the *most significant bit* and the rightmost the *least significant bit*. These terms are abbreviated as MSB and LSB, respectively. In the number 1101.001, the MSB has a value of 1 and weight 2^{+3} equals 8. On the other hand, the least significant bit is 1 with a weight of 2^{-3} or $\frac{1}{8}$.

EXAMPLE 2.2

Find the value and weight of the MSB and LSB in the binary number 11000.1.

SOLUTION:

The most significant bit is 1 and has a weight of 2^{+4} equals 16. The LSB is 1 with a weight of 2^{-1} or $\frac{1}{2}$.

2.4 CONVERSION BETWEEN DECIMAL AND BINARY NUMBERS

All numbers can be expressed in both the decimal and binary number systems. In Fig. 2.3 the decimal numbers 0 through 15 are listed along with their equivalent binary numbers. From this list, we find that the binary equivalent of decimal number 0 is just binary 0. On the other hand, decimal 15 is written in binary form as 1111.

Decimal number	Binary number
0	0
1	1
2	10
3	11
4	100
5	101
6	110
7	111
8	1000
9	1001
10	1010
11	1011
12	1100
13	1101
14	1110
15	1111

Figure 2.3 Equivalent decimal and binary numbers.

For digital circuit operation, information organization, and programming digital equipment, it is important to be able to quickly convert between decimal and binary number forms.

Decimal Equivalent of a Binary Number

We begin our study of number system conversions with the method for changing a decimal number to binary form. To find the decimal equivalent of a binary number, we multiply the value in each bit of the number by the weight of the corresponding bit. After this, the products are added to give the decimal number.

As an example, let us find the decimal number for binary 1010_2. Multiplying bit values and weights, we get

$$
\begin{aligned}
1010_2 &= 1(2^{+3}) + 0(2^{+2}) + 1(2^{+1}) + 0(2^0) \\
&= 1(8) + 0(4) + 1(2) + 0(1) \\
&= 8 + 2 \\
1010_2 &= 10_{10}
\end{aligned}
$$

This shows that 1010_2 is the binary equivalent of decimal number 10. Looking in the table of Fig. 2.3, we see that our result is correct.

EXAMPLE 2.3

Evaluate the decimal equivalent of binary number 101010.

SOLUTION:

$$
\begin{aligned}
101010 &= 1(2^5) + 0(2^4) + 1(2^3) + 0(2^2) + 1(2^1) + 0(2^0) \\
&= 1(32) + 0(16) + 1(8) + 0(4) + 1(2) + 0(1) \\
&= 32 + 8 + 2 \\
101010 &= 42
\end{aligned}
$$

Binary Equivalent of a Decimal Number

The other conversion we must be able to perform is to express a decimal number in binary form. The binary equivalent of a decimal number is found by a method known as the *double dabble process*.

With the double dabble process, the decimal number is repeatedly divided by the base of 2, and the remainders of these divisions are used to form the binary number. When doing this, the decimal number is divided by 2, the quotient brought down, and the remainder written to the right. This procedure is repeated until the quotient is zero. Now, we must use the remainders to form the binary number. The least significant bit of the binary number is the remainder of the first division or original number. Each of the remainders that follow gives the bits up to the last remainder, which gives the most significant bit.

To illustrate this method, let us convert the decimal number 10_{10} to binary form. Dividing by 2 gives the results that follow:

2	10 ⟶	0	LSB
2	5 ⟶	1	
2	2 ⟶	0	
2	1 ⟶	1	MSB
	0		

$10_{10} = 1010_2$

Here we see that dividing 10 by 2 gives a quotient of 5 with 0 remainder. The quotient is brought down and the remainder written on the right. This 0 is the LSB of the binary number for decimal 10. Now the 5 is divided once again by 2 to give a quotient of 2 and a remainder of 1. Dividing twice more, we end up with a quotient of 0 and two more remainders 0 and 1 in that order. The last remainder is the MSB of the binary answer.

At this point all remainders are known; it remains to form a binary number for 10_{10}. To do this, we start at the MSB remainder and work back toward the LSB to give 1010_2. This binary value is the same as that listed for 10 in the table of Fig. 2.3.

EXAMPLE 2.4

Convert the decimal number 32 to binary form.

SOLUTION:

2	32 →	0	LSB
2	16 →	0	
2	8 →	0	
2	4 →	0	
2	2 →	0	
2	1 →	1	MSB
	0		

32 = 100000

2.5 THE OCTAL NUMBER SYSTEM

The *octal number system* is another number system important to the study of digital electronic equipment. In fact, many minicomputer systems are programmed using instructions written in octal form.

The base of the octal number system is 8, and it uses numerical symbols 0 through 7. These symbols are listed in Fig. 2.4(a). To make a useful number, the basic octal symbols must be written in digit notation. Here the weights of the separate digits are the base 8 raised to a power. Typical octal weights are given in Fig. 2.4(b).

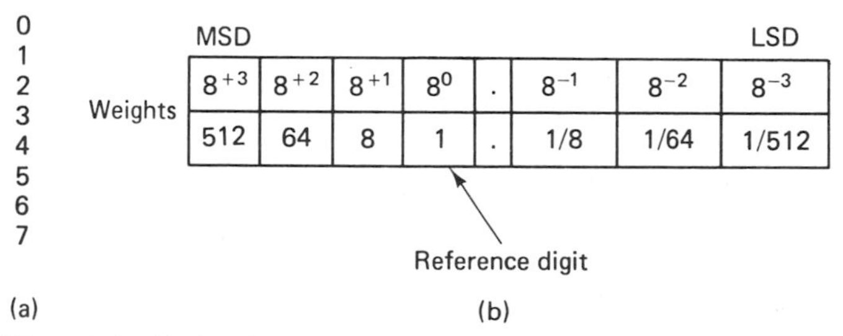

Figure 2.4 (a) Octal number system symbols; (b) Octal digit notation and weights.

From Fig. 2.4(b), we find that the weight of the reference digit is 8^0 or 1. On the other hand, the most significant digit and least significant digit locations shown have weights of 8^{+3} and 8^{-3}, respectively. Rewriting these weights in decimal form, we get 512 for the MSD and $\frac{1}{512}$ for the LSD.

EXAMPLE 2.5

What are the symbols and weights of the MSD and LSD in the octal number 721.05_8?

SOLUTION:

The number in the MSD is 7 and the weight of the digit is 8^{+2} equals 64. The symbol in the LSD is 5 and its weight is 8^{-2} or $\frac{1}{64}$.

2.6 CONVERSION BETWEEN OCTAL AND BINARY NUMBERS

The importance of using octal numbers with computers and other digital equipment is that they can be used to rewrite digital information or instructions in a compact way. For instance, a multibit binary number can be expressed with just a few octal digits. For this reason, it is important to learn how to directly convert between binary and octal number forms.

In Fig. 2.5(a), we have listed all 3-bit binary numbers and their octal equivalents. Here a 3-bit binary zero is the same as a one-digit octal zero. Each binary number that follows up through 111 is the same as one of the octal numbers from 1 through 7. In this way, we see that 3 binary bits give a single octal digit. This fact forms the basis for converting between binary and octal number forms.

Binary number	Octal number
000	0
001	1
010	2
011	3
100	4
101	5
110	6
111	7

(a)

	MSB			LSB	
	$2^{11}\ 2^{10}\ 2^9$	$2^8\ 2^7\ 2^6$	$2^5\ 2^4\ 2^3$	$2^2\ 2^1\ 2^0$	Bits
	8^3	8^2	8^1	8^0	Digits
	MSD			LSD	

(b)

Figure 2.5 (a) Equivalent binary and octal numbers; (b) Binary bits and octal digits.

The diagram in Fig. 2.5(b) shows how bits of a binary number are grouped to make digits of an octal number. From this illustration, we see the three least significant bits 2^0, 2^1, and 2^2 give the octal least significant digit 8^0. This is followed by groups of three binary bits for the 8^1 and 8^2 octal digits. The MSD 8^3 of the octal number is formed from the three MSBs 2^9 through 2^{11} of the binary number. In this way, a 12-bit binary number is rewritten with just four octal digits.

Octal Equivalent of a Binary Number

The first conversion we shall take is to rewrite a binary number in octal form. To do this, we start at the rightmost bit of the binary number and separate into groups with 3 bits. After this, we replace each group of bits by its equivalent octal number.

For example, let us change the binary number 110000100101_2 to octal form. Making groups of 3 bits and replacing with octal equivalents from Fig. 2.5(a), we get

$$110 \vdots 000 \vdots 100 \vdots 101_2$$
$$6 \vdots 0 \vdots 4 \vdots 5$$
$$110000100101_2 = 6045_8$$

EXAMPLE 2.6

What is the octal equivalent of the binary number 111001?

SOLUTION:

$$111 \vdots 001$$
$$7 \vdots 1$$
$$111001_2 = 71_8$$

Binary Equivalent of an Octal Number

To find a binary number from its octal equivalent, we reverse the method just used. Now, the number in each octal digit is replaced with its equivalent 3-bit binary number.

As an illustration of this method, we can convert the number 6045_8 into a binary number:

$$6 \vdots 0 \vdots 4 \vdots 5$$
$$110 \vdots 000 \vdots 100 \vdots 101$$
$$6045_8 = 110000100101_2$$

EXAMPLE 2.7

Evaluate the binary equivalent of the octal number 1705.

SOLUTION:

$$1 \vdots 7 \vdots 0 \vdots 5$$
$$001 \vdots 111 \vdots 000 \vdots 101$$
$$1705_8 = 1111000101_2$$

2.7 THE HEXADECIMAL NUMBER SYSTEM

The last type of numbers we shall look into are those of the *hexadecimal number system*. Hexadecimal numbers are widely used to describe microcomputer operation and programming. This is the base 16 number system. Figure 2.6(a)

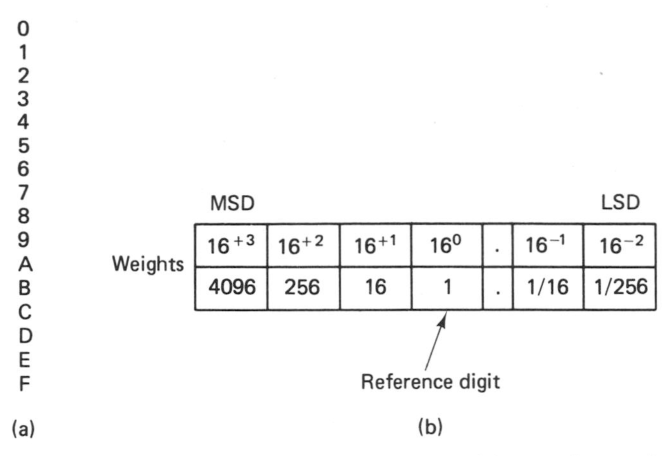

Figure 2.6 (a) Hexadecimal symbols; (b) Digit notation and weights.

shows the 16 symbols of the hexadecimal number system. Here, we see the numbers 0 through 9 followed by letters A through F. The letters A through F stand for numerical values equal to decimal numbers 10 through 15, respectively.

The weights of some hexadecimal digits are indicated in Fig. 2.6(b). These weights are formed by raising the base 16 to a + or − exponent value corresponding to the digit location. For the digits shown, the MSD weight is 16^{+3} or 4096 and the LSD weight is 16^{-2} or $\frac{1}{256}$.

EXAMPLE 2.8

What are the weights and values of the MSD and LSD in the hexadecimal number $A3F0.1B_{16}$?

SOLUTION:

The symbol in the MSD of the number is A, and this is the 16^{+3} digit. So its decimal value is 10 and weight 4096. For the LSD, we get a value of B or 11 and weight of 16^{-2} equals $\frac{1}{256}$.

2.8 CONVERSION BETWEEN HEXADECIMAL AND BINARY NUMBERS

Like octal numbers, hexadecimal notation can be used to write digital information in a more compact way. In Fig. 2.7(a), we see that four binary bits equal one hexadecimal digit. Some examples are $0000_2 = 0_{16}$, $1010_2 = A_{16}$, and $1111_2 = F_{16}$.

Using hexadecimal representation, a large binary number can be reduced to a couple of digits. For instance, in Fig. 2.7(b) we have grouped 12 binary bits to make three hexadecimal digits.

Binary number	Hexadecimal number
0000	0
0001	1
0010	2
0011	3
0100	4
0101	5
0110	6
0111	7
1000	8
1001	9
1010	A
1011	B
1100	C
1101	D
1110	E
1111	F

(a)

MSB		LSB	
$2^{11}\,2^{10}\,2^{9}\,2^{8}$	$2^{7}\,2^{6}\,2^{5}\,2^{4}$	$2^{3}\,2^{2}\,2^{1}\,2^{0}$	Bits
16^2	16^1	16^0	Digits
MSD		LSD	

(b)

Figure 2.7 (a) Equivalent binary and hexadecimal numbers; (b) Binary bits and hexadecimal digits.

Hexadecimal Equivalent of a Binary Number

The conversion of a binary number to hexadecimal form is done in a similar way as the earlier binary to octal conversion. However, now we begin by separating the bits into groups of four instead of three. The next step is to replace each group with its equivalent hexadecimal number from the table of Fig. 2.7(a). In this way, a multidigit hexadecimal number is obtained.

EXAMPLE 2.9

Express the binary number 10100101 as a hexadecimal number.

SOLUTION:

1010 : 0101
A : 5

$$10100101_2 = A5_{16}$$

EXAMPLE 2.10

What is the hexadecimal equivalent of the number 011011110001_2?

SOLUTION:

0110 : 1111 : 0001
6 : F : 1

$$011011110001_2 = 6F1_{16}$$

Binary Equivalent of a Hexadecimal Number

When a hexadecimal number is to be written in binary form, the method we just used must be reversed. In this case, the value in each hexadecimal digit is replaced by its equivalent 4-bit binary number.

EXAMPLE 2.11

Convert the hexadecimal number A5 to its binary equivalent.

SOLUTION:

A : 5

1010:0101

$A5_{16} = 10100101_2$

EXAMPLE 2.12

Rewrite the number $3C51_{16}$ in binary form.

SOLUTION:

3 : C : 5 : 1

0011:1100:0101:0001

$3C51_{16} = 11110001010001_2$

2.9 WORDS AND WORD ORGANIZATION

Having completed our introduction to number systems, we shall turn our interest to how information is organized in digital systems.

In digital equipment, data, codes, and instructions are represented with binary numbers. The processing of this information is done by electronic circuitry, and the results are normally described in binary form. However, these results are often rewritten in octal or hexadecimal notation for compactness.

Words and Word Length

Most electronic systems handle information in a fixed length group of binary bits. A fixed length group of bits is called a *word*. The number of bits in a word is known as its *word length*. In computers and other digital equipment, we find words of various length. Some typical lengths are 4 bits, 8 bits, 12 bits, and 16 bits; 4-, 8-, and 12-bit words are used in many microcomputers. On the other hand, most minicomputers operate off 16-bit words.

In Fig. 2.8 all 16 4-bit binary words are listed. Besides this, equivalent decimal and hexadecimal numbers are provided in this table. Looking under the binary number column, we see that each word must be written with the same number of bits. For instance, the binary number for 2 is written as 0010 instead of just 10. When using word notation, most significant 0s cannot be left out.

EXAMPLE 2.13

What is the word length of the binary word 111100110000? Express the word in hexadecimal form.

Decimal number	Binary number	Hexadecimal number
0	0000	0
1	0001	1
2	0010	2
3	0011	3
4	0100	4
5	0101	5
6	0110	6
7	0111	7
8	1000	8
9	1001	9
10	1010	A
11	1011	B
12	1100	C
13	1101	D
14	1110	E
15	1111	F

Figure 2.8 Four-bit binary words.

SOLUTION:

There are 12 bits in the binary word. Therefore, its word length is 12 bits. Changing the word to hexadecimal form, we get

1111 ⋮ 0011 ⋮ 0000
F ⋮ 3 ⋮ 0

$F30_{16}$

The Byte

In some applications, a word is processed in pieces. The term used to indicate a piece of a word is *byte*. For example, a 16-bit word can be processed as four 4-bit bytes. For some electronic devices and equipment, only 8 bits are considered as 1 byte.

EXAMPLE 2.14

Break the 9-bit word 001101000 into 3-bit bytes and express in octal form.

SOLUTION:

Starting with the LSB, we get the 3 bytes that follow:

$$000 = 0_8$$

$$101 = 5_8$$

$$001 = 1_8$$

Rewriting the word in octal form results in

$$150_8$$

Word Format

The way in which information is organized in a word is called its *format*. Many different formats are used in digital equipment. Two simple 8-bit word formats are shown in Figs. 2.9(a) and (b). They are formats for a *data word* and *instruction word*, respectively.

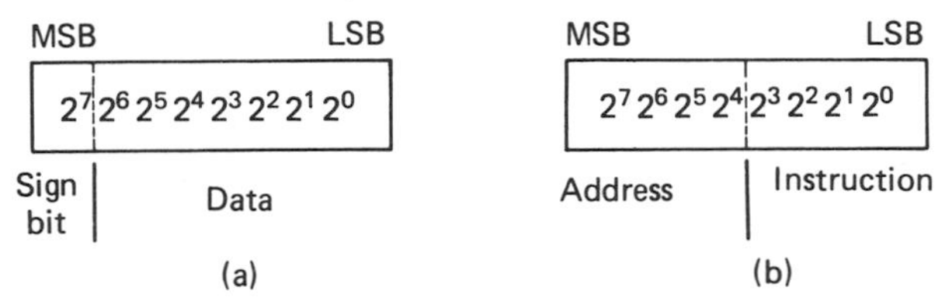

Figure 2.9 (a) Data word format; (b) Instruction word format.

A data word is used to indicate a signed number. In the diagram of Fig. 2.9(a), we see that the 8-bit data word is divided into two parts. The first part is just the MSB, and it is known as the *sign bit*. When this sign bit is 0, the number is positive, and a 1 in this location means a negative number.

The other 7 bits of the data word are used to identify the value of the numerical information. With an 8-bit data word length, we can represent plus or minus numbers over the range $0000000_2 = 0_{10}$ to $1111111_2 = 127_{10}$. By using a larger word length, such as 16 bits, we obtain a wider range of numbers.

EXAMPLE 2.15

Find the sign and value of the 8-bit data word 01000011.

SOLUTION:

$$0 \vdots 1000011$$
$$+ \vdots 1(2^6) + 0(2^5) + 0(2^4) + 0(2^3) + 0(2^2) + 1(2^1) + 1(2^0)$$
$$+ \vdots 64 + 2 + 1$$
$$01000011 = +67$$

The instruction word in Fig. 2.9(b) uses a different format. Here the 8-bit word is split into two 4-bit bytes. The four LSBs are a binary code used to indicate an instruction.

On the other hand, the MSBs represent a 4-bit binary address. This address is a code which selects the device the instruction is to be sent to.

EXAMPLE 2.16

Rewrite the instruction word 01101111 in hexadecimal form. What are the hexadecimal address and instruction codes?

SOLUTION:

$$0110 \vdots 1111$$
$$6 \vdots F$$
$$01101111 = 6F$$

In the hexadecimal word, the address is 6, and the code for the instruction is F.

ASSIGNMENTS

Section 2.2

1. What is the base of the decimal number system?

2. Write the symbols used to represent numerical values in the decimal system.

3. What term is used to refer to the positional value of a digit?

4. What do the abbreviations MSD and LSD stand for?

5. What are the symbols and weights of the MSD and LSD in the decimal number 9057.238?

Section 2.3

6. Write the two basic symbols used in the binary number system.

7. What is the name given to a binary digit?

8. How are most significant bit and least significant bit abbreviated?

9. What is the weight of the second bit to the left of the binary point?

10. Find the symbols and weights of the most significant bit and least significant bit in the number 100111.0101_2.

Section 2.4

11. Make a list of the binary equivalents of decimal numbers from 0 through 9.

12. Evaluate the decimal equivalent for each of the following binary numbers.

(a) 110 (b) 1011 (c) 010101 (d) 1111111

13. Find the decimal value of the number 1000111_2.

14. Convert the decimal numbers that follow to binary form.

(a) 5 (b) 9 (c) 42 (d) 100

15. What is the binary number for decimal 500?

Section 2.5

16. What is the base of the octal number system?

17. List the basic symbols used in the octal number system.

18. What is the weight of the fifth octal digit to the left of the point?

19. Write the octal number that comes after 7.

20. Find the symbols and weights of the MSD and LSD in the number 17.0235_8.

Section 2.6

21. Express each of the binary numbers that follow in octal form.

(a) 001100 (b) 101110 (c) 111000010 (d) 111010101111

22. What is the octal value of the number 000110011_2?

23. Convert the following octal numbers to binary form.
(a) 32 (b) 47 (c) 210 (d) 6760

24. Write the binary value for the number 345_8.

25. List the octal values for the decimal numbers 0 through 9.

Section 2.7

26. Write the 16 basic symbols of the hexadecimal number system.

27. What is the weight of the second hexadecimal digit to the left of the point?

28. Find the symbol and weight of the MSD in the hexadecimal number C8B.

Section 2.8

29. Convert the binary numbers that follow to hexadecimal form.
(a) 00111001 (b) 11100010 (c) 10011010 (d) 001110100000

30. What is the hexadecimal equivalent of the number 11110000_2?

31. Evaluate the binary equivalent of each hexadecimal number that follows.
(a) 6B (b) F3 (c) 45 (d) 2B0

32. Write the binary number for $A1B_{16}$.

33. Make a list of the hexadecimal numbers for the decimal numbers 0 through 20.

Section 2.9

34. What is the word length for each of the following binary words?
(a) 01011100 (b) 111000 (c) 0000111000000001

35. Break the word in Problem 34(b) into 3-bit bytes and express in octal form.

36. Divide the word in Problem 34(c) into 4-bit bytes and write in hexadecimal form.

37. Each of the following binary data words has the format shown in Fig. 2.9(a). What is the sign and value of the data word?
(a) 10010101 (b) 01100000

38. The instruction words that follow have the same format as shown in Fig. 2.9(b). Write the 4-bit address byte and instruction byte in hexadecimal form.
(a) 10000001 (b) 11111001

The Basic Logic Gates 3

3.1 INTRODUCTION

The *logic gate* is a building block of digital electronic circuitry. In this chapter we begin our study of digital circuitry. Here, we shall introduce the basic logic gates and learn to describe their operation in both mathematical and graphical form.

To study the operation of logic circuits, we use logic symbols, Boolean equations, and truth tables, which are covered in the early part of this chapter. The rest of the chapter is devoted to a study of standard logic gates. Here, we shall cover the topics that follow:

1. The logic gate
2. Boolean variables, operators, and equations
3. Truth tables and symbols of standard logic gates
4. Operation of the standard logic gates

3.2 THE LOGIC GATE

The first question we may ask when beginning the study of digital electronics is how a logic gate is defined. A logic gate is an electronic circuit in which the logic level of the output signal depends on the logic state of one or several input

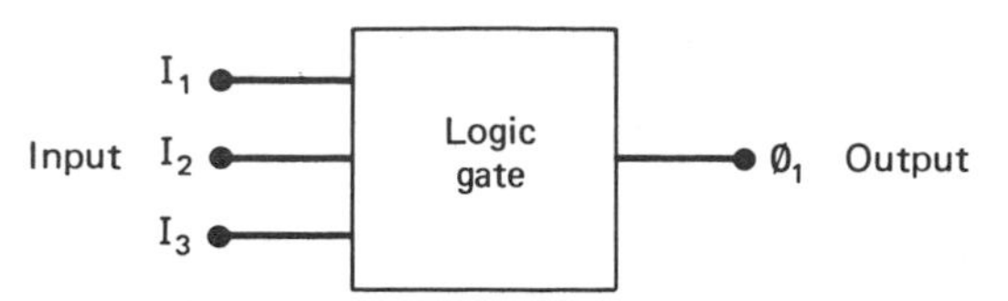

Figure 3.1 General logic gate block diagram.

signals. Figure 3.1 shows a general block diagram of a logic gate circuit. On this block, we find three input signals marked I_1, I_2, and I_3. On the other hand, it has a single output lead labeled $Ø_1$.

In a logic gate circuit, the input and output signals can only take on one of two logic levels or states. These logic levels are indicated by the binary numbers 0 and 1. The logic levels at the input of a logic gate determine the state of the output. For example, making the inputs all binary 1, $I_1 = I_2 = I_3 = 1$, the output $Ø_1$ could be logic 0. However, by setting at least one input to the 0 logic state the output changes to the 1 logic level.

There are five basic logic gate circuits; they are the *AND gate*, *OR gate*, *NOT gate*, *NAND gate*, and *NOR gate*. It is standard practice to write the name of gate circuits with capital letters.

Each of these standard logic gates can be made using different types of digital circuitry. For instance, a NAND gate can be constructed with RTL, DTL, TTL, and CMOS circuitry as well as other types. A listing of different types of integrated digital circuitry is provided in Fig. 1.8.

Voltage and Current Mode Logic

Besides using different types of circuitry, gates can be made to work in different ways or modes. For instance, gate circuits can use either *voltage mode logic* or *current mode logic*. By voltage mode logic, we mean that the 0 or 1 logic states at the inputs and outputs of the circuit are voltage levels. For example, the voltage level representing a binary 1 could be +5 V dc and binary 0 equals 0 V or ground. This voltage mode logic condition is illustrated with the waveform in Fig. 3.2(a). Here, we see the voltage signal starts at a 1 logic level of +5 V and switches to a logic 0 of 0 V.

Most types of logic circuitry operate off voltage mode logic signals. Some examples are RTL, DTL, TTL, and CMOS integrated circuits.

Another type of circuitry has signals that are currents. That is, the change from the binary 0 to 1 state is indicated with current levels. Circuits that work this way are called current mode logic gates. ECL integrated logic gates are the most widely used current mode logic devices.

Current mode logic circuits operate faster than all types of voltage mode devices. For this reason, ECL integrated devices are used in some high-speed applications. However, in modern digital equipment TTL and CMOS voltage mode IC devices are the most popular.

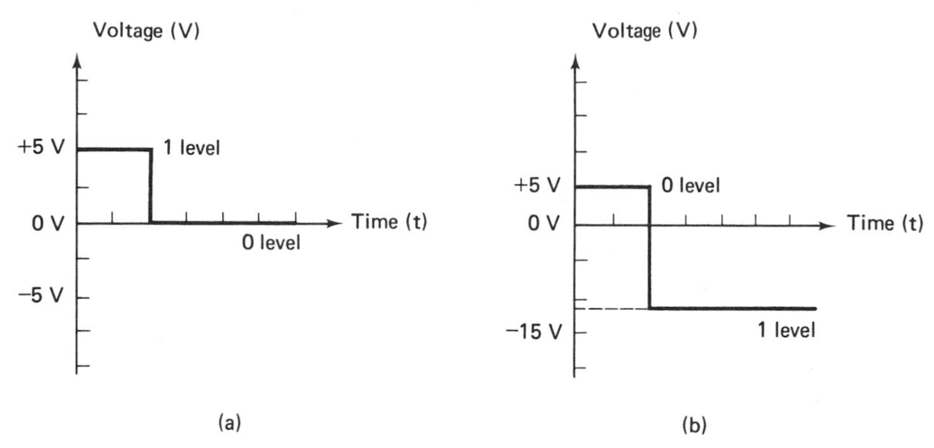

Figure 3.2 (a) Voltage mode positive logic; (b) Voltage mode negative logic.

Positive and Negative Logic

The operation of a logic gate circuit can be further described using *positive* and *negative logic*. Which type of logic is in use is decided by the way we select voltage or current levels to indicate the 0 and 1 binary states.

A gate is said to work off positive logic levels when its 1 state voltage or current has a more positive value than the 0 state level. Some integrated circuits have a 1 logic level of +5 V and 0 logic level equal to 0 V. These voltage levels correspond to positive logic. A signal like this is shown in Fig. 3.2(a).

Other kinds of logic circuits have logic levels such as 1 equals −12 V and a 0 of +5 V. Figure 3.2(b) gives an example of this type of signal. Looking at this signal, we notice that the 0 state logic level is more positive than the 1 state. This condition represents a waveform that uses negative logic.

EXAMPLE 3.1

The logic levels of a gate circuit are 1 = 0 V and 0 = −5 V. What type of logic is this?

SOLUTION:

Since voltages are used to specify the 1 and 0 logic levels, the circuit uses voltage mode logic. Moreover, the 1 logic level of 0 V is more positive than the −5-V logic 0 level. Therefore, it also operates off positive logic.

3.3 TRUTH TABLES

The relationship between input and output logic levels of a gate circuit is graphically illustrated with what is known as a *truth table*. For instance, the logic block in Fig. 3.3(a) has two inputs I_1 and I_2. Its output is labeled $Ø_1$. The operation of this circuit can be described with a truth table like that shown in Fig. 3.3(b).

From this table, we see that a truth table lists the input and output logic

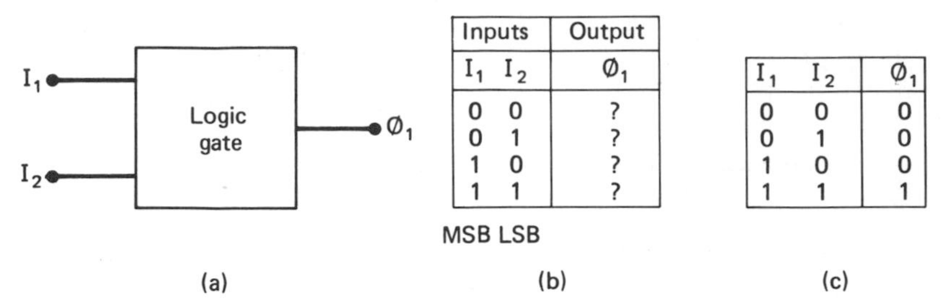

Inputs		Output
I_1	I_2	$Ø_1$
0	0	?
0	1	?
1	0	?
1	1	?

I_1	I_2	$Ø_1$
0	0	0
0	1	0
1	0	0
1	1	1

Figure 3.3 (a) General two-input logic gate; (b) General two-input gate truth table; (c) Two-input AND gate truth table.

states of a circuit. The columns on the left are used for inputs to the circuit, and the output is the last column on the right. The most significant bit of the input is normally the leftmost column in the table.

The operation of a logic gate is indicated with 0 and 1 binary states for inputs and output. Under the input columns of the table, all 2-bit binary combinations of I_1 and I_2 are written. They are the four 2-bit binary numbers corresponding to decimal numbers 0 through 3 and are listed down the input columns in correct numerical order.

Each of the five basic logic gate circuits has a unique input/output relation for these same four binary inputs. In the output column of the table, the 0 and 1 logic level produced by the circuit is listed next to the corresponding input combination. However, in the general table of Fig. 3.3(b) the output states are not listed. Instead, each is replaced by a question mark to indicate an unknown state.

A complete truth table of a two-input AND gate is shown in Fig. 3.3(c). Here, we see both input and output logic levels. The truth tables of OR gate, NAND gate, and NOR gate operation are similar; however, the logic levels in the output column change. The truth table of a NOT gate is somewhat different, because the NOT gate circuit has just one input and one output.

EXAMPLE 3.2

A logic gate has three inputs I_1, I_2, and I_3. How many binary number combinations must be entered into the truth table? Set up a three-input truth table and enter 3-bit binary input words. But do not list outputs for the gate.

SOLUTION:

For three inputs, the truth table must have eight binary words. These are the 3-bit equivalent words for decimal numbers 0 through 7. The result is the truth table shown in Fig. 3.4.

3.4 BOOLEAN TERMS AND EQUATIONS

In the previous section, we have shown how a truth table can be used to describe the operation of a gate circuit. A second way of indicating gate operation is with *Boolean algebra*. This is a mathematical method of describing circuit operation instead of a tabular approach like the truth table.

I_1	I_2	I_3	$\emptyset_1$
0	0	0	?
0	0	1	?
0	1	0	?
0	1	1	?
1	0	0	?
1	0	1	?
1	1	0	?
1	1	1	?

Figure 3.4 General three-input truth table.

The mathematical theory of Boolean algebra was developed by George Boole in the mid-1800s. However, it was not until 1938 that Boolean algebra was used on digital circuits. The first use was to simplify complex switching circuits. But at present Boolean techniques are widely used in the design of digital logic networks.

To describe the operation of a logic gate, we must write an equation that relates the output of the circuit to its inputs. This *Boolean equation* has three elements: *input variables, output variables,* and *Boolean operators.* Here we shall introduce these concepts and how to use them to write Boolean equations for the operation of the five basic logic gates.

Boolean Variables

The input and output terminals of a logic gate are marked with Boolean variables. The symbols used for input variables on a gate circuit are *A, B, C,* and *D.* On the other hand, we shall normally label the output *F,* for *Boolean function.* Each of these variables can only take on the binary values 0 or 1 during circuit operation.

In Fig. 3.5, a general gate block is shown with its inputs and output marked using this notation.

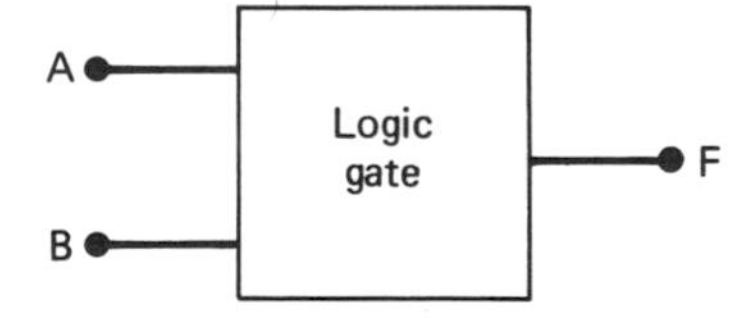

Figure 3.5 General gate marked with Boolean variables.

Boolean Operators

Three logic operations are used in Boolean algebra. They are the *AND function, OR function, and NOT function.* Each of these Boolean functions is indicated with a different mathematical symbol. The symbols for the AND, OR, and

NOT operators are $\cdot$, $+$, and $^-$ respectively:

$$\cdot = \text{AND function operator}$$
$$+ = \text{OR function operator}$$
$$^- = \text{NOT function operator}$$

These three symbols are the logical connectors used along with Boolean variables to make a Boolean equation.

Boolean Equations

By combining the Boolean variables at the inputs of a circuit with Boolean operators, we form a Boolean expression. Normally, the Boolean expression is equated to the output function F of the circuit.

As an example, let us write the AND operation of input variables A and B. This results in the equation

$$F = A \cdot B$$

This Boolean equation can be read several ways. For instance, we can say F equals A AND B. A second way is F equals A dot B.

A shorthand notation for the AND operation is to leave out the dot between the variables. Then, we get

$$F = AB$$

Here the AND operator is understood, and it can be read as just F equals AB. Both of these equations mathematically describe the operation at output F of an AND gate circuit with two inputs marked A and B.

As another example, we shall write the NOT operation of the variable A. This gives the equation

$$F = \bar{A}$$

Again, the function can be described a couple of ways. One is to say F is equal to NOT A. Another way of expressing this equation is F equals A bar.

A last description of this equation is to say F equals the *complement* of A. In this way, we find that the term complement means the same as NOT function. These two terms can be used interchangeable. Therefore, we can talk about the expression $\bar{B}$ as NOT B or the complement of B.

EXAMPLE 3.3

Write an equation for the OR function of variables A, B, and C.

SOLUTION:

The Boolean function will be called F and the OR function indicated by $+$. So the equation for the OR of input variables A, B, and C is written as

$$F = A + B + C$$

EXAMPLE 3.4

How is the Boolean equation $F = A + B + C + D$ read?

SOLUTION:

This equation is the OR function of the four variables A, B, C, and D. One way of reading this expression is F equals A OR B OR C OR D. A second way is A plus B plus C plus D.

NAND and NOR Functions

Up to this point, all Boolean equations have been formed using just one type of Boolean operator. But this is not true for the *NAND* and *NOR functions*. Both NAND and NOR are formed using two Boolean operators. For instance, the NAND function is a contraction of NOT-AND. Therefore, it is formed by using the NOT operator along with the AND operator. The NOR function stands for NOT-OR.

For an example, let us write the NAND function of variables A and B. This gives the Boolean equation

$$F = \overline{A \cdot B} = \overline{AB}$$

The NAND operation above can be read as F equals NOT, A AND B. Another way is to say F equals A AND B, bar. This equation describes the operation of a two-input NAND gate circuit.

EXAMPLE 3.5

What is the equation for the NOR function of variables A, B, and C?

SOLUTION:

To write this equation, we set up the OR function of variables A, B, and C. Then a bar must be put over the top of the complete OR function to give the NOR expression. This results in

$$F = \overline{A + B + C}$$

3.5 STANDARD LOGIC SYMBOLS

To illustrate a logic gate circuit in a schematic diagram of a digital system, a *logic symbol* is used instead of drawing its circuitry. The logic symbols of the AND gate, OR gate, NOT gate, NAND gate, and NOR gate circuits are shown in Figs. 3.6(a) through (e), respectively.

Here we see that the AND, OR, NAND, and NOR gate symbols are all drawn with two-inputs. On the other hand, the NOT gate has just one input. On each logic symbol, inputs are labeled with Boolean variables, and the output is marked with a Boolean equation of the gate's operation. Letters A and B are used to indicate inputs, and the output function is F.

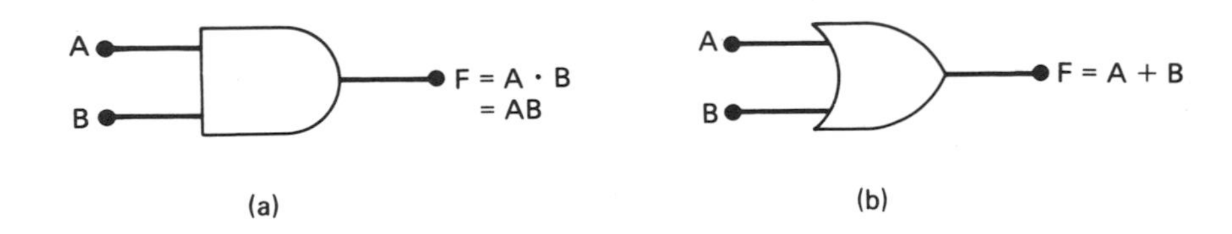

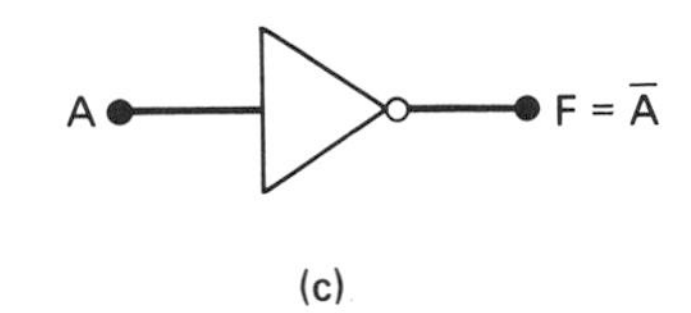

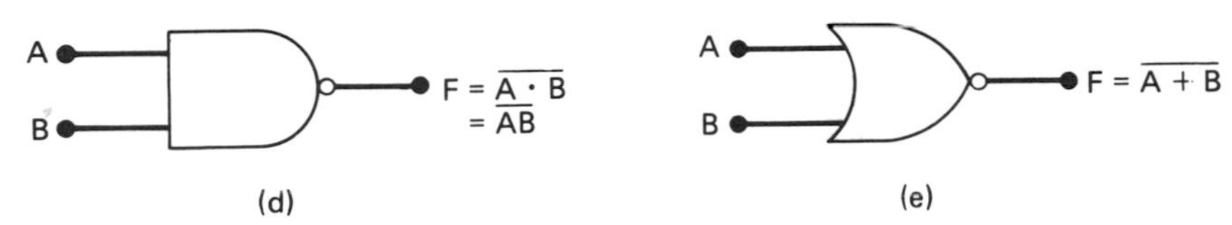

Figure 3.6 (a) AND gate symbol; (b) OR gate symbol; (c) NOT gate symbol; (d) NAND gate symbol; (e) NOR gate symbol.

In these diagrams, we have shown circuits with just two inputs. However, logic gates with more than two inputs are also in use. Some other common numbers of inputs are three, four, and eight inputs.

Besides this, we notice the NOT gate, NAND gate, and NOR gate symbols all have a dot at the output lead. This dot identifies the NOT logic function. For this reason, the NAND and NOR gate symbols actually indicate NOT-AND and NOT-OR circuit operations, respectively.

EXAMPLE 3.6

Draw the logic symbol of a three-input AND gate. The inputs are to be called A, B, and C. Mark the inputs and label the output with the Boolean equation for the circuit's operation.

SOLUTION:

The symbol of a three-input AND gate is drawn in Fig. 3.7(a). Its inputs are labeled A, B, and C. The output function F is equal to $A \cdot B \cdot C$ or just ABC:

$$F = A \cdot B \cdot C = ABC$$

EXAMPLE 3.7

Show the logic symbol for the Boolean equation $F = \overline{A + B + C + D}$.

SOLUTION:

Looking at the given Boolean expression, we see it describes a four-input NOR gate. The inputs of the circuit are A, B, C, and D. This gate symbol is drawn in Fig. 3.7(b), where its inputs and output are labeled.

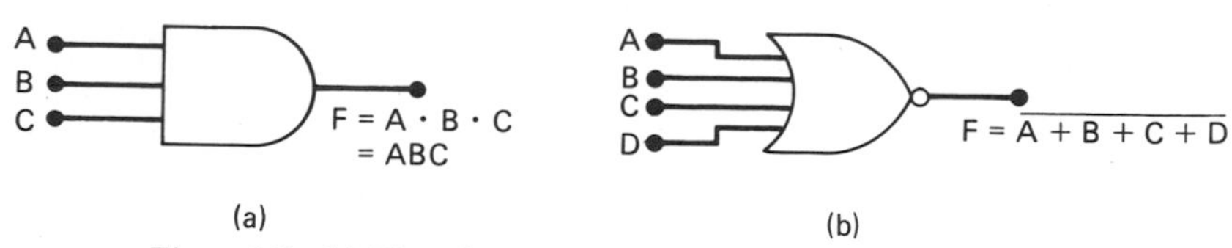

(a) (b)

Figure 3.7 (a) Three-input AND gate; (b) Four-input NOR gate.

3.6 AND GATE OPERATION

In this section, we shall begin our study of operation for the five basic logic gates. The first gate to be discussed is the AND gate.

Boolean Equation and Truth Table Operation

The logic diagram of a two-input AND gate is shown in Fig. 3.8(a). Here we see that its inputs are A and B. The output F of the circuit is given by the Boolean AND function:

$$F = A \cdot B = AB$$

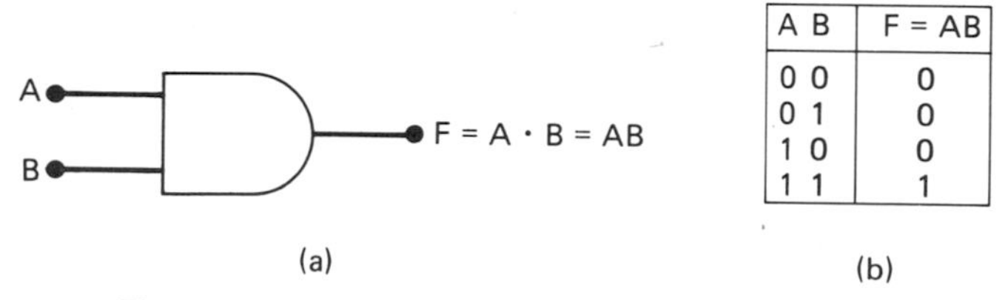

A B	F = AB
0 0	0
0 1	0
1 0	0
1 1	1

(a) (b)

Figure 3.8 (a) Two-input AND gate; (b) Truth table.

This equation gives a mathematical description of AND gate operation.

The truth table of AND gate operation is shown in Fig. 3.8(b). This tells us that the logic state of output F on an AND gate is binary 1 when inputs A and B are 1 together. Unless both inputs are 1, the output F remains at the 0 logic level. For instance, making the inputs $A = 0$ and $B = 1$ gives the second condition under the input columns of the truth table. Looking at the output column, we find that the corresponding output of the AND gate is F equals binary 0.

A three-input AND gate like that shown in Fig. 3.9(a) operates in the same way as the two-input circuit we just described. Its output F stays at the 0 logic level until all inputs A, B, and C are logic 1. This operation is shown in the truth table of Fig. 3.9(b).

EXAMPLE 3.8

In Fig. 3.10(a), an AND gate is shown with inputs $A = 1$ and $B = 1$. What is the logic level at output F?

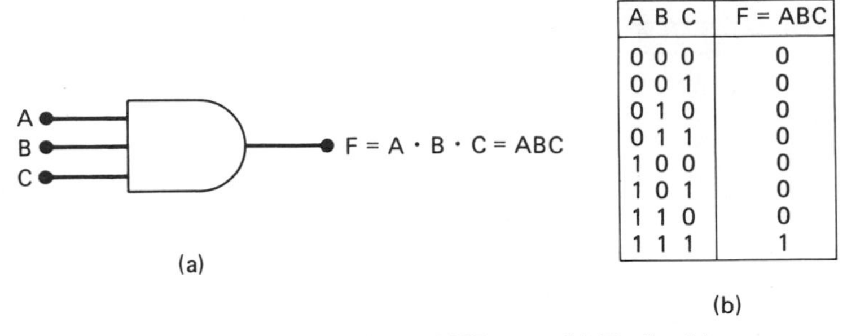

A B C	F = ABC
0 0 0	0
0 0 1	0
0 1 0	0
0 1 1	0
1 0 0	0
1 0 1	0
1 1 0	0
1 1 1	1

Figure 3.9 (a) Three-input AND gate; (b) Truth table.

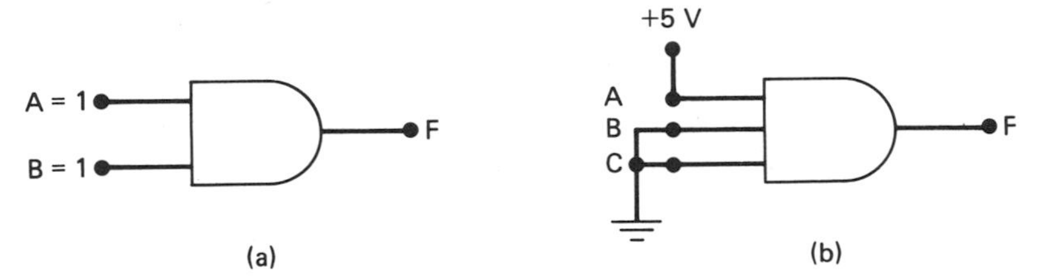

Figure 3.10 (a) Two-input AND gate with known inputs; (b) Three-input AND gate with known input voltages.

SOLUTION:

The truth table in Fig. 3.8(b) tells the output of the AND gate for all combinations of inputs A and B. Here, we find that output F is 1 when A and B are both 1:

$$AB = 11$$

$$F = 1$$

EXAMPLE 3.9

A three-input AND gate is shown in Fig. 3.10(b). If the circuit works off positive voltage mode logic with logic levels $1 = +5$ V and $0 = 0$ V, what is the binary state of the output?

SOLUTION:

Using positive logic, the input logic levels are

$$A = 1$$

$$B = 0$$

$$C = 0$$

This makes the binary input $ABC = 100$. Since all inputs are not at the 1 logic level, the output is in the 0 state:

$$F = 0$$

AND Gate Switch Circuit

To give an illustration of AND gate circuit operation, let us use the simple lamp circuit in Fig. 3.11(a). This circuit is made with a 5-V voltage source, two switches, and a lamp connected in series. The two switches represent the Boolean variables A and B, while the lamp indicates the output F.

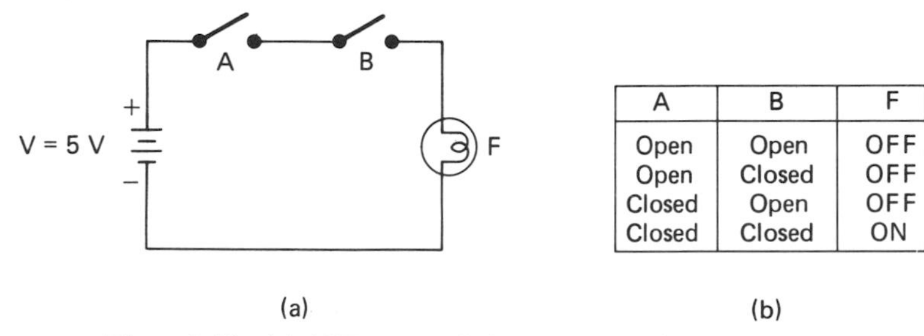

A	B	F
Open	Open	OFF
Open	Closed	OFF
Closed	Open	OFF
Closed	Closed	ON

(a) (b)

Figure 3.11 (a) AND gate switch connection; (b) Truth table.

The operation of the circuit will be described in terms of binary numbers. An open switch represents a 0 and a closed switch 1. For the output, a lighted lamp means binary 1 and an OFF lamp binary 0.

When both switches in the circuit are open, the binary inputs are $A = 0$ and $B = 0$. For this condition, the path between battery and lamp is an open circuit, and current will not flow through the lamp. Therefore, the lamp is OFF, and the output F is binary 0 or 0 V.

Closing either switch A or B still leaves the series path open and the light stays OFF. When B is closed, we get the second binary case $AB = 01$, and the output stays at the 0 logic level of 0 V because of the open A switch. The third state is $AB = 10$ and corresponds to the A switch closed and B open. Again, the circuit path is open, and the output is in the 0 logic state.

If both switches are closed, the path between battery and lamp is completed, current flows, and the lamp goes ON. In this state, the input is $AB = 11$, and output is also 1. This makes the output voltage +5 V.

In this way, we find that the lamp does not go ON unless both switches are closed. This operation is summarized by the truth table in Fig. 3.11(b) and is consistent with the operation of the AND gate shown in the table of Fig. 3.8(b).

3.7 OR GATE OPERATION

Having described the operation of the AND gate, we shall turn our interest to a second gate—the OR gate. In this section, we describe OR gate operation with a Boolean equation and truth table. Furthermore, its operation is demonstrated with an OR gate circuit constructed with switches.

A B	F = A + B
0 0	0
0 1	1
1 0	1
1 1	1

Figure 3.12 (a) Two-input OR gate; (b) Truth table.

Boolean Equation and Truth Table Operation

In Fig. 3.12(a), the symbol of a two-input OR gate is shown. The inputs to the circuit are the signals labeled A and B. A Boolean equation can be written to describe the output operation of this circuit. This gives the OR gate equation

$$F = A + B$$

The truth table of OR gate operation is shown in Fig. 3.12(b). Here we see that output F is at the binary 1 logic level if input A, input B, or both A and B are in the 1 logic state. When both inputs A and B are binary 0, we get the one condition where the output switches to the 0 logic level.

As an example, let us make the inputs of an OR gate $A = 1$ and $B = 0$. These inputs give the third case in the truth table of Fig. 3.12(b). Looking at the table, we find that the gate has a logic 1 at the output.

The truth table of a three-input OR gate is given in Fig. 3.13. Again, the table shows that output F is 1 if any combination of inputs A, B, and C is 1.

A B C	F = A + B + C
0 0 0	0
0 0 1	1
0 1 0	1
0 1 1	1
1 0 0	1
1 0 1	1
1 1 0	1
1 1 1	1

Figure 3.13 Three-input OR gate table.

EXAMPLE 3.10

In Fig. 3.14, the inputs to a two-input OR gate are shown. This gate works off positive voltage mode logic. The logic levels are $1 = +5$ V and $0 = 0$ V. Is the output in the binary 0 or 1 state?

SOLUTION:

For positive logic, the input to the logic gate is

$$AB = 01$$

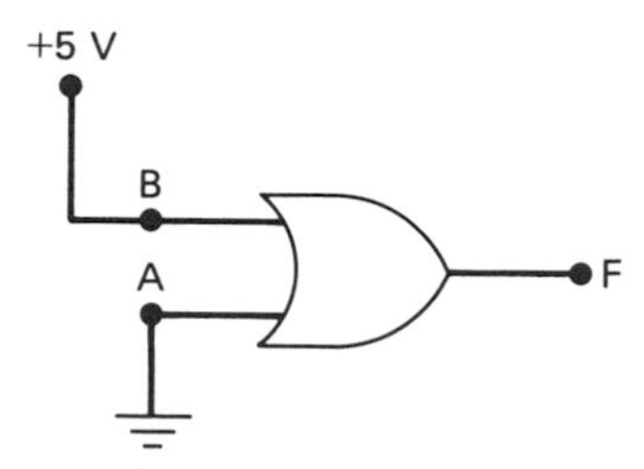

Figure 3.14 Two-input OR gate with known input voltages.

Therefore, the output F is logic 1:

$$F = 1$$

EXAMPLE 3.11

The input to a four-input OR gate is $ABCD = 1010$. What is the output F of the logic gate?

SOLUTION:

Since both the A and C inputs are 1, the output F is in the 1 logic state:

$$F = 1$$

OR Gate Switch Circuit

As an example of OR gate operation, let us look at the lamp circuit in Fig. 3.15(a). This circuit is made with a 5-V voltage supply, two parallel connected switches, and a lamp.

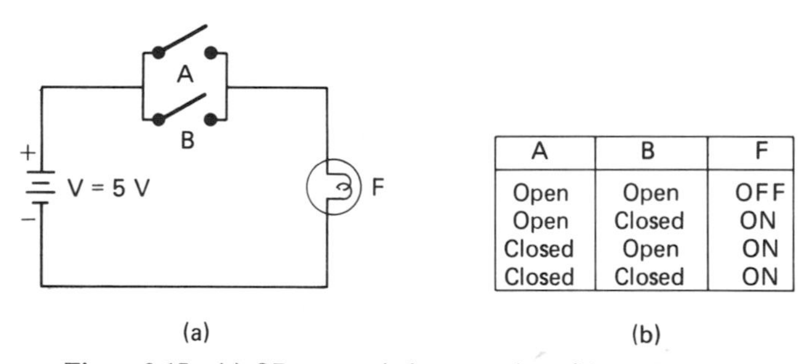

A	B	F
Open	Open	OFF
Open	Closed	ON
Closed	Open	ON
Closed	Closed	ON

Figure 3.15 (a) OR gate switch connection; (b) Truth table.

To describe the operation of the circuit with binary numbers, we begin by defining the states of switches A and B as 1 equal to closed and 0 for open. The lamp is the output, and indicates F equals 1 when ON and 0 for OFF.

The lamp in this circuit is ON unless both switches are open. This operation is given by the table in Fig. 3.15(b). The way in which this circuit works is consistent with two-input OR gate operation. If another switch called C is wired in parallel with those in Fig. 3.15(a), a three-input OR gate is formed.

3.8 NOT GATE OPERATION

The third gate we shall consider is the NOT gate. The circuit used to make a NOT gate is often called an *inverter*. In Fig. 3.16(a), the logic symbol of a NOT gate or inverter is shown.

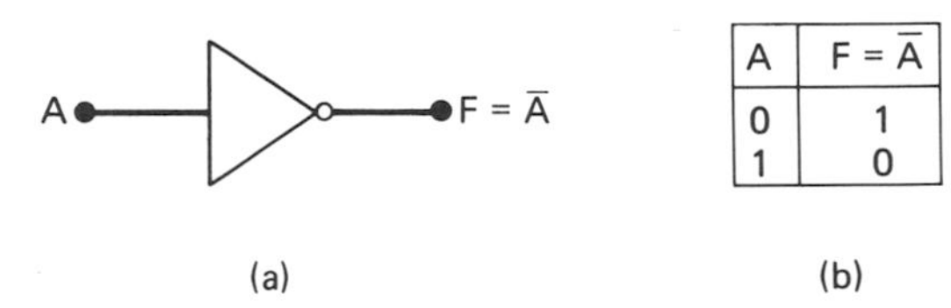

A	$F = \bar{A}$
0	1
1	0

Figure 3.16 (a) NOT gate; (b) Truth table.

Boolean Equation and Truth Table Operation

The input/output operation of a NOT gate can be described two ways. One way is to write a Boolean equation of its output. For the NOT gate circuit in Fig. 3.16(a), the input is A, and its output is given by the equation

$$F = \bar{A}$$

Earlier, we pointed out that this expression is sometimes read as F equals the complement of A.

Another description of NOT gate operation is obtained by constructing a truth table for the circuit. Figure 3.16(b) gives this truth table. Here, we notice that a binary 0 applied to the input gives a 1 at the output. Similarly, a 1 state input makes the output logic 0. So we say a NOT gate or inverter circuit changes the logic level of a signal between input and output.

EXAMPLE 3.12

The input A of an inverter circuit is -5 V. If the circuit works off positive voltage mode logic and logic levels are $1 = 0$ V and $0 = -5$ V, what is the logic level at output F?

SOLUTION:

The -5-V input is the same as a binary 0 logic level:

$$A = 0$$

A 0 into an inverter gives a 1 logic level at the output:

$$F = 1$$

3.9 NAND GATE OPERATION

One of the most widely used logic circuits is the NAND gate. The logic symbol of a two-input NAND gate is given in Fig. 3.17(a). Here, we shall describe the operation of a NAND logic gate. Besides this, we shall compare its operation to the AND gate.

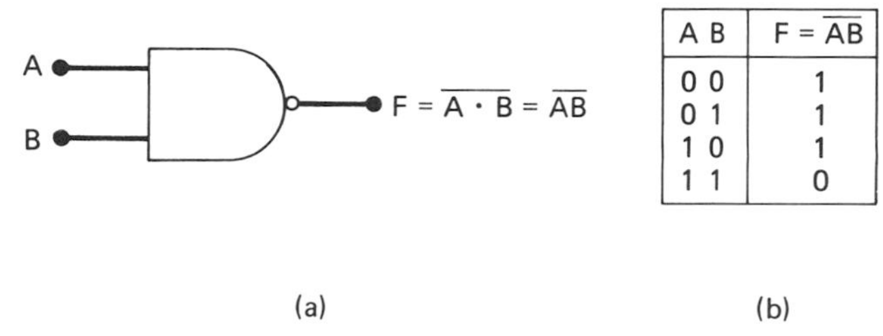

A B	F = $\overline{AB}$
0 0	1
0 1	1
1 0	1
1 1	0

Figure 3.17 (a) Two-input NAND gate; (b) Truth table.

Boolean Equation and Truth Table Operation

The inputs to the NAND gate in Fig. 3.17(a) are marked A and B. Its output is labeled F. A Boolean equation of operation for this NAND gate is

$$F = \overline{A \cdot B} = \overline{AB}$$

In Fig. 3.17(b), the truth table of two-input NAND gate operation is shown. From this table, we see that the output is binary 0 when both inputs A and B are at the 1 logic level. On the other hand, the output is in the 1 logic state if at least one input is 0.

For an example, let us take the input $AB = 11$ to a two-input NAND gate. This gives the fourth input condition in the truth table of Fig. 3.17(b). For this reason, output F of the gate is binary 0.

Comparing the NAND gate truth table to that of the AND gate in Fig. 3.8(b), we see that the NAND output is the complement of the AND function. That is, changing the logic level of each output state in the AND gate table gives the table for a NAND gate. So a NAND gate works like an AND gate with a NOT gate at the output. This circuit connection is given in Fig. 3.18.

NAND gates are also available with more than two inputs. For instance, the table of four-input gate operation is shown in Fig. 3.19. Here again we find that the output is logic 0 when all inputs are 1.

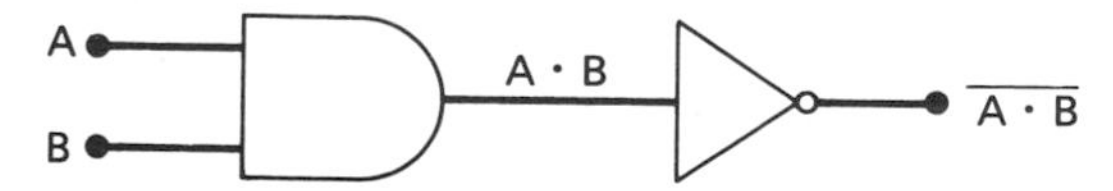

Figure 3.18 AND-NOT equivalent of the NAND gate.

A B C D	F = ABCD
0 0 0 0	1
0 0 0 1	1
0 0 1 0	1
0 0 1 1	1
0 1 0 0	1
0 1 0 1	1
0 1 1 0	1
0 1 1 1	1
1 0 0 0	1
1 0 0 1	1
1 0 1 0	1
1 0 1 1	1
1 1 0 0	1
1 1 0 1	1
1 1 1 0	1
1 1 1 1	0

Figure 3.19 Four-input NAND truth table.

EXAMPLE 3.13

A two-input NAND gate works off positive voltage mode logic with logic levels 1 = +5 V and 0 = 0 V. If the input voltage level at A is 0 V and that of B equals +5 V, is the output in the binary 0 or 1 state?

SOLUTION:

For positive logic, the input logic states are

$$A = 0$$
$$B = 1$$

These inputs correspond to the second input condition in the truth table of Fig. 3.17(b). From this table, we see that the output logic level is 1:

$$F = 1$$

EXAMPLE 3.14

The inputs to a three-input NAND gate are $A = 1$, $B = 1$, and $C = 0$. Find the logic level of output F.

SOLUTION:

Since all inputs are not in the binary 1 state, output F is at the 1 logic level:

$$F = 1$$

EXAMPLE 3.15

The A and B waveforms in Fig. 3.20 are the input signals of a two-input NAND gate. Draw the output waveform.

SOLUTION:

In time interval 0 to 1, the input logic levels are

$$A = 1$$
$$B = 0$$

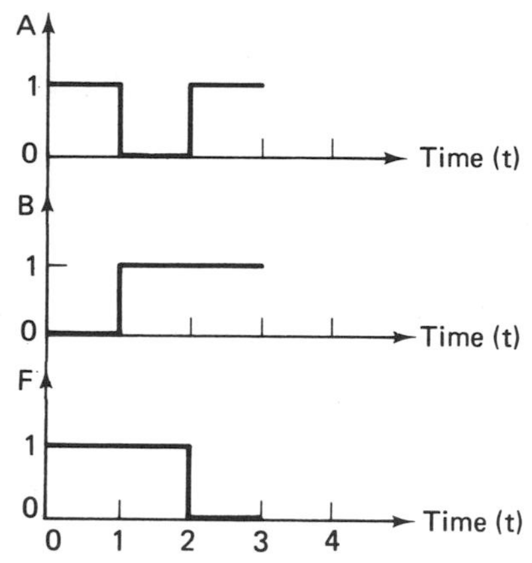

Figure 3.20 Input/output waveform of a two-input NAND gate.

This gives the third condition in the NAND gate table of Fig. 3.17(b), and the output F is in the 1 state. This is shown in the waveforms of Fig. 3.20:

$$F = 1$$

For the time interval 1 to 2, the input states are

$$A = 0$$

$$B = 1$$

These inputs are the same as condition 2 in the NAND truth table [Fig. 3.17(b)]. Therefore, the output F remains logic 1 as shown:

$$F = 1$$

The third region has both inputs A and B equal to 1:

$$A = 1$$

$$B = 1$$

For this case, the truth table tells us that the output switches to the binary 0 level:

$$F = 0$$

This level completes the output waveform F in Fig. 3.20.

3.10 NOR GATE OPERATION

The last of the basic logic gates is the NOR gate. Figure 3.21(a) shows the logic symbol of a two-input NOR gate circuit. This section describes the operation of the NOR gate.

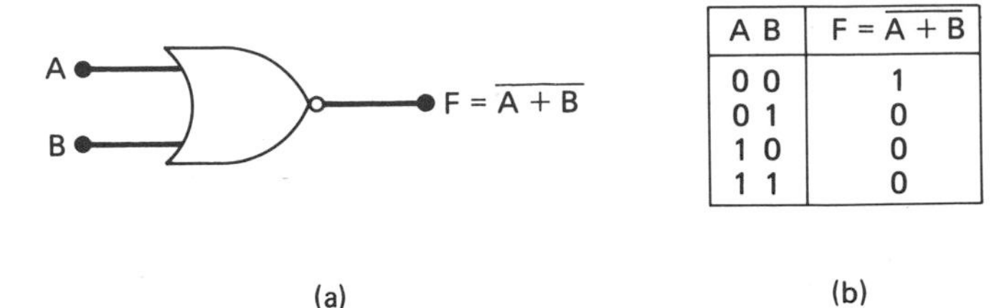

A B	F = $\overline{A + B}$
0 0	1
0 1	0
1 0	0
1 1	0

(a) (b)

Figure 3.21 (a) NOR gate; (b) Truth table.

Boolean Equation and Truth Table Operation

The Boolean equation for the NOR gate output F is written using input variables A and B. This gives the expression

$$F = \overline{A + B}$$

A truth table for NOR gate operation is in Fig. 3.21(b). Looking at this table, we see that the output of a NOR gate is binary 1 for the condition of both inputs logic 0 together. Furthermore, the output is 0 if at least one input is at the 1 logic level.

From the truth table, we find that the NOR gate output is the complement of the OR gate function in Fig. 3.12(b). For this reason, the NOR gate operates like an OR gate with a NOT gate connected at its output.

EXAMPLE 3.16

The two inputs, A and B, of a NOR gate are measured to be 0 V. If the circuit operates off positive voltage mode logic with binary levels $1 = +5$ V and $0 = 0$ V, find the logic level of the output.

SOLUTION:

Using positive logic, the binary inputs are

$$A = 0$$
$$B = 0$$

This input is $AB = 00$ and the first case in the truth table of Fig. 3.21(b). Therefore, the output is at the 1 logic level:

$$F = 1$$

EXAMPLE 3.17

The 4-bit binary word $ABCD = 1010$ is applied to the input of a four-input NOR gate. What is the output F?

SOLUTION:

Since all inputs are not logic 0, the output F is at the 0 logic level:

$$F = 0$$

EXAMPLE 3.18

In Fig. 3.22 the waveforms at the A and B inputs of a two-input NOR gate are shown. Construct the output waveform.

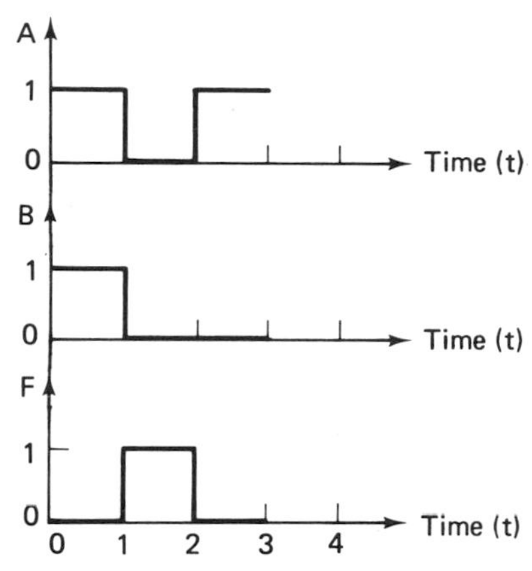

Figure 3.22 Input/output waveforms.

SOLUTION:

During time interval 0 through 1, the input levels are

$$A = 1$$
$$B = 1$$

This is the fourth case in the NOR gate table of Fig. 3.21(b), and the output is 0, as shown in the F waveform of Fig. 3.22:

$$F = 0$$

For the second time region, 1 to 2, the A and B signals are

$$A = 0$$
$$B = 0$$

Both inputs equal to 0 gives the first state in the NOR gate table. This gives a 1 output:

$$F = 1$$

The last interval has $A = 1$ and $B = 0$. These inputs make the output return to the 0 logic level:

$$A = 1$$
$$B = 0$$
$$F = 0$$

The output waveform F is drawn in Fig. 3.22.

ASSIGNMENTS

Section 3.2

1. Describe the input/output operation of a logic gate in general.

2. List the names of the five basic logic gates.

3. What is meant by the term *voltage mode logic*? *Current mode logic*?

4. Define the terms *positive logic* and *negative logic*.

5. The binary logic levels of a gate circuit are $1 = +12$ V and $0 = 0$ V. What type logic is used?

Section 3.3

6. What type of illustration of gate operation is a truth table?

7. Are the inputs listed on the right or left in a truth table?

8. Is the MSB of the input the leftmost or rightmost input column in a truth table?

9. For a logic gate with four inputs I_1, I_2, I_3, and I_4, how many number combinations must be listed in the truth table? Construct a four-input truth table and enter the input words. Leave the outputs undefined.

Section 3.4

10. What type of description of gate operation is a Boolean equation?

11. A logic gate is to have three inputs and an output. Give Boolean variables to indicate the inputs and output.

12. List the three Boolean operators.

13. Write the Boolean equation for the OR function of inputs A and B.

14. What is the Boolean equation for the NOT function of input D?

15. Write the AND function of variables A, B, and C.

16. How is the equation $F = ABCD$ expressed in words?

17. Describe the equation $F = \bar{C}$ in words.

18. Write a Boolean equation for the NOR function of variables A and B.

19. What is the NAND function equation for inputs A, B, C, and D?

20. How is the equation $F = \overline{A + B + C + D}$ read?

Section 3.5

21. Draw the logic symbol for a three-input OR gate. Label its inputs and output.

22. Construct the logic symbol of a four-input AND gate. Mark the inputs and output.

23. What is the logic symbol of a four-input NAND gate?

24. Draw the logic symbol for a three-input NOR gate.

25. Draw the logic symbol of an inverter with input D and output F.

Section 3.6

26. Describe the operation of an AND gate in terms of its output in the binary 1 state.

27. The inputs of a two-input AND gate are $A = 0$ and $B = 0$. Find the binary level of the output.

28. The inputs of an AND gate are $A = 0$ V and $B = -5$ V. If the logic levels are $0 = 0$ V and $1 = -5$ V, is the output binary 0 or 1?

29. The 4-bit data word sent into an AND gate is $ABCD = 1111$. What is the logic level of the output?

Section 3.7

30. Describe the input/output relationship of an OR gate in terms of its output equal to 1.

31. What is the binary output of an OR gate if its two inputs are $A = 0$ and $B = 0$?

32. The voltage levels at the input of an OR gate are $A = +5$ V and $B = +5$ V. Find the output logic level.

33. The input to a three-input OR gate is the binary equivalent of octal 5. What is the output logic level?

Section 3.8

34. Describe the relation between the input and output of a NOT gate.

35. What is the output of a NOT gate when its input $C = 0$?

36. The input signal A to a NOT gate is $+5$ V, and its logic levels are $1 = +5$ V and $0 = 0$ V. Find the output logic level.

Section 3.9

37. Describe the operation of a NAND gate in terms of its output equal to 0.

38. The inputs of a NAND gate are $A = 0$ and $B = 0$. What is the binary output?

39. If the inputs of a NAND gate are $A = 0$ V and $B = -5$ V, what is the output logic level for positive logic?

40. The data word into a 4-bit NAND gate is the binary equivalent of the hexadecimal number F. Find the binary level of the output.

41. The waveforms at the A and B inputs of a NAND gate are shown in Fig. 3.23(a). Construct an output waveform.

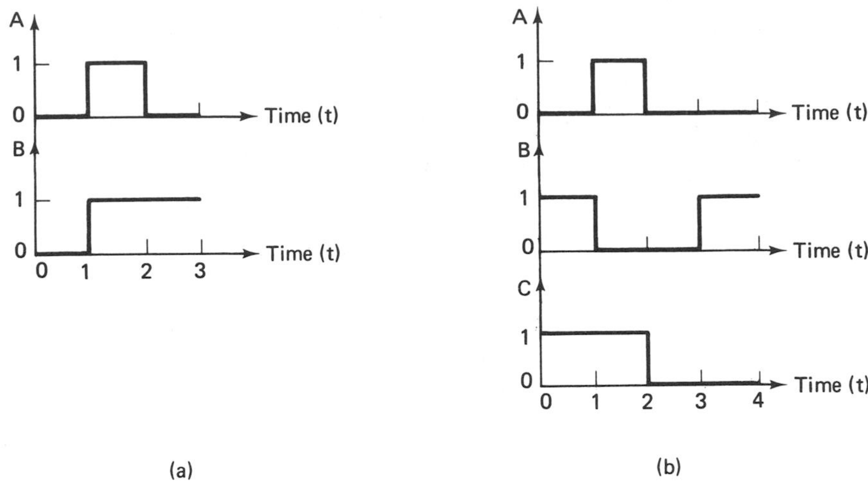

Figure 3.23 (a) *A* and *B* waveforms; (b) *A*, *B*, and *C* waveforms.

Section 3.10

42. Describe the operation of a NOR gate in terms of its output equal to 1.

43. What is the output of a NOR gate when its inputs are $A = 1$ and $B = 1$?

44. The input to a NOR gate is the 3-bit binary number for octal 3. Is the binary output 0 or 1?

45. Draw the output waveform of a NOR gate if its inputs are the *A* and *B* waveforms in Fig. 3.23(a).

46. The inputs *A*, *B*, and *C* of a NOR gate are the waveforms in Fig. 3.23(b). Construct the output waveform.

TTL Logic Gate Integrated Circuits

4

4.1 INTRODUCTION

In Chapter 3, the five basic logic gates were introduced. For each, we showed its logic symbol and described operation with both a truth table and Boolean equation.

Here, the integrated circuits or ICs that contain the electronic circuits for each of these gates will be described. In this section, we shall study the logic gates in the 7400 line of transistor-transistor logic (TTL) ICs.

Besides this, the electrical characteristics used to describe the operation and ratings of integrated devices will be defined in this chapter. These electrical properties are as follows:

1. Logic levels
2. Fan in and fan out
3. Propagation delay
4. Noise immunity

4.2 THE TTL LOGIC GATE

The 7400 line of ICs is one of the most widely used types of integrated devices. Some of the standard logic gate integrated circuits and their type circuits are

7400	Two-input NAND gate
7402	Two-input NOR gate
7404	Inverter gate
7408	Two-input AND gate
7432	Two-input OR gate

This list contains one device for each of the basic logic functions.

All of these 7400 devices are available in both the flat package and dual-in-line package. However, the DIP package is far more popular. For this reason, devices included in this chapter and those that follow will be illustrated with DIP cases.

Power Supply Ratings

TTL integrated circuits are designed to operate off a 5-volt (5-V) dc power supply. This power supply ranges from $+V_{cc}$ equals 5 V to a 0-V ground. The value of supply voltage can vary by about $\pm.5$ V before the device will fail to work correctly. The same devices can be operated off a power supply with voltage values of $+V_{cc}$ equals 0 V and a -5-V ground.

Integrated circuits are small devices; however, this does not mean they use a small amount of power from the 5-V supply. In fact, TTL circuits consume a substantial amount of power. For example, the power dissipations of the 7400 IC devices in the earlier list range from 40 milliwatts (40 mW) per device to as high as 100 mW.

In this way, we find that a single logic gate IC can draw from about 8 milliamperes (8 mA) to 20 mA of current. A system with about 100 of these integrated devices can draw as much as 2 A of current from the power supply. Due to this high-value current drain, digital systems made with TTL logic circuitry generally cannot operate off a battery power supply.

4.3 THE 7400 NAND GATE

The 7400 integrated circuit is a quad two-input NAND gate. In Fig. 4.1(a), the pin layout of the device is shown. Here we find that this IC contains four two-input NAND gate circuits each with separate inputs and output. The 7400 device is in a 14-pin dual-in-line package.

Pin Layout

The pin layout in Fig. 4.1(a) indicates the location of the inputs and outputs of the four NAND gates within the IC package. From this diagram, we find that a first NAND gate uses pin 1 and 2 for its A_1 and B_1 inputs, respectively. The output of this gate circuit is F_1 at pin 3.

A second gate has input A_2 at pin 4 and input B_2 at pin 5. Its output F_2 is at pin 6. The remaining two logic gates are tied to pins on the other side of the IC package.

Another way of expressing the pin layout of a 7400 NAND gate is with a

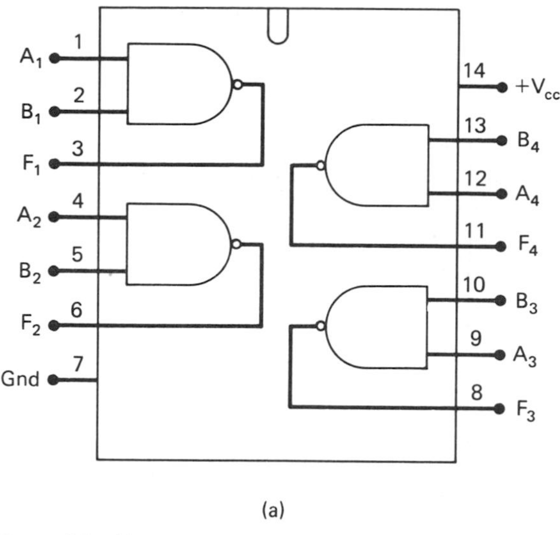

Gate	A	B	F	$+V_{cc}$	Gnd
1	1	2	3	14	7
2	4	5	6		
3	9	10	8		
4	12	13	11		

(b)

Figure 4.1 (a) 7400 pin layout; (b) Circuit layout chart.

chart, as in Fig. 4.1(b). With this information, we can again locate NAND gate 1. It has the *A* input at pin 1 and *B* input at pin 2. Besides this, the *F* column tells us that the output pin of the gate circuit is 3.

In both Figs. 4.1(a) and (b), the power supply leads of the 7400 IC are denoted by $+V_{cc}$ and gnd. The $+V_{cc}$ side of the voltage source is at pin 14 on the package of the device. It is this lead that is supplied with +5 V. On the other hand, the gnd end of the supply goes to pin 7.

All four circuits on the 7400 NAND gate IC are powered off the same set of power supply leads. However, each circuit operates as an independent gate from a logic point of view.

TTL NAND Gate Circuit[1]

The actual circuit used to make the NAND gates in a 7400 integrated circuit device is drawn in Fig. 4.2. A 7400 device has four of these circuits inside its package, one for each of the logic gates.

The input transistor to the circuit is Q_1. This device is called a *multi-emitter transistor*, and it has one emitter lead for each input *A* and *B*. Moreover, we see that the output circuit consists of two transistors Q_3 and Q_4. The hookup of these transistors is called a *totem pole output circuit*. The last transistor Q_2 is used as a driver for transistors Q_3 and Q_4 of the output circuit.

In this totem pole circuit, transistor Q_3 is used to switch the output terminal *F* to the +5-V power supply. For this reason, Q_3 is called a *pull-up transistor*, and when it goes ON the output *F* takes on the 1 logic level.

[1]A detailed description of two-input NAND gate operation is given in Appendix 1.

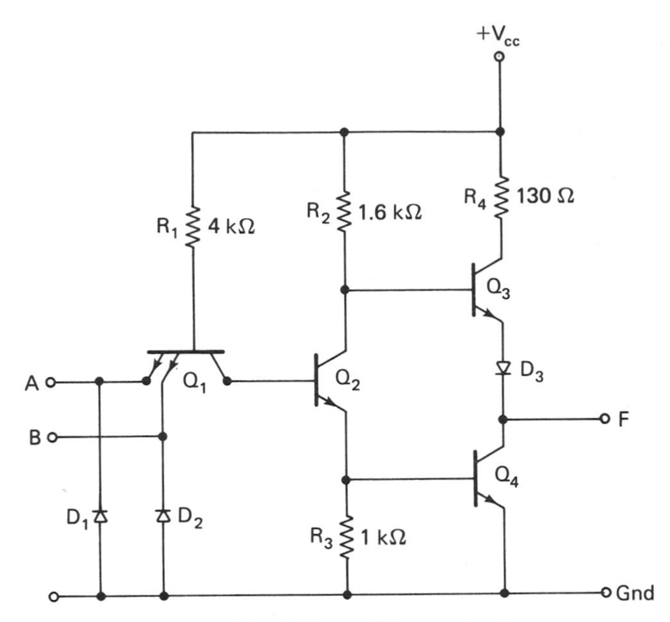

Figure 4.2 TTL NAND gate circuit.

On the other hand, the lower transistor Q_4 switches F to ground. It is called a *pull-down transistor* and makes the output logic state 0 as it goes ON.

The pull-up transistor Q_3 is also called the *current sourcing transistor*, because it supplies current to the output when turned ON. But the pull-down transistor Q_4 is known as the *current sinking transistor*. By current sinking, we mean that it completes the current return path to ground as it goes ON.

Typical resistor values are listed in the transistor circuit diagram of Fig. 4.2.

EXAMPLE 4.1

The 7400 IC in Fig. 4.1 has +5 V applied to pin 14 and 0 V at pin 7. If input A_1 at pin 1 is tied to the +5-V supply (pin 14) and B_1 at pin 2 to the 0-V ground at pin 7, what is the logic state of output F_1 at pin 3?

SOLUTION:

The IC is powered ON by the +5 V at pin 14 and ground at pin 7. By connecting pin 1 to +5 V, the A_1 input is made logic 1, and tying pin 2 to ground makes input B_1 equal 0. Repeating the input logic levels, we get

$$A_1 = 1$$
$$B_1 = 0$$

This makes the output F_1 of the NAND gate a logic 1:

$$F_1 = 1$$

4.4 ELECTRICAL CHARACTERISTICS OF THE 7400 NAND GATE

In the last section, the pin layout and circuitry of the 7400 NAND gate IC was introduced. Here we shall look into some of its important electrical characteristics. In this way, a better understanding of its operation and limitations is obtained.

Input and Output Logic Levels

The first set of electrical ratings we shall consider for the 7400 device is its *logic levels*. These logic level ratings tell the maximum 0 state voltage and minimum 1 voltage for which the circuit will correctly operate. Different values are specified at the output and input of the TTL circuit.

On the output side, the minimum 1 or high-level voltage is indicated with the notation V_{OH}. The value of this voltage is 2.4 V. If the output of a gate falls below 2.4 V, circuits connected at the output may not read the level as a logic 1. This condition could result in an error or malfunction in the digital circuit.

The same is true of an 0 or low-output logic level. Its rating is labeled V_{OL} and has a maximum value of .4 V. When this voltage level is exceeded, the output will no longer indicate a satisfactory 0 logic level. Both output levels are listed in the table of Fig. 4.3.

Characteristic	Rating	Value	Condition
Input and output logic levels	V_{OH}	2.4 V	Minimum
	V_{OL}	0.4 V	Maximum
	V_{IH}	2.0 V	Minimum
	V_{IL}	0.8 V	Maximum
Fan in and fan out	Fan in	2	Maximum
	Fan out	10	Maximum
Propagation delays	t_{PLH}	22 ns	Maximum
	t_{PHL}	15 ns	Maximum
Noise immunity	NI	0.4 V	Minimum

Figure 4.3 7400 electrical characteristics.

Sometimes typical logic level values instead of maximum and minimum ratings are used at the output of a logic gate. For the 7400 NAND gate, the typical 1 level voltage is 3.4 V and the 0 level value is .2 V.

At the input of a NAND gate, no typical values are specified for the 0 and 1 logic levels. However, the minimum 1 or high state voltage and maximum 0 level values are given.

The minimum 1 voltage level at a gate input is marked V_{IH} and has a value of 2 V. On the other hand, the maximum 0 input level value V_{IL} equals .8 V. These input voltage levels are listed in the table of Fig. 4.3.

From the values in this table, we see that the input and output logic levels of a TTL gate differ substantially from the high and low power supply values of +5 and 0 V.

EXAMPLE 4.2

The input voltages V_A and V_B at pins 4 and 5 of a 7400 NAND gate are measured to be 3.7 and 2.9 V, respectively. What voltage would be measured at the output F at pin 6? Give both the typical and maximum values.

SOLUTION:

Both input voltages are above the minimum 1 state input level of 2 V. So both inputs are binary 1:

$$A = 1$$

$$B = 1$$

This input condition makes the output switch to the 0 logic state:

$$F = 0$$

Using the typical value of a low-output voltage, we get

$$V_F = .2\text{ V}$$

But the maximum value is .4 V.

Fan In and Fan Out

Two other important properties of the 7400 NAND gate are *fan in* and *fan out*. The fan in rating corresponds to the number of inputs on the gate. Each input gives a fan in of 1. For example, the four two-input gates on a 7400 IC each have a fan in of 2.

On the other hand, fan out tells how many TTL inputs can be supplied by the output of a logic gate. A *TTL input* is defined as a load that requires the sourcing of 40 μA in the 1 state and the sinking of 1.6 mA for the 0 state. The fan out of each gate on a 7400 IC device is 10. This means the output terminal F can be connected to 10 separate inputs on other TTL logic gates without experiencing overloading. This idea is illustrated in Fig. 4.4.

Let us assume that an eleventh gate is tied to the F output in Fig. 4.4. This would overload the output and cause the output voltage to drop below the V_{OH} minimum value of 2.4 V.

However, in many applications, the output of a logic gate will drive just one or two inputs. In these cases, no fan out problem could occur. The fan in and fan out properties of the 7400 logic gate are tabulated in Fig. 4.3.

Propagation Delay

In the operation of an electronic circuit, a small amount of time is needed for the electronic devices within the circuit to change logic levels. This effect leads to the gate characteristic called *propagation delay*.

Propagation delay tells the amount of time it will take before the output of a

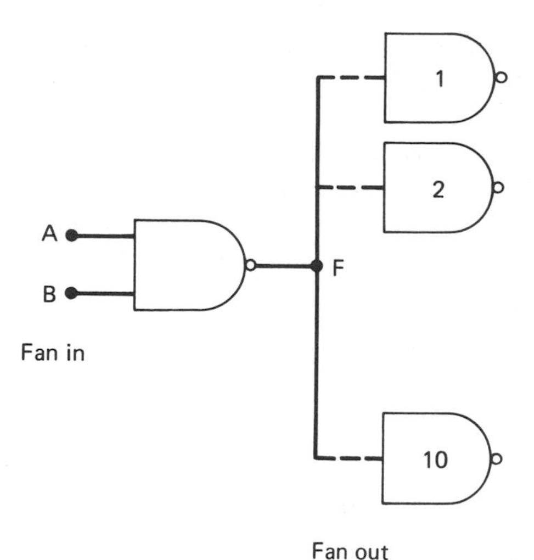

Figure 4.4 Fan in and fan out.

gate switches logic levels after the input logic levels are set. For this switching property, there are two different values defined. The first propagation delay is for the change from the 0 to 1 logic level and the other is for the return from the 1 level to 0. These two values are not the same because the devices working in the circuit are different for each case.

When the output is made to switch from 0 to 1, the propagation delay is denoted by t_{PLH}. The maximum value of the t_{PLH} rating of a 7400 NAND gate is specified to be 22 nanoseconds (22 ns).

The waveforms in Fig. 4.5 illustrate this logic gate switching characteristic.

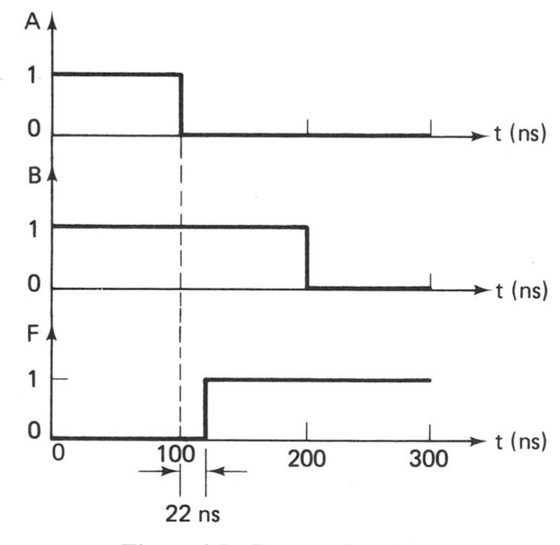

Figure 4.5 Propagation delay.

In this diagram, we find that the inputs *A* and *B* are first both 1. This gives a 0 output at *F*. But at the 100-ns instant, the *A* input is set to 0. Now the inputs are *A* equals 0 and a *B* of 1. So we would expect output *F* to immediately switch to the 1 state. Instead, a propagation delay occurs and a 22-ns interval elapses before the output responds by going to the 1 logic level.

The time needed for the output to change from logic 1 to 0 is shorter than from 0 to 1. This propagation delay is labeled t_{PHL} and has a maximum value of 15 ns.

Generally, the maximum propagation delay value is specified, because it indicates the worst-case switching speed, and all 7400 NAND devices will work in 22 ns or faster.

Both propagation delay values are marked in the table of Fig. 4.3. Looking at these values, we see that TTL gates operate very fast. In fact, TTL ICs are considered high-speed switching devices. Besides this, we realize that in circuit applications where gates are required to operate at speeds slower than 10 microseconds (10 μs) the propagation delay properties can often be neglected.

EXAMPLE 4.3

Let us assume that the output of a 7400 NAND gate was in the 1 state and then switches to 0. How long would be the maximum propagation delay that could occur in changing to the 0 logic level?

SOLUTION:

This transition in output logic level is from high to low. In this case, the propagation delay is given by t_{PHL}, and its value is 15 ns:

$$t_{PHL} = 15 \text{ ns}$$

Noise Immunity

A problem that sometimes occurs in gate circuitry is noise at an input. This noise can cause a false change in output logic level. The ability of a TTL logic gate to withstand input noise is called *noise immunity* or *noise margin*.

The noise immunity characteristic of the 7400 NAND gate is illustrated by the diagram in Fig. 4.6. Here, the inputs of gate 1 are both at the 1 logic level and its output is 0. Furthermore, the maximum 0 state voltage value V_{OL} equals .4 V is indicated at this output.

The output of gate 1 is used as an input to gate 2. The maximum 0 voltage

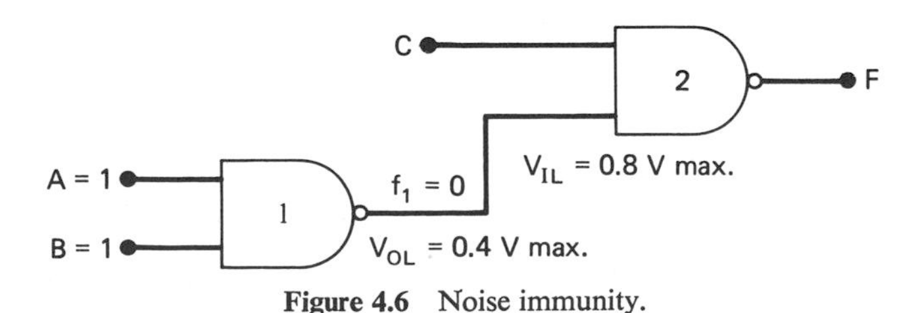

Figure 4.6 Noise immunity.

level allowed at this input is V_{IL} and equals .8 V. In this way, we find that the input of NAND gate 2 could withstand .4 V of noise before its .8-V logic 0 level is exceeded.

This tells us that the minimum noise immunity of a TTL gate is found as the difference in its maximum 0 state logic levels:

$$\text{NI} = V_{IL} - V_{OL} = .8\text{ V} - .4\text{ V}$$
$$\text{NI} = .4\text{ V}$$

Noise immunity can also be defined based on the minimum 1 input and output logic levels. However, their difference is also .4 V, and the same result is obtained.

The last property listed in Fig. 4.3 is noise immunity. A noise immunity value of .4 V is low. For this reason, TTL circuitry is considered to be sensitive to random noise.

EXAMPLE 4.4

The 0 state output of a gate circuit is measured and found to be the .2-V typical value. How much noise voltage could the output receive without causing a false 1 to the circuits tied at its output?

SOLUTION:

The maximum 0 voltage at the input of the next gate is V_{IL} equals .8 V and the measured V_{OL} of this gate is .2 V. Noise immunity is the difference of these two levels:

$$\text{NI} = V_{IL} - V_{OL} = .8\text{ V} - .2\text{ V}$$
$$\text{NI} = .6\text{ V}$$

4.5 OTHER TTL NAND GATES

The 7400 is not the only TTL NAND gate integrated circuit; devices containing three-input, four-input, and eight-input gates are also available. At this point, we shall look into other standard NAND gate ICs.

Three-Input NAND Gate IC

Figure 4.7(a) gives the pin layout of a 7410 NAND gate device. In this diagram, we find that the integrated circuit contains three independent three-input NAND gate circuits.

The circuit used in this device is the same as the TTL NAND gate in Fig. 4.2. However, it has one emitter added on transistor Q_1 to accommodate the C input.

Since the circuit in the 7410 IC is the same as in the 7400 two-input NAND gate, their electrical characteristics are mainly the same. The input and output logic levels, fan in and fan out, propagation delay, and noise immunity values are listed in Fig. 4.7(b). From this list, we see that the one property that changes is fan in, and its value is now 3 due to the C input.

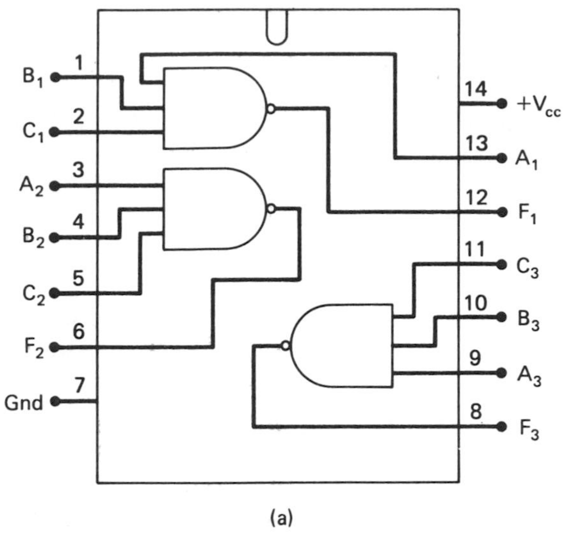

(a)

Rating	Value	Condition
V_{OH}	2.4 V	Minimum
V_{OL}	0.4 V	Maximum
V_{IH}	2.0 V	Minimum
V_{IL}	0.8 V	Maximum
Fan in	3	Maximum
Fan out	10	Maximum
t_{PLH}	22 ns	Maximum
t_{PHL}	15 ns	Maximum
NI	0.4 V	Minimum

(b)

Figure 4.7 (a) 7410 pin layout; (b) Electrical characteristics.

EXAMPLE 4.5

Inputs A and B of a three-input NAND gate are measured to be steady at 3.8 and 4.1 V, respectively. The third input C is observed to switch from a level of 4 V to a new value of .1 V. What will be the new logic level and typical voltage at output F?

SOLUTION:

The minimum input voltage level for a logic 1 is 2 V. This says that inputs A and B are at logic 1:

$$A = 1$$

$$B = 1$$

The C input is now .1 V and is below the .8-V maximum 0 level:

$$C = 0$$

Therefore, output F is 1:

$$F = 1$$

The typical value of output voltage is 3.4 V.

Four-Input and Eight-Input NAND Gate ICs

Two other NAND gate devices are shown in Figs. 4.8(a) and (b). The pin layout in Fig. 4.8(a) is that of a 7420 dual four-input NAND gate IC. In Fig. 4.8(b), a 7430 single eight-input NAND device is shown.

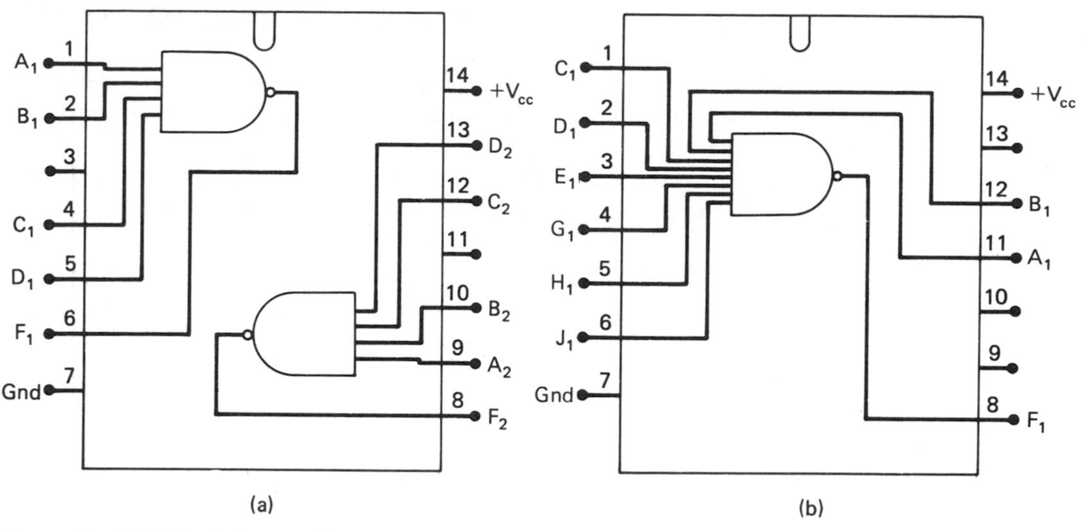

Figure 4.8 (a) 7420 pin layout; (b) 7430 pin layout.

4.6 THE 7402 NOR GATE

One of the most widely used NOR gate integrated circuits is the 7402 device. Figure 4.9(a) shows that the layout of the four gates in a 7402 package is different from the 7400 NAND gate device. Here, the A_1 input of NOR gate 1 is at pin 2, and the other input B_1 is at pin 3. The output F_1 of this gate is at pin 1.

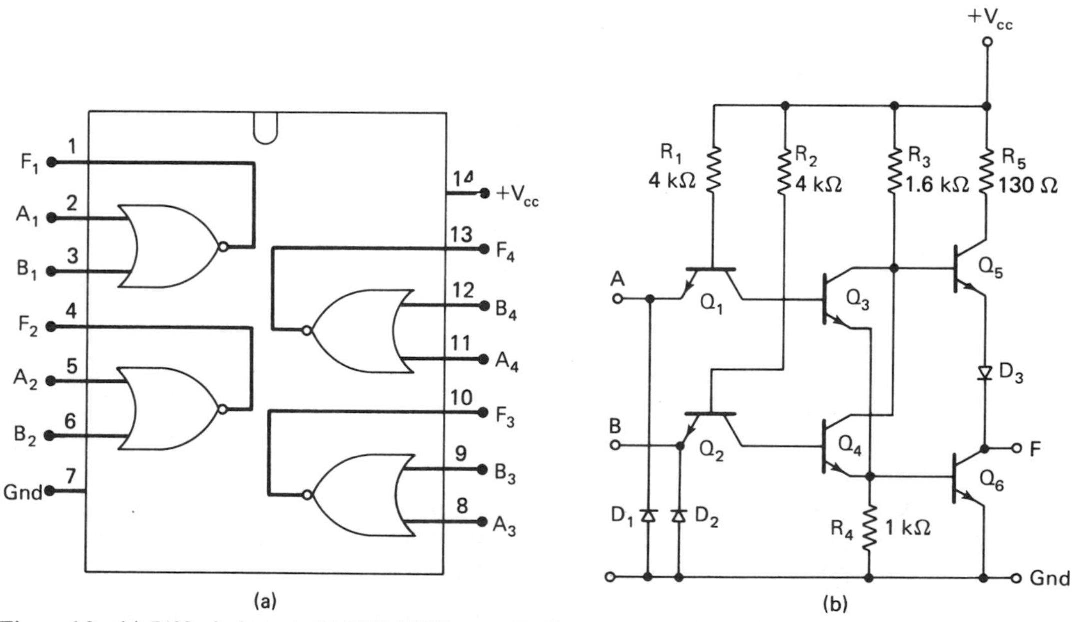

Figure 4.9 (a) 7402 pin layout; (b) TTL NOR gate circuit.

The second NOR gate has inputs A_2 and B_2 at pin 5 and 6, respectively, with output F_2 at pin 4. In addition to these two gates, there are two more at pins on the other side of the device.

As with all gate devices described to this point, the power supply is indicated by $+V_{cc}$ and gnd. Furthermore, $+V_{cc}$ is again at pin 14 and gnd at pin 7.

TTL NOR Gate Circuit[2]

The circuit used to make the TTL NOR gate is shown in Fig. 4.9(b). In this circuit transistors Q_5 and Q_6 form the same totem pole output configuration used in the TTL NAND gate of Fig. 4.2. On the other hand, the input circuit is completely different. Here two separate transistors Q_1 and Q_2 are used for the A and B inputs, respectively. Each input transistor has its own output drive transistor, Q_3 and Q_4, respectively. Drive transistors Q_3 and Q_4 are connected in parallel.

The resistor values marked in the circuit diagram are essentially the same as those in the NAND gate. However, one more resistor is needed in this circuit.

Electrical Characteristics

The electrical characteristics of a 7402 NOR gate are listed in Fig. 4.10. From this table, we see that the values for input and output voltage levels, fan in and fan out, propagation delays, and noise immunity are exactly the same as those of the 7400 NAND gate device.

Rating	Value	Condition
V_{OH}	2.4 V	Minimum
V_{OL}	0.4 V	Maximum
V_{IH}	2.0 V	Minimum
V_{IL}	0.8 V	Maximum
Fan in	2	Maximum
Fan out	10	Maximum
t_{PLH}	22 ns	Maximum
t_{PHL}	15 ns	Maximum
NI	0.4 V	Minimum

Figure 4.10 7402 electrical characteristics.

Three-Input NOR Gate

Other integrated NOR gate devices are available. The pin layout of a 7427 three-input NOR gate integrated circuit is drawn in Fig. 4.11. This device is constructed using the same basic circuit as in the 7402 NOR gate, and, except fan in, their electrical characteristics are alike.

EXAMPLE 4.6

The voltage at the A input pin 2 of a 7402 NOR gate is observed to be constant at .1 V. However, the B input at pin 3 is observed to switch from 3.9 V down

[2]In Appendix 1, a detailed description of two-input NOR gate operation is given.

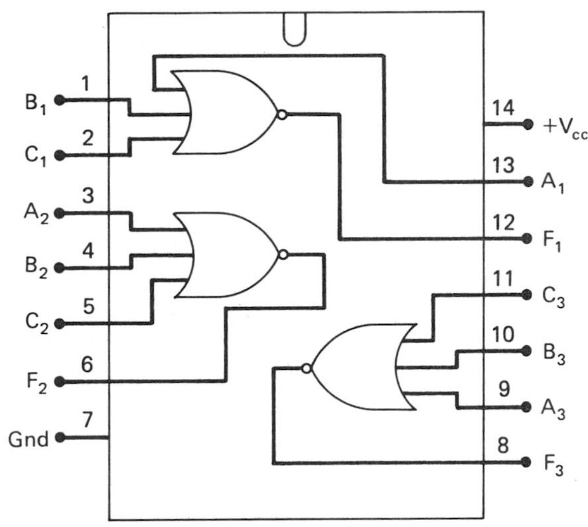

Figure 4.11 7427 pin layout.

to .15 V. What will be the new logic level and typical voltage value of output F at pin 1? How long will it take the output to get into this logic state?

SOLUTION:

After the B input switches low, both inputs are below the maximum value for an 0. Now, both inputs are in the 0 logic state:

$$A = 0$$
$$B = 0$$

Therefore, the output will switch to the 1 logic level:

$$F = 1$$

A typical 1 at output F is 3.4 V. However, the circuit will experience a delay between switching at the input and the occurrence of the new output logic state. This time is given by the propagation delay t_{PLH}:

$$t_{PLH} = 22 \text{ ns}$$

4.7 THE 7404 INVERTER

Another important integrated circuit is the 7404 hex-inverter. The pin layout of a 7404 device is illustrated in Fig. 4.12(a). In this diagram, we see that the hex-inverter has six independent NOT gates or inverter circuits.

The pin layout in Fig. 4.12(a) gives the location of the six inverter circuits in the 7404 device. For instance, the first NOT gate has input A_1 at pin 1 and the output F_1 at pin 2. Again, the power supply is between pins 14 and 7.

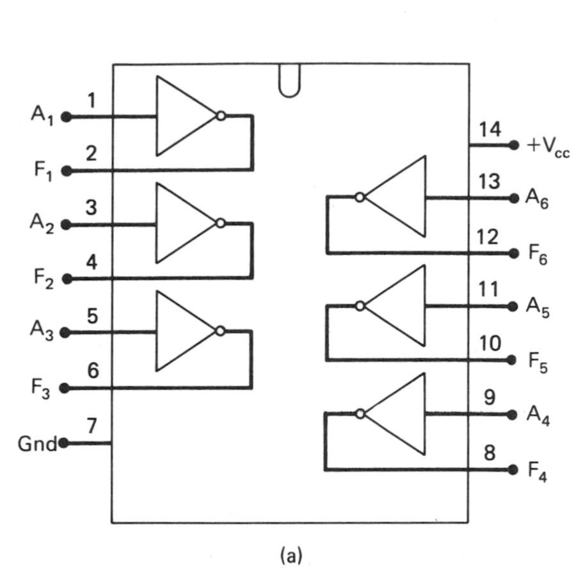

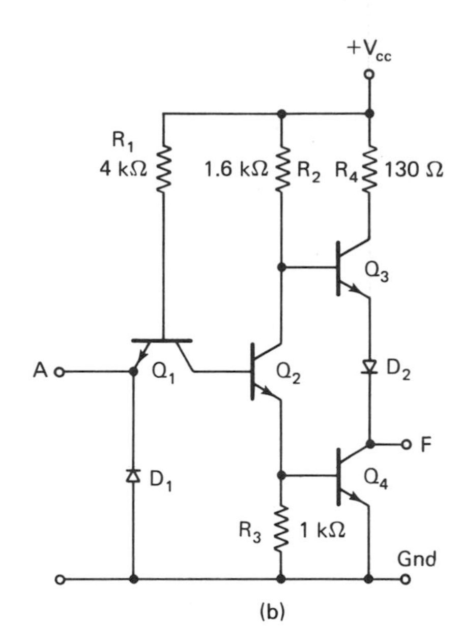

Figure 4.12 (a) 7404 pin layout; (b) TTL inverter circuit.

TTL Inverter Circuit and Electrical Characteristics

The circuit used to make a NOT gate in the 7404 device is shown in Fig. 4.12(b). Here we see that the circuit is similar to the one used in the 7400 NAND gate. However, the inverter circuit has just one emitter on the input transistor Q_1. The connection of driver transistor Q_2 to the totem pole pull-up transistor Q_3 and pull-down transistor Q_4 is exactly the same as in the NAND gate circuit.

Electrical characteristics of the 7404 are similar to those given for the 7400 and 7402 gates. These values are listed in Fig. 4.3 and repeated in Fig. 4.10. The one difference is fan in, which is now 1 instead of 2.

EXAMPLE 4.7

The input signal at pin 1 of a 7404 inverter is observed with an oscilloscope. The input voltage is steady at .15 V and then suddenly switches to 4.2 V. What is the new logic level of the output? How long is the propagation delay to this state?

SOLUTION:

After the input switches to a 4.2-V value, it is above the input logic 1 voltage level V_{IH}. The input is in the 1 logic state, and the output goes to the 0 logic level:

$$A = 1$$
$$F = 0$$

The change in output logic level is delayed by the propagation delay t_{PHL}:

$$t_{PHL} = 15 \text{ ns}$$

4.8 THE 7408 AND GATE AND 7432 OR GATE

To this point, we have described integrated circuits for three of the five basic logic gates. These are the NAND gate, NOR gate, and inverter. In this section, AND gate and OR gate ICs will be introduced.

TTL AND Gate IC

One of the most widely used AND gate integrated circuits is the 7408. This device is a quad two-input AND gate. The pin layout of a 7408 is shown in Fig. 4.13. From this diagram, we see that the four AND gates are located at the same pins as the NAND gates in a 7400 device.

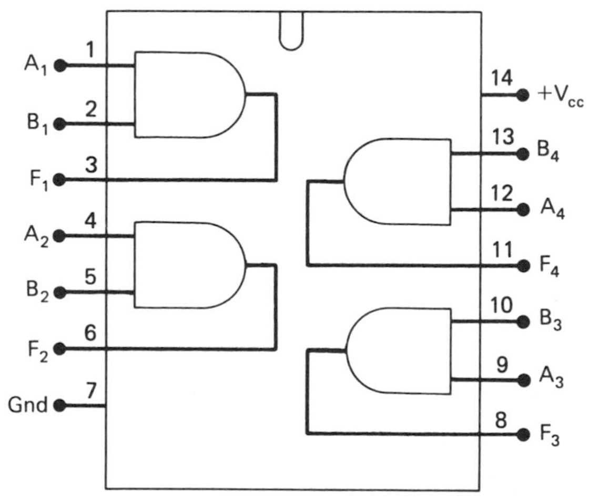

Figure 4.13 7408 pin layout.

A three-input AND gate device is also available. The 7411 is a triple three-input AND gate integrated circuit, and its pin layout is identical to the 7410.

TTL OR Gate IC

The 7432 is a TTL OR gate device. A pin layout of this integrated circuit is shown in Fig. 4.14. Here we see that the device contains four two-input OR gates. Again, the layout of the gates within the package is the same as for the 7400 IC.

Electrical Characteristics

Up to now, all TTL integrated circuits had identical electrical characteristics. This is also true for the 7432 OR gate device. However, the 7408 AND gate has

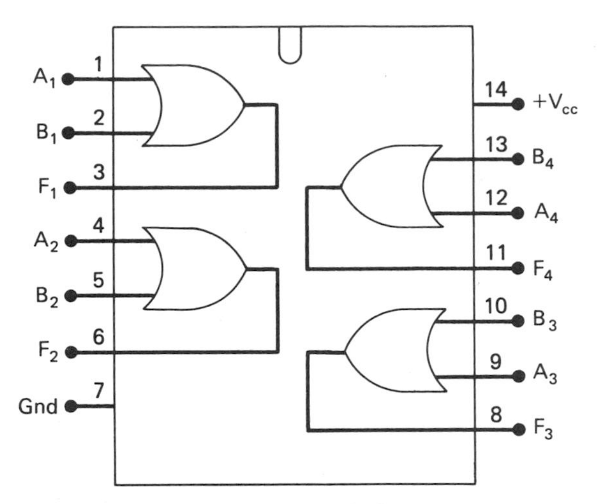

Figure 4.14 7432 pin layout.

slower switching properties. Its propagation delay t_{PLH} equals 27 ns, and t_{PHL} is 19 ns. All other 7408 characteristics remain the same.

EXAMPLE 4.8

The pin 1 input on a 7408 AND gate is held at 4.1 V. At pin 2, the input is measured to switch from .1 to 3.9 V. Find the new output logic level. What is the propagation delay for this change in logic level?

SOLUTION:

When pin 2 switches to the 3.9-V level, both inputs of the AND gate are above the V_{IH} voltage level and are in the 1 logic state:

$$A = 1$$
$$B = 1$$

In this way, we find that the output logic level is also 1:

$$F = 1$$

For the output to change from 0 to 1, the propagation delay is t_{PLH} and equals 27 ns:

$$t_{PLH} = 27 \text{ ns}$$

4.9 GATES WITH OPEN-COLLECTOR AND THREE-STATE OUTPUTS

Some of the basic logic gates are available with special types of output circuits. One of the common types contains what is known as an *open-collector output*. A second device recently in use has what is called a *three-state output*. Both of these types of gates are needed in certain special digital applications.

The Open-Collector Output Gate

The circuit in Fig. 4.15(a) is a two-input NAND gate with open-collector output. Here, we see that the pull-up transistor part of the circuit is not put in the device. So the pull-down transistor Q_3 is left with an open at its collector.

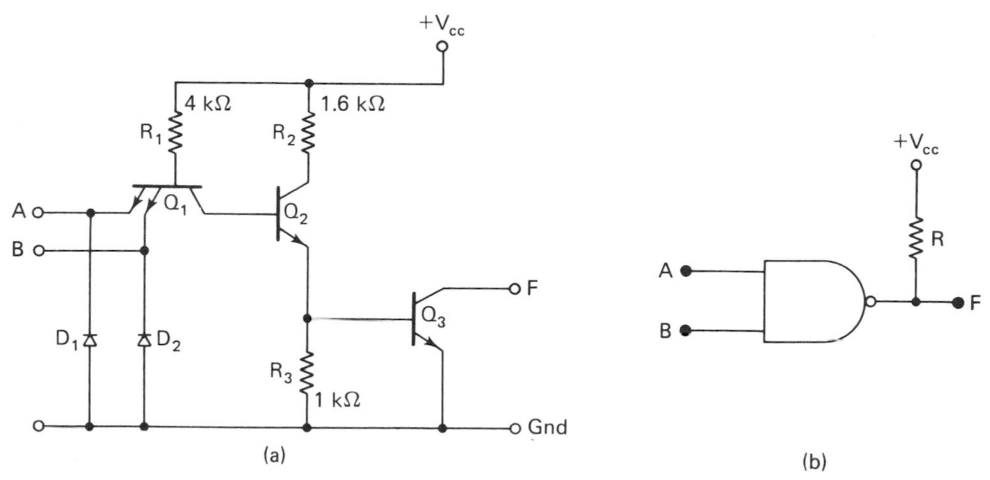

Figure 4.15 (a) Open-collector NAND gate circuit; (b) Open-collector gate with resistive load.

This TTL NAND gate circuit is employed in the 7401 quad two-input NAND gate integrated circuit. A pin layout of the 7401 device is identical to the 7402 NOR gate in Fig. 4.9(a). AND gates, NOR gates, and inverter circuit ICs are also available with open-collector outputs.

To make this gate operate, a load must be supplied at the output. This load could be a transistor or just a resistor. The connection of a resistor to the open-collector NAND gate output is shown in Fig. 4.15(b). In this diagram, the external resistor is returned to the +5-V supply and provides pull-up.

Normally, the outputs of TTL gates cannot be wired together. However, open-collector devices can have their outputs connected. A diagram of two parallel-connected open-collector NAND gates is shown in Fig. 4.16. From this circuit, we notice that a single pull-up resistor is needed for both gates. This parallel connection of gates is used in some important applications. One of the most popular open-collector connections is called the wire-AND circuit.

Three-State Output

The three-state output gate is a newly developed device. A logic symbol for a three-state inverter is shown in Fig. 4.17. Here, we observe that the device has an extra lead marked C, which is a control lead for the output.

This circuit has three different output states. Two are the normal 0 and 1

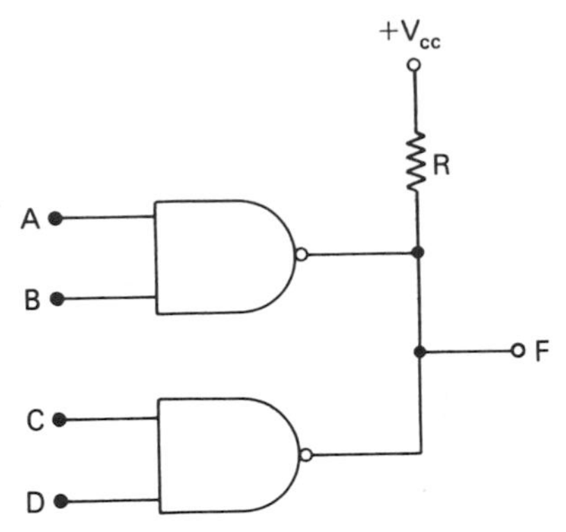

Figure 4.16 Paralleled open-collector outputs.

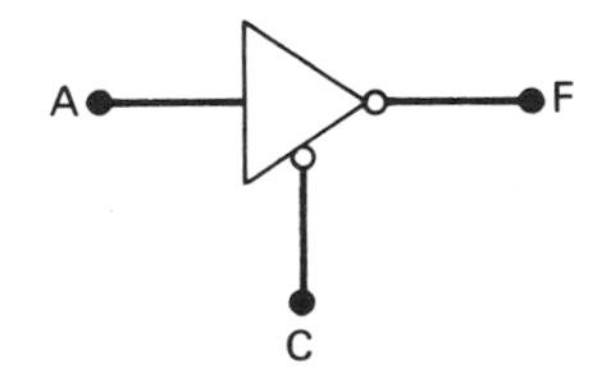

Figure 4.17 Inverter with three-state output.

binary logic levels. However, the third output state is called *hi-Z* for *high impedance*. In the hi-Z state the output is electrically like an open circuit.

The 8096 TTL integrated circuit is a hex-inverter with three-state outputs. This device has two control leads C_1 and C_2. Together, C_1 and C_2 determine the operation of all six gates. When at least one of the control inputs is at logic 1, all outputs are in the hi-Z third state. To operate like normal inverters, both control leads must be made logic 0.

Using a three-state output gate, the outputs can be directly connected in parallel. In fact, up to 128 outputs can be tied in parallel. When working in this way, just one of the gates can operate as an inverter at a time. All other gates must be held in the hi-Z output state by the control signals.

4.10 THE BUFFER AND GATES WITH BUFFERED OUTPUTS

In some applications, the output of a circuit must be buffered. By *buffer*, we mean that the output of a gate has its ratings changed. For instance, the output current rating can be increased to give a higher fan out. Another reason for buffering is to change the voltage level at the output of the circuit.

Buffer Gate Operation

The logic symbol used to indicate a buffer gate is in Fig. 4.18. This symbol is a lot like the inverter, but there is no dot at the output. Looking at Fig. 4.18, we notice that the output of a buffer is the same logic level as the input. The truth table tells

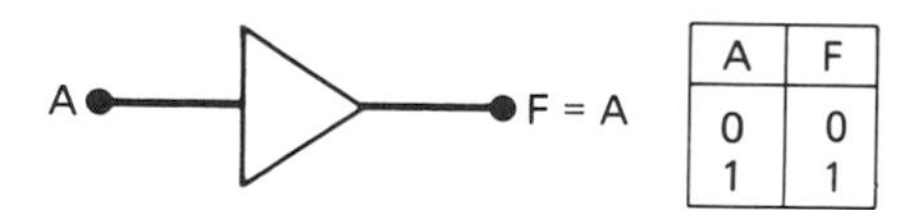

A	F
0	0
1	1

Figure 4.18 Buffer symbol and truth table.

us that a 0 input gives a 0 output and that 1 at the input makes the output 1. However, a change in fan out or voltage rating has been achieved. The Boolean equation of buffer operation is

$$F = A$$

TTL Buffer IC

The 7407 is an integrated circuit containing buffer gates. This device has six high-voltage buffers. Each of the circuits has an output voltage rating of 30 V. The pin layout of this device is the same as that shown for the 7404 inverter in Fig. 4.12(a).

The output of this device is open-collector and can be returned with a resistor to a power supply as high as 30 V. In this way, the circuit can be used to increase the logic level of the output above the typical value of 3.4 V. The buffering of logic levels is important when interconnnecting different types of logic circuitry. For example, to go from TTL circuitry to CMOS devices each output must have its logic level raised before it can be applied to the CMOS input.

Three-State Buffer

Buffer devices are also available with a three-state output. The 74125 is a quad three-state buffer gate. This integrated circuit contains four buffer gates, and each has its own output control lead.

The pin layout of the 74125 is the same as the 7400 NAND gate. However, the control lead C is in the position of the A inputs of the NAND gates. To make the output go into the hi-Z state, the control input C must be made logic 1.

This device can be used to directly parallel the output of circuits. The advantage of using a buffer in this application is that the logic levels are not changed when signals pass from input to output.

Gates with Buffered Outputs

The basic logic gates are available with buffered outputs. Some of the common devices are listed in Fig. 4.19.

Device	Gate	Buffering
7406	Hex-inverter	30 V rating
7426	Quad 2-input NAND	15 V rating
7428	Quad 2-input NOR	Fan out of 30
7437	Quad 2-input NAND	Fan out of 30
7440	Dual 4-input NAND	Fan out of 30

Figure 4.19 Common gates with buffered outputs.

ASSIGNMENTS

Section 4.2

1. Give the number and the type of the device for three integrated circuits in the 7400 line.

2. What is the most popular package for 7400 line integrated circuit devices?

3. How much voltage is needed for a TTL IC power supply?

4 Are TTL devices high- or low-current integrated circuits?

Section 4.3

5. List the pins of the inputs A_3 and B_3 and the output F_3 of NAND gate 3 on a 7400 IC.

6. What two names are used to refer to the upper transistor Q_3 in the TTL NAND gate output circuit of Fig. 4.2? The lower transistor Q_4?

7. The A_2 input at pin 4 of a 7400 NAND gate is +5 V, and the other input B_2 at pin 5 is also +5 V. What are the logic levels of the inputs and output F_2?

Section 4.4

8. What are the typical values of the binary 1 and 0 at the output of a 7400 IC?

9. The inputs of a 7400 NAND gate are V_A equals .2 V and V_B equals .35 V. Find the logic level of the output. What would be the minimum voltage that could be measured at F?

10. What is the current definition of a TTL input?

11. How much is the fan out of a 7400 NAND gate?

12. Both inputs of a 7400 NAND gate are 1, and its output is 0. When one input switches 0, what is the output logic state? How long is the propagation delay until this output state occurs?

13. The 0 state at the output of a 7400 NAND gate is .1 V. How much noise voltage could this output withstand before it would indicate a false 1 to the circuits supplied by its output?

Section 4.5

14. How long would it take the output of the 7410 NAND gate in Example 4.5 to switch to the 1 logic level?

15. The inputs of a 7420 four-input NAND gate are $V_A = 3.9$ V, $V_B = 4.0$ V, $V_C = 4.1$ V, and $V_D = 3.7$ V. Find the output logic level. What is the typical value of output voltage?

Section 4.6

16. Make a pin layout chart like that in Fig. 4.1(b) for a 7402 NOR gate IC.

17. In the TTL NOR gate circuit of Fig. 4.9(b), is output transistor Q_5 or Q_6 the current sinking transistor?

18. The A_2 input at pin 5 of a 7402 NOR gate is measured to be .2 V. The B_2 input at pin 6 is observed to switch from .15 to 3.6 V. What is the new binary state of the output F_2? How long does it take the output to get to this state?

19. A 7427 three-input NOR gate has input voltages $V_A = 3.4$ V, $V_B = 3.7$ V, and $V_C = 4.1$ V. What is the logic state of the output F? Give the typical value for the output voltage V_F.

Section 4.7

20. For the 7404 inverter IC, make a pin layout chart like the one in Fig. 4.1(b).

21. The TTL inverter circuit is illustrated in Fig. 4.12(b). Is output transistor Q_3 or Q_4 the current sourcing transistor?

22. The input A_2 at pin 3 of a 7404 IC is held at 4.1 V. This input is then switched to .1 V. Find the new output logic level. What is the propagation delay to this state?

Section 4.8

23. The three inputs on a 7411 AND gate IC are measured to be $V_A = 3.8$ V, $V_B = 4.1$ V, and $V_C = 4.3$ V. What is the binary level of the output F?

24. One input on a 7408 two-input AND gate switches from .1 to 4.2 V every 100 μs. Describe the output F if the other input is held constant at .2 V. What would the output look like if the constant input is switched to 4 V and held?

25. The input A_1 at pin 1 of a 7432 OR gate is .2 V. What is the output F_1 at pin 3 after the B_1 input switches from .25 to 4.3 V?

Section 4.9

26. In the open-collector TTL NAND gate circuit of Fig. 4.15(a), is the current sinking or current sourcing transistor left out?

27. Is the external load resistor on a 7401 open-collector NAND gate connected from the output terminal to ground or to the +5-V supply?

28. How does the output of a three-state inverter function electrically when it is in the hi-*Z* state?

29. The control signals C_1 and C_2 on a 8096 three-state inverter are both made 0, and the *A* input is logic 1. Is the output *F* in the 0, 1, or hi-*Z* state?

Section 4.10

30. Give two functions of a buffer gate.

31. What is the output voltage rating of each gate on the 7407 hex-buffer IC?

32. The control input at pin 1 of a 74125 three-state buffer is binary 0, and the input signal *A* at pin 2 is logic 1. What is the state of the output *F* at pin 3?

33. For what purpose is the output of a 7440 dual four-input NAND gate buffered?

Combinational Logic Networks

5

5.1 INTRODUCTION

Having introduced number systems, the basic logic functions, and their corresponding TTL integrated circuits, we shall turn our interest to the study of logic networks. For this, the topics that follow are included in this chapter:

1. The combinational logic network
2. Levels of logic
3. Network Boolean equation
4. Network truth table
5. Truth table analysis

5.2 THE COMBINATIONAL LOGIC NETWORK

When logic gates are interconnected and provide a useful function, a logic network is formed. For instance, the gate combination in Fig. 5.1 is a simple AND-OR-NOT network. This group of logic gates is called a *combinational logic network*, and the techniques used to analyze its operation are known as *combinational theory*.

Combinational networks are used to make logical decisions and to control the operation of circuits in digital electronic systems. They can be designed three

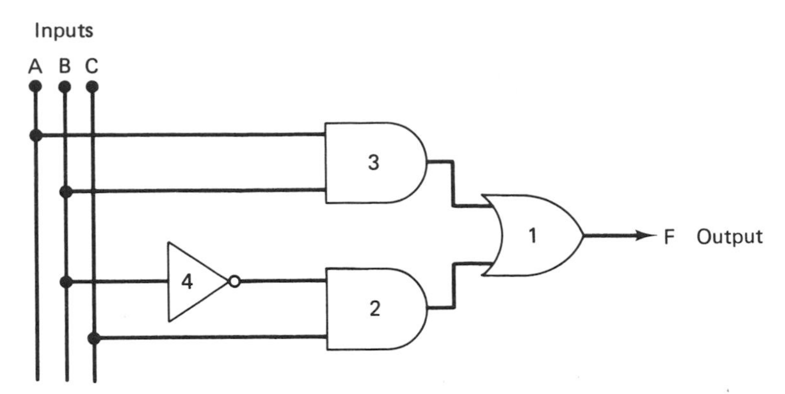

Figure 5.1 Combinational logic network.

different ways. First, AND-OR-NOT circuitry can be used as in the network of Fig. 5.1. However, identical operation can be obtained using NAND-NOT logic or NOR-NOT circuitry.

Certain logic gate combinations are very widely used in digital equipment. For this reason, they are given names and are available as standard integrated circuits. Some examples are as follows:

The exclusive-OR gate
Magnitude comparators
Decoders
Multiplexers and demultiplexers
Adders

Moreover, many nonstandard logic networks are designed using the basic logic gates.

In this chapter, we shall study the operation of nonstandard logic gate networks. But in later chapters standard logic networks and their integrated circuits will be treated.

5.3 LEVELS OF LOGIC, BOOLEAN EQUATIONS, AND TRUTH TABLES

The operation of a combinational logic network gives the relationship between its inputs A, B, and C, and the output F. We can describe the operation of a logic network in either of two ways. One way is with a *network Boolean equation.* This gives a mathematical description of circuit operation and is important for the design of digital circuitry. The second approach is to set up a *truth table for the network*. In this way, circuit operation is indicated by a table showing the 0 and 1 logic levels at the output of each gate in the network.

The first step in writing the Boolean equation or making the truth table of a logic network is to assign levels of logic to the separate logic gates.

Levels of Logic

An important property of a logic network we must recognize is the level of its logic gates. The levels in a logic gate circuit are defined from output toward the input. In the example logic network of Fig. 5.2(a), the output is *F*, and the inputs are *A*, *B*, and *C*. Here the logic circuit has two levels of gate circuitry. The first level has a single OR gate labeled 1, and there are two AND gates, gate 2 and gate 3, in the second level. So the level of a logic gate in a combinational logic network is determined by the number of gates its inputs must pass through to get to the output.

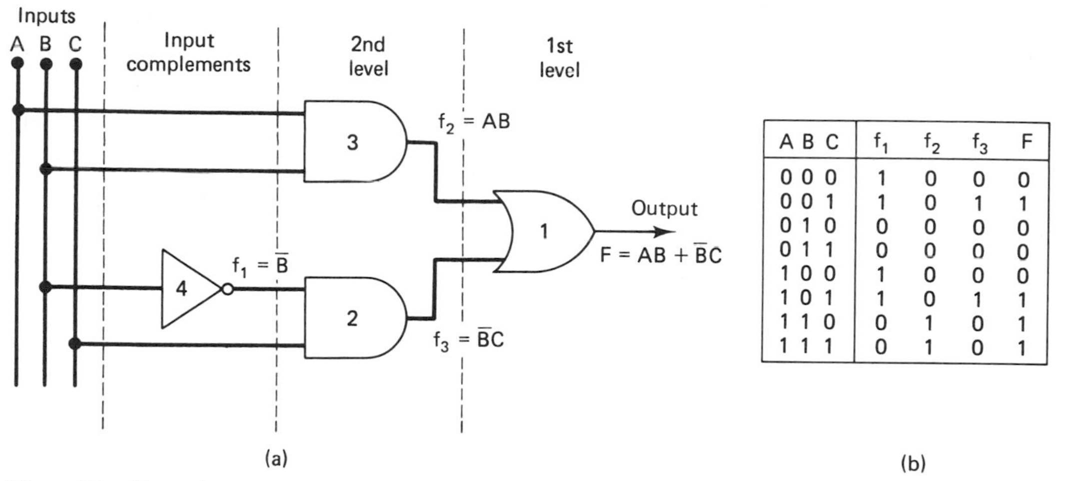

A B C	f_1	f_2	f_3	F
0 0 0	1	0	0	0
0 0 1	1	0	1	1
0 1 0	0	0	0	0
0 1 1	0	0	0	0
1 0 0	1	0	0	0
1 0 1	1	0	1	1
1 1 0	0	1	0	1
1 1 1	0	1	0	1

Figure 5.2 (a) Logic levels and Boolean equations; (b) Truth table.

Inverters at the input are not considered as a new logic level. Instead, they just provide the complement of an input signal. Therefore, the inverter, gate 4, in our logic circuit gives the complement of input *B*, which is $\bar{B}$.

Network Boolean Equation

To write the Boolean equation of a logic network, we must indicate the output of each gate as a Boolean function of its inputs. We begin at the inputs of the network and work through the complemented inputs and levels of logic toward the output.

For the logic circuitry in Fig. 5.2(a), the Boolean output of each gate is marked into the logic diagram. In this network, we start with the inverter gate. Its output is indicated by the Boolean function f_1 and is the complement of input B:

$$f_1 = \bar{B}$$

Now the Boolean functions f_2 and f_3 of the second level logic gates are written. They are given by the AND operation of their input signals:

$$f_2 = AB$$
$$f_3 = f_1 C = \bar{B}C$$

Finally, the output F of the network is found from the inputs of the OR gate. This gives

$$F = f_2 + f_3 = AB + \bar{B}C$$

So the network Boolean equation F is the output of the first level logic gate.

Network Truth Table

The truth table of a logic network represents the logic state for each input, the complemented inputs, and the output of each logic gate. The complete truth table for the logic diagram of Fig. 5.2(a) is shown in Fig. 5.2(b). In this table, the first columns of binary numbers are for the circuit's inputs A, B, and C. They are written as 3-bit binary words and arranged with the most significant bit first. The numbers are listed in numerical order down the columns.

The inputs are followed in the table by columns for the complemented inputs, second level logic gates, and logic network output. In the logic diagram and truth table of Fig. 5.2, the complemented input is f_1, the second level gate outputs f_2 and f_3, and the circuit output F. So they are listed in this order across the columns of the truth table in Fig. 5.2(b).

In this way, we see that the organization of logic information in the truth table depends on recognizing the levels of logic in a network.

The actual 0 and 1 logic states for the gate outputs in the truth table are found by a technique called *truth table analysis*.

5.4 TRUTH TABLE ANALYSIS

By truth table analysis, we can form the table showing the logic operation of a combinational logic network. To do this, we use the truth tables of the fundamental logic gates. Truth tables of AND, OR, NOT, NAND, and NOR gate operation are repeated in Figs. 5.3(a) through (e).

A B	AB
0 0	0
0 1	0
1 0	0
1 1	1

(a)

A B	A + B
0 0	0
0 1	1
1 0	1
1 1	1

(b)

A	$\bar{A}$
0	1
1	0

(c)

A B	$\overline{AB}$
0 0	1
0 1	1
1 0	1
1 1	0

(d)

A B	$\overline{A + B}$
0 0	1
0 1	0
1 0	0
1 1	0

(e)

Figure 5.3 (a) AND gate; (b) OR gate; (c) Inverter; (d) NAND gate; (e) NOR gate.

The procedure that follows can be used to form the truth table of a gate network:

1. Mark the inputs and outputs of the logic circuit.
2. Indicate the levels of logic gates.
3. Construct the truth table and label its columns.
4. Write in the binary words for the inputs.
5. Use the AND, OR, NOT, NAND, and NOR gate truth tables to evaluate the logic states for the complemented input columns and then the gates in each level until the output is found.

EXAMPLE 5.1

Check the operation of the logic diagram in Fig. 5.2(a) by truth table analysis. The table of circuit operation is shown in Fig. 5.2(b).

SOLUTION:

The inputs, logic levels, and outputs are already marked into the logic diagram. So the analysis begins with the inverter, gate 4. Its output is the complement of input B and is 1 for $B = 0$ or 0 when $B = 1$. Comparing the f_1 column in the table to the B column, we see that this is correct.

Now, the output of the second level AND gate 3 will be found. Here the output f_2 is 1 if both inputs A and B are 1. Otherwise it is 0. Checking the A and B columns shows that the output f_2 is also correct.

Next, we shall take the other second level AND gate. For gate 2 the output is 1 when inputs f_1 and C are 1 together. Looking at columns C and f_1 in the table, we see that the output is again correct.

Finally, the output of the network is found. To do this, the output of the first level OR gate is determined. The output of gate 1 is in the 1 state if either of its inputs f_2 and f_3 is 1. From the f_2, f_3, and F columns of the truth table, we see that the output of the network is correctly indicated.

5.5 AND-OR-NOT GATE EXAMPLES

Here we shall use the truth table analysis technique on several other AND-OR-NOT logic circuits.

EXAMPLE 5.2

Perform a truth table analysis of the logic diagram in Fig. 5.4(a).

SOLUTION:

The output function f_1 of the inverter gate is the complement of input A:

$$f_1 = \bar{A}$$

So its logic state is 1 when the A input is 0. The A input is the first column of the truth table in Fig. 5.4(b) and f_1 the third column.

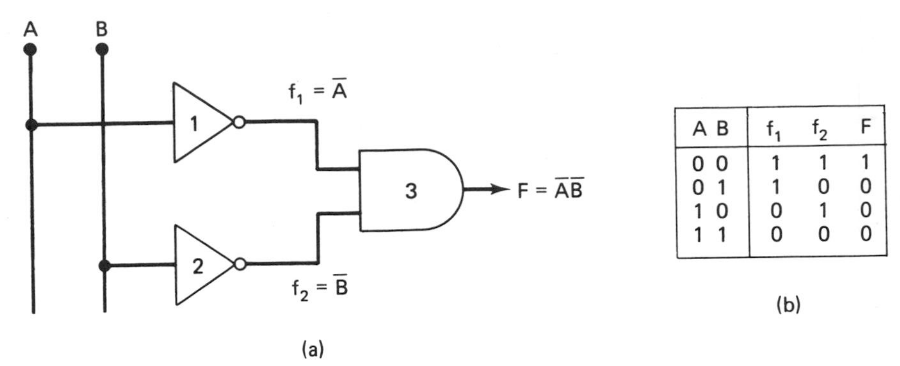

A B	f_1	f_2	F
0 0	1	1	1
0 1	1	0	0
1 0	0	1	0
1 1	0	0	0

Figure 5.4 (a) Logic diagram; (b) Truth table.

The second output f_2 is input B inverted:

$$f_2 = \bar{B}$$

Its logic value is 1 for each 0 at the B input. The B input and f_2 output are in the second and fourth columns of the truth table, respectively.

The output of the logic network is from the AND gate. Its inputs are f_1 and f_2. Therefore, the output is given by the expression

$$F = \bar{A}\bar{B}$$

and is 1 for both f_1 and f_2 equal to 1. This is shown in the last column of the truth table.

From the truth table of circuit operation, we see that this combinational logic circuit works like a single NOR gate.

EXAMPLE 5.3

Form a truth table for the combinational logic network in Fig. 5.5(a).

SOLUTION:

The output of gate 1 is $f_1 = \bar{B}$, and its value in the truth table of Fig. 5.5(b) is 1 when the B input is 0. Gate 2 inverts the C input and gives $f_2 = \bar{C}$. Therefore, the f_2 output in the table is 1 for C equal to 0.

Now, the third level OR gate is considered. The output of gate 3 is $f_3 = A + C$ and is 0 for both A and C equal to 0 together.

Then, the output of gate 4 in the second logic level is evaluated. The expression for f_4 is f_3B or $(A + C)B$. This output is 1 when f_3 and B are both 1.

Next, the AND gate output f_5 is to be found. This is given by the expression $f_5 = f_1 f_2$ or $f_5 = \bar{B}\bar{C}$ and is 1 if both f_1 and f_2 are equal to 1.

Finally, the output F is found from the first level OR gate. This gives the network Boolean equation that follows:

$$F = f_4 + f_5 = (A + C)B + \bar{B}\bar{C}$$

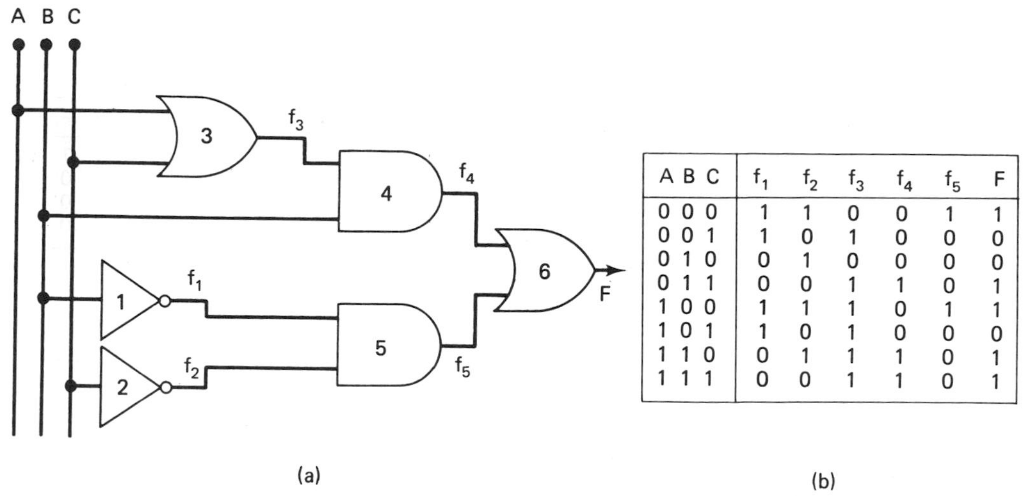

A B C	f_1	f_2	f_3	f_4	f_5	F
0 0 0	1	1	0	0	1	1
0 0 1	1	0	1	0	0	0
0 1 0	0	1	0	0	0	0
0 1 1	0	0	1	1	0	1
1 0 0	1	1	1	0	1	1
1 0 1	1	0	1	0	0	0
1 1 0	0	1	1	1	0	1
1 1 1	0	0	1	1	0	1

Figure 5.5 (a) Logic diagram; (b) Truth table.

So the circuit output is 1 if f_4 or f_5 is in the 1 logic state. This completes the truth table of Fig. 5.5(b).

5.6 NAND-NOT GATE EXAMPLES

Having shown how the truth table analysis technique is used to find the operation of an AND-OR-NOT logic circuit, we shall extend its use to NAND-NOT circuitry.

EXAMPLE 5.4

Use truth table analysis to find the operation of the NAND-NOT logic circuitry in Fig. 5.6(a).

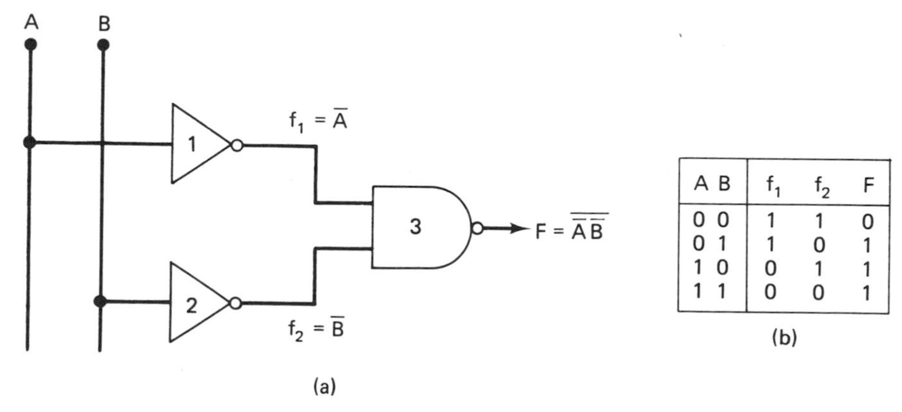

A B	f_1	f_2	F
0 0	1	1	0
0 1	1	0	1
1 0	0	1	1
1 1	0	0	1

Figure 5.6 (a) Logic diagram; (b) Truth table.

SOLUTION:

The f_1 output is the NOT function of the A input:

$$f_1 = \bar{A}$$

Its value is 1 when A is 0. This is shown in the third column of the table of Fig. 5.6(b).

The output of the other inverter is the complement of the B input:

$$f_2 = \bar{B}$$

So its logic state is 1 for $B = 0$ in the truth table.

The network output is given by the NAND expression

$$F = \overline{f_1 f_2} = \overline{\bar{A}\bar{B}}$$

and the circuit output is 0 when both the f_1 and f_2 inputs are in the 1 state together.

Looking at the truth table, we see that a NAND gate with both inputs inverted results in OR gate operation.

EXAMPLE 5.5

Form the truth table of circuit operation for the gate network in Fig. 5.7(a). How does the operation of this circuit compare to that in Fig. 5.2(a)?

SOLUTION:

First the complemented input $f_1 = \bar{B}$ is found to be 1 when input $B = 0$. This is entered in the f_1 column of the truth table in Fig. 5.7(b).

Next, output $f_2 = \overline{AB}$ is found from the A and B inputs. When A and B are in the 1 logic state together, f_2 is 0, as shown in the table.

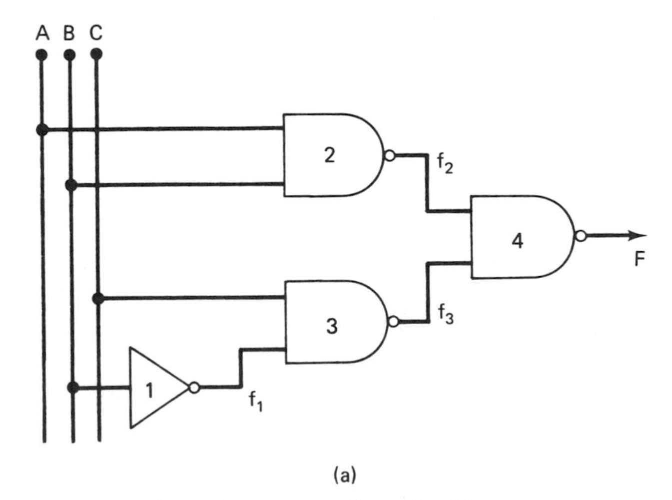

A B C	f_1	f_2	f_3	F
0 0 0	1	1	1	0
0 0 1	1	1	0	1
0 1 0	0	1	1	0
0 1 1	0	1	1	0
1 0 0	1	1	1	0
1 0 1	1	1	0	1
1 1 0	0	0	1	1
1 1 1	0	0	1	1

(b)

Figure 5.7 (a) Logic diagram; (b) Truth table.

Then f_3 is given by the expression $f_3 = \overline{f_1 C}$ or $\overline{\bar{B}C}$. So it is 0 for both $f_1 = 1$ and $C = 1$.

In the same way, the network output is given by the expression

$$F = \overline{f_2 f_3} = \overline{(\overline{AB})(\overline{\bar{B}C})}$$

This makes output $F = 0$ when both f_2 and f_3 are 1, as shown in the table.

Comparing the output of this circuit to that of the AND-OR-NOT circuit in Fig. 5.2(a), we see that the output F is the same for both. However, the logic states of all gates within the network are not the same.

EXAMPLE 5.6

In Fig. 5.8 a NAND-NOT logic network is shown, and input logic levels are indicated. Find the logic level at gate outputs f_1, f_2, and f_3. What is the network output F?

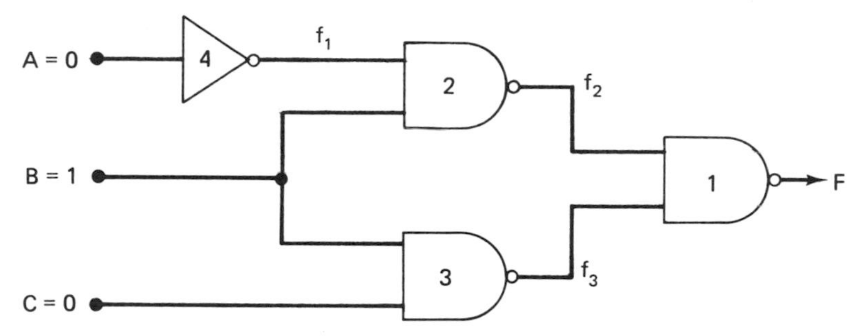

Figure 5.8 Logic network with known inputs.

SOLUTION:

The output f_1 is the A input complemented. Since the A input is at logic 0, the f_1 output is binary 1:

$$f_1 = 1$$

Now the inputs to gate 2 are known. B and f_1 are its inputs, and both are logic 1. This makes the output f_2 binary 0:

$$f_2 = 0$$

The other gate in the second level has inputs B and C. Here, the B input is 1, and C equals 0. Therefore, output f_3 is binary 1:

$$f_3 = 1$$

The output F of the network is determined by the logic level of inputs f_2 and f_3 on gate 1. Input f_2 is logic 0, and f_3 is logic 1. For this reason, output F is 1:

$$F = 1$$

5.7 NOR-NOT GATE EXAMPLES

The last type of circuitry that can be analyzed by the truth table method is NOR-NOT combinational logic networks. Here again examples will be used to illustrate the technique.

EXAMPLE 5.7

By truth table analysis, find the operation of the NOR-NOT circuitry in Fig. 5.9(a). How does this logic combination work compared to the AND-OR-NOT and NAND-NOT logic networks of Figs. 5.2(a) and 5.7(a), respectively.

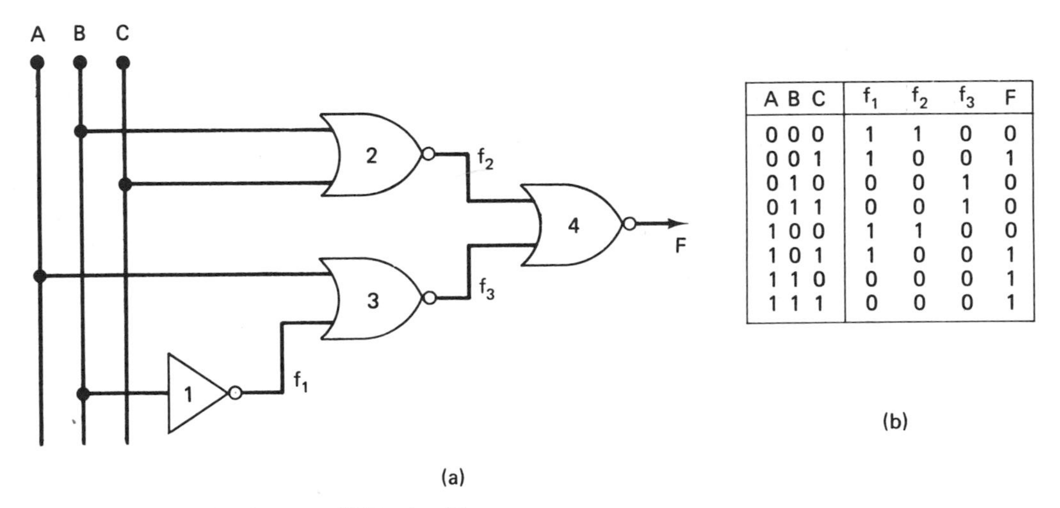

A B C	f_1	f_2	f_3	F
0 0 0	1	1	0	0
0 0 1	1	0	0	1
0 1 0	0	0	1	0
0 1 1	0	0	1	0
1 0 0	1	1	0	0
1 0 1	1	0	0	1
1 1 0	0	0	0	1
1 1 1	0	0	0	1

Figure 5.9 (a) Logic diagram; (b) Truth table.

SOLUTION:

The output of the inverter is $f_1 = \bar{B}$ and is 1 when the B input is 0. This is recorded in the truth table of Fig. 5.9(b).

Now the output of NOR gate 2 is found as $f_2 = \overline{B + C}$ and is in the 1 state for both B and C equal 0.

For gate 3, the output is $f_3 = \overline{A + f_1}$ or $\overline{A + \bar{B}}$. So its output is 1 when A and f_1 are both 0.

Finally, the F output is found from gate 4:

$$F = \overline{f_2 + f_3} = \overline{\overline{(B + C)} + \overline{(A + \bar{B})}}$$

In this way, the output is found to be 1 if f_2 and f_3 are both equal to 0.

Comparing the output of the NOR-NOT network to that of the earlier AND-OR-NOT and NAND-NOT combinations, we see that the output F is the same in each.

EXAMPLE 5.8

Assume that the A, B, and C waveforms in Fig. 5.10 are input signals to the NOR-NOT logic network in Fig. 5.9(a). Use the network truth table in Fig. 5.9(b) to determine the waveform at output F.

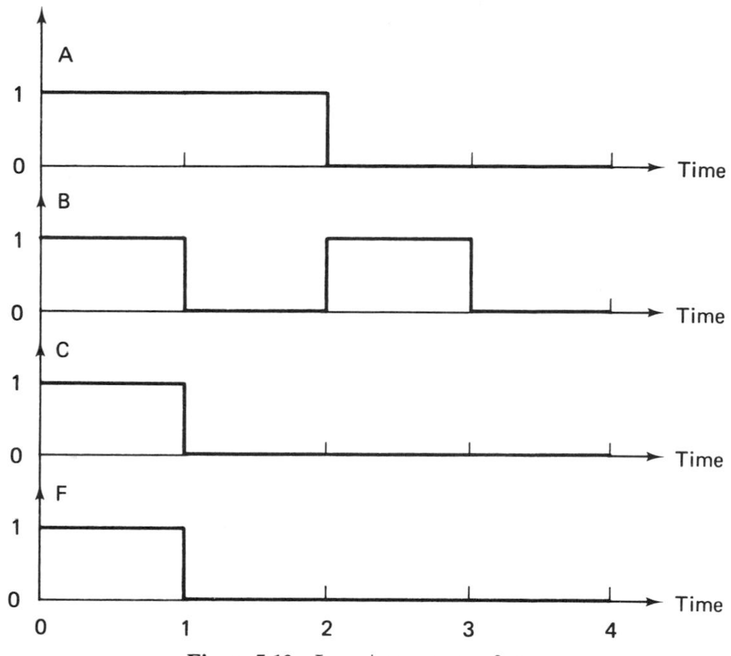

Figure 5.10 Input/output waveforms.

SOLUTION:

During the time interval from 0 to 1, all inputs are logic 1:

$$ABC = 111$$

Looking at the truth table in Fig. 5.9(b), we see that the output F is 1 for this input condition:

$$F = 1$$

In the next interval, 1–2, the A input is 1, and both B and C are 0. This input state is

$$ABC = 100$$

The output in the table corresponding to this input condition is a logic 0:

$$F = 0$$

The third set of inputs is $ABC = 010$, and the output under this condition is again 0:

$$ABC = 010$$
$$F = 0$$

For the last time interval, all inputs are 0, and the output stays at the 0 logic level:

$$ABC = 000$$
$$F = 0$$

These output states are used to form the F waveform shown in Fig. 5.10.

5.8 INTEGRATED CIRCUIT COMBINATIONAL LOGIC NETWORK

The gate circuitry for a combinational logic network can be constructed using the basic TTL logic gates introduced in Chapter 4. For example, the NAND-NOT logic circuit in Fig. 5.11(a) can be built using just two integrated circuits. To do this, we need one inverter gate and three two-input NAND gates. So the complete circuit can be made with one 7404 hex-inverter device and a quadruple two-input NAND gate like the 7400.

However, the circuitry can be further simplified by making the inverter from a NAND gate. This hookup is shown in the logic diagram of Fig. 5.11(b). Here the inputs of one NAND gate are shorted together to make an inverter. In this way, the complete logic network can be constructed with one 7400 IC.

5.9 WIRE-AND CONNECTION

A widely used gate network made with open-collector gates is the *wire-AND* connection. In Fig. 5.12(a), a simple wire-AND gate combination is shown. Here we see that the outputs of two open-collector AND gates are tied to a single pull-up resistor R.

The output of this circuit is the AND operation of the four inputs A, B, C, and D. It is given by the equation

$$F = (AB) \cdot (CD) = ABCD$$

In this way, we find that wiring the outputs together produces the AND operation of the gates connected together. One advantage of this gate combination is that it can be used to create a multi-input AND gate.

By wiring the two outputs together, point F cannot switch to logic 1 until both outputs are logic 1. For instance, if the input is $ABCD = 1110$, the output of gate 1 should be 1 and that of gate 2, 0. However, the 0 output of gate 2 holds F at logic 0.

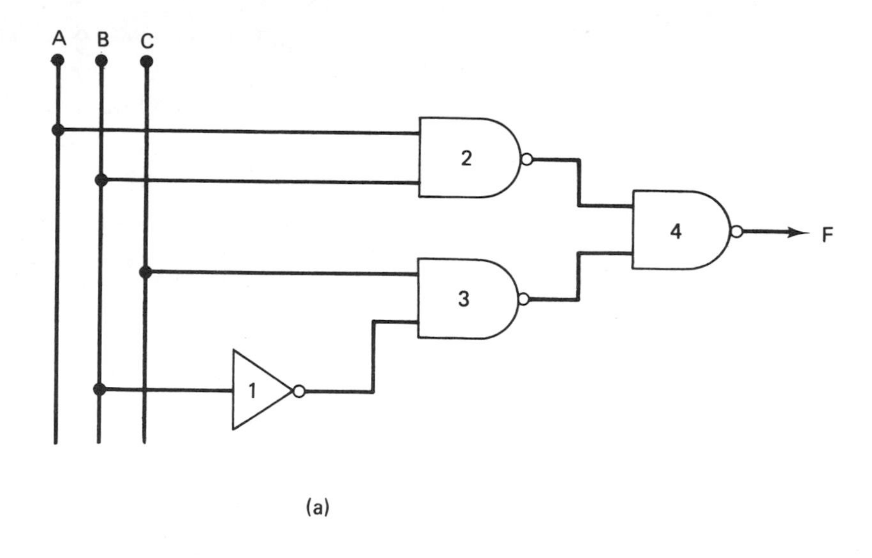

(a)

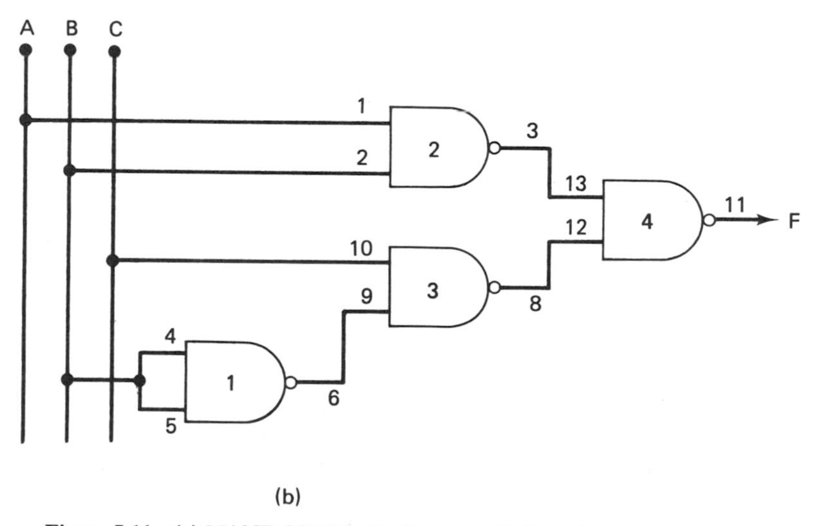

(b)

Figure 5.11 (a) NAND-NOT logic diagram; (b) Complete NAND circuit.

For output F to be logic 1, the input must be $ABCD = 1111$. In this case, the outputs of gate 1 and gate 2 both switch to logic 1. Therefore, the network output F is also 1. This is the one input combination for which the output equals 1 and operation is consistent with that of a four-input AND gate.

In a schematic or logic diagram, the symbol of Fig. 5.12(b) is used to indicate a wire-AND connection.

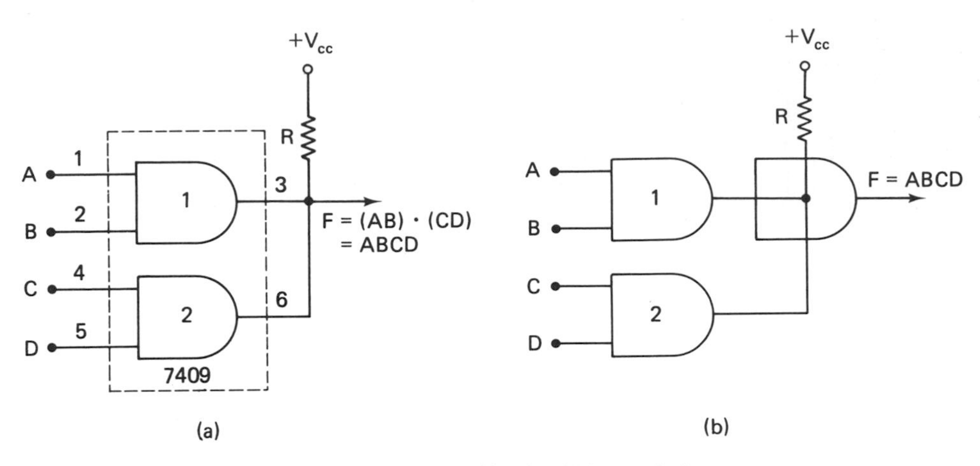

Figure 5.12 (a) Wire-AND gate connection; (b) Wire-AND symbol.

ASSIGNMENTS

Section 5.2

1. Define a combinational logic network.
2. What three types of logic gate circuitry can be used to design combinational logic circuitry?
3. Give the names of four standard combinational logic circuits available as integrated circuits.

Section 5.3

4. Mark the levels of logic into the logic gate diagram of Fig. 5.13(a). Then

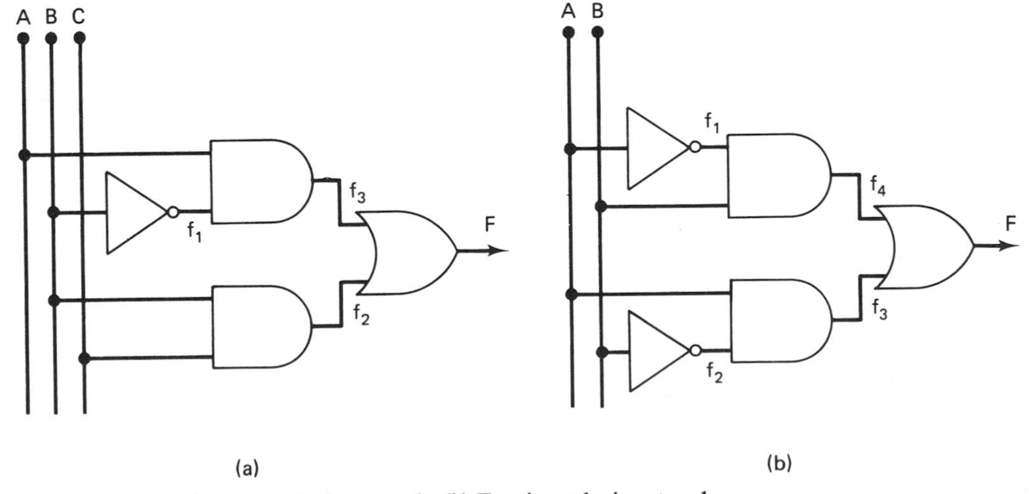

Figure 5.13 (a) Three-input logic network; (b) Two-input logic network.

write the Boolean equation for outputs f_1, f_2, f_3, and F in terms of the inputs A, B, and C and their complements.

5. For the logic diagram in Fig. 5.13(b), indicate the different levels of logic and write a Boolean expression for the output of each gate.

Sections 5.4 and 5.5

6. Use truth table analysis to form a truth table of circuit operation for the logic gate connection in Fig. 5.14. Describe the operation of the network.

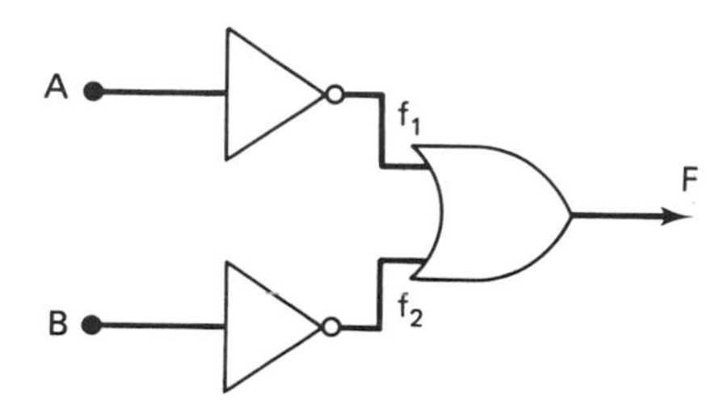

Figure 5.14 OR-NOT logic diagram.

7. Perform a truth table analysis for the logic diagram of Fig. 5.13(a).

8. Form a truth table of the logic circuitry in Fig. 5.13(b).

Section 5.6

9. Construct a truth table for the NAND-NOT logic circuit in Fig. 5.15(a).

10. What is the logic level at points f_1, f_2, f_3, f_4, and F in the network of Fig. 5.15(b)? The input is $ABCD = 1000$.

Section 5.7

11. Perform a truth table analysis of the NOR-NOT logic diagram in Fig. 5.16(a). Describe the operation of the gate combination.

12. Find the outputs f_1, f_2, f_3, and F in the logic network of Fig. 5.16(b). The input is $ABC = 101$.

13. The A and B waveforms in Fig. 5.17 are applied to inputs of the network in Fig. 5.16(a). Draw the output waveform.

Section 5.8

14. Redraw the logic diagram in Fig. 5.15(b) using only NAND gates.

15. Redraw the logic diagram of Fig. 5.16(a) after replacing the inverters with NOR gates connected as inverters.

16. Redraw the logic network in Fig. 5.16(b) using only NOR gates.

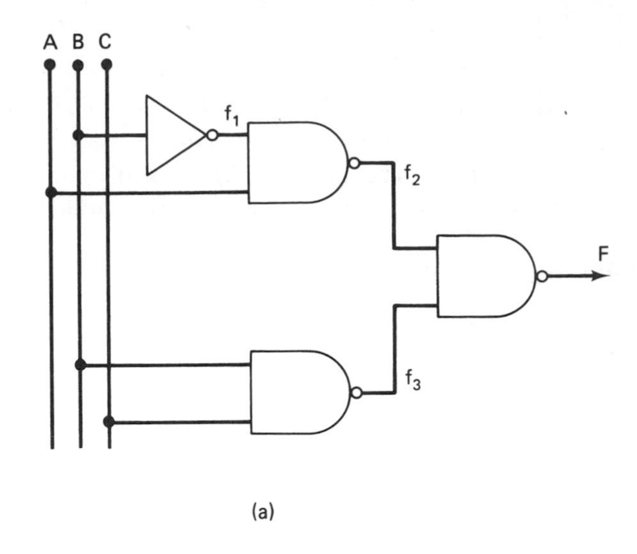

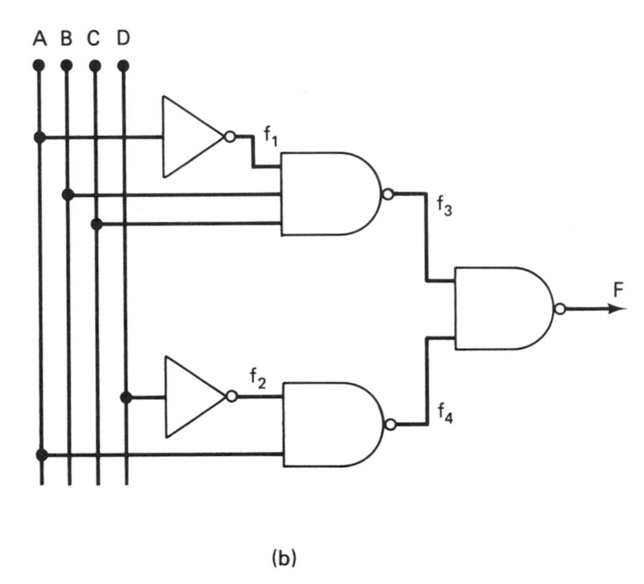

Figure 5.15 (a) Three-input NAND-NOT logic network; (b) Four-input NAND-NOT logic network.

Section 5.9

17. Draw the wire-AND connection of two four-input AND gates.

18. In Fig. 5.18, two two-input NAND gates are wired together in the wire-AND configuration. Make a truth table of the circuit's operation.

19. Redraw the circuit in Fig. 5.18 using the wire-AND symbol at the output.

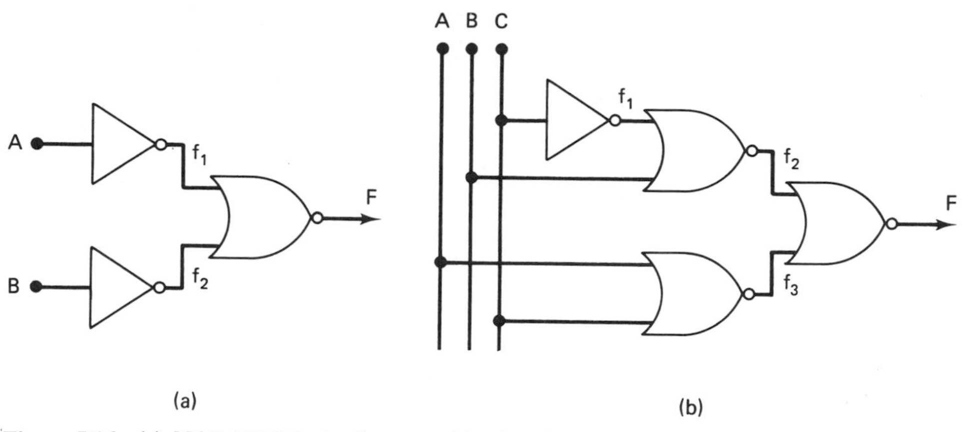

Figure 5.16 (a) NOR-NOT logic diagram; (b) Three-input NOR-NOT logic diagram.

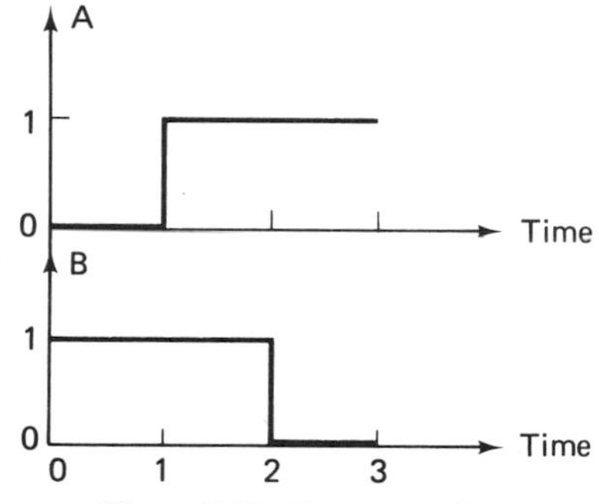

Figure 5.17 Input waveforms.

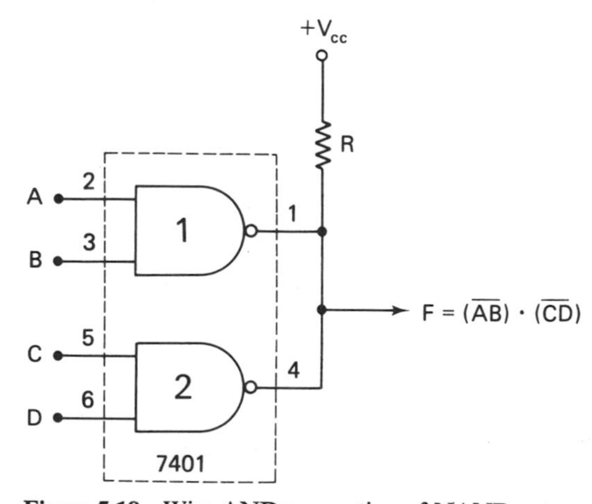

Figure 5.18 Wire-AND connection of NAND gates.

Implementing Logic Circuitry

6

6.1 INTRODUCTION

In Chapter 5, the truth table analysis technique was used to find the output operation of a logic network. Moreover, we illustrated how the identical output function *F* could be obtained with three different types of logic circuits: the AND-OR-NOT network, the NAND-NOT network, and the NOR-NOT network.

Here we shall do the reverse of the approach introduced in Chapter 5. Using the truth table of a logic function, we shall write a Boolean equation describing the operation. From this equation, we can make logic diagrams showing how to set up the network with different types of logic circuitry.

For this purpose, the topics listed below are included in this chapter:

1. The sum-of-products and product-of-sums equations
2. Implementing with AND-OR-NOT logic
3. Implementing with NAND-NOT logic
4. Implementing with NOR-NOT logic
5. The AND-OR-invert gate integrated circuit

6.2 THE TRUTH TABLE OF A LOGIC NETWORK

The first step needed to implement a logic network is to define its operation with a truth table. To do this, we begin by identifying the inputs of the circuit and label them with Boolean variables such as *A*, *B*, *C*, and *D*.

After this, the output function F is defined for each binary combination of the inputs. To tell how F will respond to the different sets of inputs, the operation of the logic network must be described.

As an illustration of this idea, let us look at the block diagram in Fig. 6.1(a). Here we see three digital circuits. The upper circuit produces a signal called A, and the lower circuit makes a second signal marked B. These signals are used as the A and B inputs of a logic network. The output of the logic network is called F.

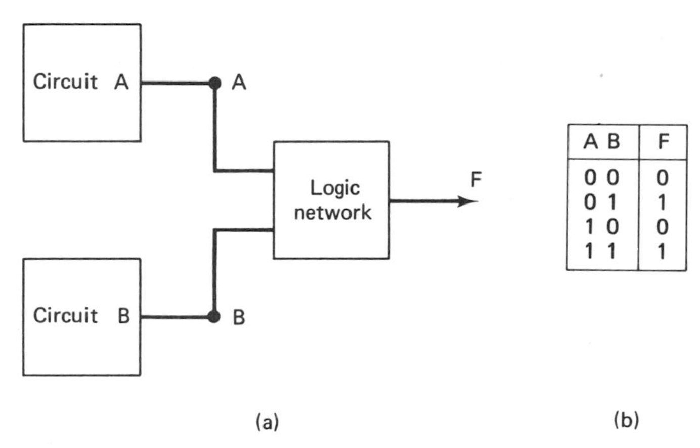

A B	F
0 0	0
0 1	1
1 0	0
1 1	1

Figure 6.1 (a) Block diagram; (b) Truth table.

We want to make the output of the logic network signal a logic 1 for two input binary combinations. The first condition is when just the B input is at the 1 logic level, and the other is for both A and B equal 1 together. This completes the description of operation for the network, and it remains to form a truth table.

A truth table showing the operation of our logic network is shown in Fig. 6.1(b). From this table, we see that inputs A and B are listed along with their four binary combinations. The output under column F is made 1 for the two specified conditions. The first case is the binary input $AB = 01$. This corresponds to input B at the 1 logic level alone. The other input combination that makes the output switch to logic 1 is $AB = 11$, and it corresponds to the state where both inputs are 1 at the same time. The output is in the binary 0 state for the other two input conditions.

Once this table is formed, a Boolean equation can be written to give a mathematical description of operation for the logic network. With this equation, a logic diagram can be drawn to show the circuitry needed to implement the network.

6.3 SUM-OF-PRODUCTS BOOLEAN EQUATION

In the last section, we showed how the truth table of a logic network can be formulated. With this table, we can write two different Boolean equations for the output. These expressions are called the *sum-of-products* and *product-of-sums* equations. Both of these equations correctly describe the truth table and network operation.

To implement a network with AND-OR-NOT logic circuitry, we can use either the sum-of-products or product-of-sums equations. However, the sum-of-products equation is needed when using NAND-NOT logic, and the product-of-sums equation is used for NOR-NOT gate networks.

Product Terms

A *product term* can be set up for each input binary combination in a truth table. To make a product term, we use the AND function of the input Boolean variables. If an input is binary 0, the corresponding variable in the product term is complemented.

The four product terms of a two-input truth table are listed in Fig. 6.2(a). Here, the first product term is for the condition of both A and B equal 0. Since both inputs are binary 0, A and B are complemented to give the product term $\bar{A}\bar{B}$. The result is read as A bar AND B bar.

A B	Product terms
0 0	$\bar{A}\,\bar{B}$
0 1	$\bar{A}\,B$
1 0	$A\,\bar{B}$
1 1	$A\,B$

(a)

A B	F	Product terms
0 0	0	
0 1	1	$\bar{A}\,B$
1 0	0	
1 1	1	$A\,B$

(b)

Figure 6.2 (a) All two-variable product terms; (b) Product terms of a logic function.

The next input combination is A equals 0 and B equals 1. For this case, only A is 0 and is written with a bar in the product term. On the other hand, the third term has A equals 1 and B is 0. This gives a product term made with the AND function of A and $\bar{B}$.

Last, the term for both inputs equal 1 is written. This input condition is described by the AND of variable A with B.

In these product terms, a Boolean variable expressed without a bar indicates that it has a binary value of 1. On the other hand, a variable with a bar represents a binary 0.

EXAMPLE 6.1

Make a chart of product terms like that in Fig. 6.2(a) for a three-input truth table.

SOLUTION:

For a three-input table, the inputs are marked A, B, and C. In Fig. 6.3, 3-bit binary words from 000 to 111 are listed. Next to each input state the corresponding product term is written. Each term is formed with the AND function

A B C	Product terms
0 0 0	$\bar{A}\,\bar{B}\,\bar{C}$
0 0 1	$\bar{A}\,\bar{B}\,C$
0 1 0	$\bar{A}\,B\,\bar{C}$
0 1 1	$\bar{A}\,B\,C$
1 0 0	$A\,\bar{B}\,\bar{C}$
1 0 1	$A\,\bar{B}\,C$
1 1 0	$A\,B\,\bar{C}$
1 1 1	$A\,B\,C$

Figure 6.3 Three-input product terms.

of inputs A, B, and C. Variables expressed with a bar correspond to inputs that are binary 0.

The table begins with $ABC = 000$ and the product term $\bar{A}\bar{B}\bar{C}$. For the last input combination $ABC = 111$; the product term is ABC.

The Sum-of-Products Equation

Having introduced product terms, we are now ready to write a sum-of-products equation from a truth table. To do this, the OR function is used to combine the product terms for which the output function F is binary 1.

As an example, let us write the sum-of-products equation for the truth table in Fig. 6.1(b). This table is rewritten in Fig. 6.2(b) along with the product terms for the two input conditions where output F equals logic 1.

Now, the sum-of-products Boolean equation is formed by combining these two product terms with the OR function. The result is the equation

$$F = \bar{A}B + AB$$

In this way, we see that the truth table output F is expressed as the sum of two product terms.

EXAMPLE 6.2

Write a sum-of-products Boolean equation for the logic function indicated by the truth table in Fig. 6.4.

A B C	F	Product terms
0 0 0	0	
0 0 1	1	$\bar{A}\,\bar{B}\,C$
0 1 0	0	
0 1 1	0	
1 0 0	1	$A\,\bar{B}\,\bar{C}$
1 0 1	1	$A\,\bar{B}\,C$
1 1 0	1	$A\,B\,\bar{C}$
1 1 1	1	$A\,B\,C$

Figure 6.4 Three-input logic function with product terms.

SOLUTION:

This function F has five input states for which the output is logic 1. Product terms for these five states are listed next to the output in Fig. 6.4. Combining these terms with the OR function, we get

$$F = \bar{A}\bar{B}C + A\bar{B}\bar{C} + A\bar{B}C + AB\bar{C} + ABC$$

6.4 IMPLEMENTING WITH AND-OR-NOT LOGIC

Up to this point, we have shown how to form a truth table of a logic function and write a sum-of-products equation for the output. Our next step is to draw a logic diagram of the gate circuitry needed to implement the Boolean expression.

When implementing a sum-of-products equation with AND-OR-NOT logic circuitry, just two levels of logic gates are required. The first level is always an OR gate. This gate must have one input for each product term in the Boolean equation.

On the other hand, the second level of logic has one AND gate for each product term. The number of inputs on each of these gates is the same as the number of inputs to the network. Inverters are required to provide complemented inputs.

The procedure followed to set up a logic diagram from a known Boolean equation is outlined as follows:

1. Determine the number of levels of logic.
2. How many and what type of gates are needed in each level?
3. Find the number of inputs on each gate.
4. Draw the logic diagram.

To illustrate this procedure, let us construct a logic diagram for the equation $F = \bar{A}B + AB$. This equation uses two input variables A and B. Besides this, it has two sum terms. So the first level of logic has a two-input OR gate. This gate is marked 1 in the logic diagram of Fig. 6.5 and its output is labeled F.

On the other hand, the second level of logic needs two two-input AND gates. These gates are labeled 2 and 3 in the logic diagram. Moreover, one inverter is needed to provide the $\bar{A}$ input. This complemented input is made by gate 4.

To check the network just formed, we can write its network Boolean equation. We begin with the output of the inverter gate. Its output is given by the expression

$$f_1 = \bar{A}$$

Next, equations for the second level output functions f_2 and f_3 can be written with the AND operation:

$$f_2 = \bar{A}B$$
$$f_3 = AB$$

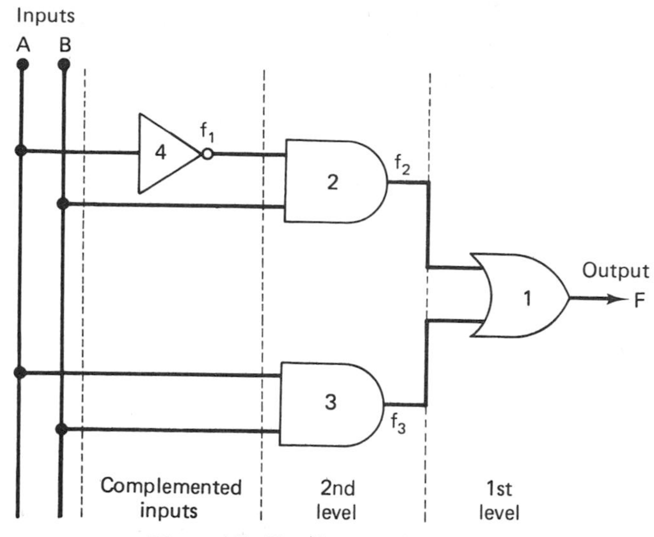

Figure 6.5 Two-input logic network.

Finally, the output F of the network is formed. Using the OR function, we get

$$F = f_2 + f_3 = \bar{A}B + AB$$

Comparing this output equation to the original sum-or-products equation, we see that they are identical. In this way, we find that the circuit in Fig. 6.5 is correct for the given Boolean function.

To determine if the logic diagram does provide the original truth table in Fig. 6.1(b), we can use the truth table analysis technique. By truth table analysis, the binary states of the output can be found and compared to the output in the original truth table.

EXAMPLE 6.3

Construct an AND-OR-NOT logic diagram for the sum-of-products equation $F = \bar{A}\bar{B}C + A\bar{B}\bar{C} + A\bar{B}C + AB\bar{C} + ABC$. The Boolean expression was written in Example 6.2.

SOLUTION:

To implement this equation, the first level of logic has one five-input OR gate. This gate is labeled 1 in the logic diagram of Fig. 6.6.

In the second level, there must be five three-input AND gates. These gates are marked 2 through 6 in the logic diagram.

Since each input appears in complemented form, three inverters are needed. The inverters are labeled gate 7 through gate 9.

EXAMPLE 6.4

The truth table of a four-input logic function is shown in Fig. 6.7(a). Write a sum-of-products Boolean equation for its output and then draw the AND-OR-NOT logic circuitry needed to implement the function.

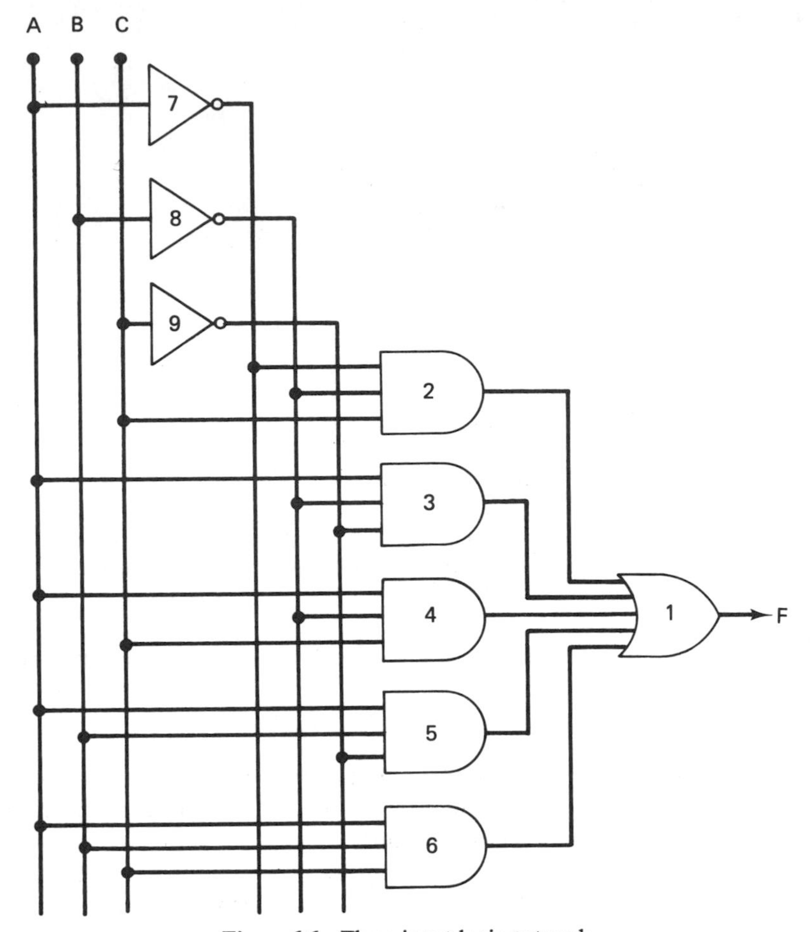

Figure 6.6 Three-input logic network.

SOLUTION:

The output F is 1 for just two logic states. These two states correspond to the product terms $A\bar{B}CD$ and $AB\bar{C}D$. Using the OR function, the sum-of-products equation is

$$F = A\bar{B}CD + AB\bar{C}D$$

To implement this logic function the first level in the logic diagram needs a two-input OR gate. However, the second level uses two four-input AND gates. Furthermore, inverters are required to make the $\bar{B}$ and $\bar{C}$ complemented inputs. The logic diagram for the equation is shown in Fig. 6.7(b).

6.5 IMPLEMENTING WITH NAND-NOT LOGIC

In the last section, we implemented a Boolean equation with AND-OR-NOT circuitry. Here, the same type of expression will be implemented with NAND-NOT logic.

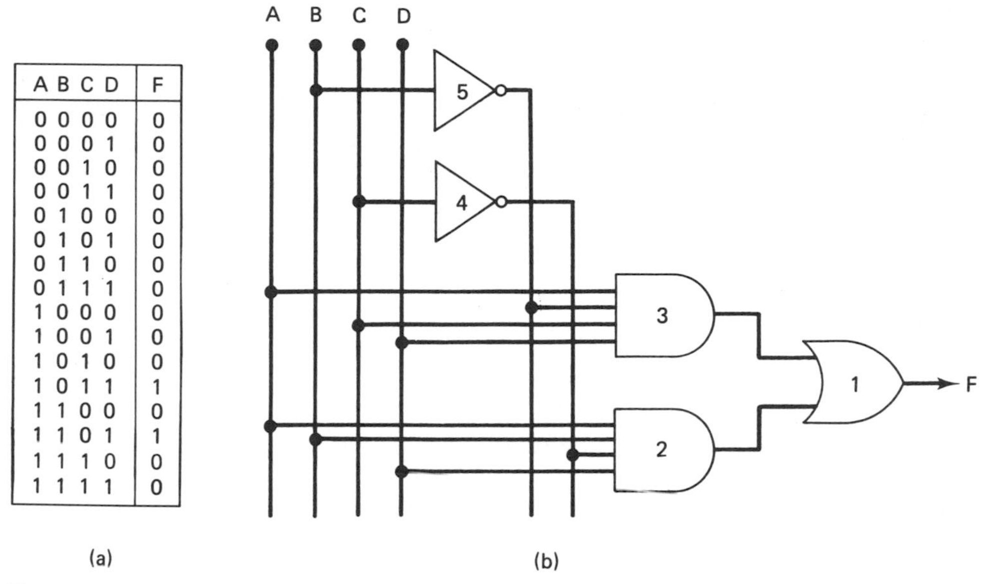

A B C D	F
0 0 0 0	0
0 0 0 1	0
0 0 1 0	0
0 0 1 1	0
0 1 0 0	0
0 1 0 1	0
0 1 1 0	0
0 1 1 1	0
1 0 0 0	0
1 0 0 1	0
1 0 1 0	0
1 0 1 1	1
1 1 0 0	0
1 1 0 1	1
1 1 1 0	0
1 1 1 1	0

Figure 6.7 (a) Four-input truth table; (b) Logic network.

To do this, the procedure introduced for AND-OR-NOT circuitry is again followed. For a standard sum-of-products equation, two levels of logic gates are needed; however, in this case NAND gates are used in both levels. Besides this, inverters are required to produce complemented inputs. These NAND and NOT gates are connected in the same way as done for AND-OR-NOT logic.

As an example, let us set up the logic circuitry needed to implement the truth table in Fig. 6.2(b). Earlier, the sum-of-products equation for output F was written for this table. The result was the expression

$$F = \bar{A}B + AB$$

From this equation, we notice that the network has two inputs A and B. Moreover, the expression has two product terms. So the first level of logic needs a two-input NAND gate. This is gate 1 in the logic diagram of Fig. 6.8(a).

In addition, the second level requires two two-input NAND gates. These gates are marked 2 and 3 in the logic diagram. Last, one inverter is needed for the complemented input $\bar{A}$.

An AND-OR-NOT logic network was constructed for the same Boolean equation in Fig. 6.5. The NAND-NOT network we just set up in Fig. 6.8(a) will produce the same output F as this earlier AND-OR-NOT logic diagram. This fact can be checked by the truth table analysis technique.

The network truth table of our NAND-NOT circuit is shown in Fig. 6.8(b). In this table, we see that the f_1 output is the opposite logic state of the A input.

Next, the output f_2 of NAND gate 2 is found from the f_1 and B columns. Signal f_2 is logic 1 unless both f_1 and B are binary 1 together.

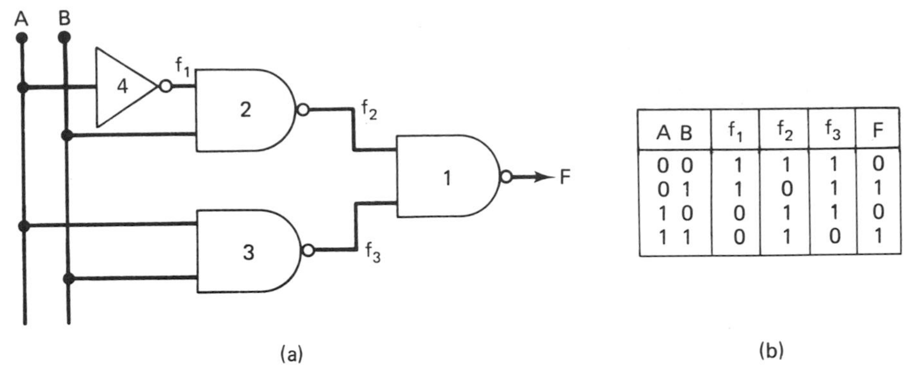

A B	f_1	f_2	f_3	F
0 0	1	1	1	0
0 1	1	0	1	1
1 0	0	1	1	0
1 1	0	1	0	1

Figure 6.8 (a) NAND-NOT logic network; (b) Network truth table.

The output of gate 3 is called f_3, and it is caused by the A and B inputs. This output is 0 when both A and B are 1 at the same time.

F is the output of the logic network. The binary state of F is found from the logic states in the f_2 and f_3 columns of the table. The F output is 0 when both f_2 and f_3 are 1. Otherwise, the output is at the 1 logic level.

To decide if the circuit is working correctly, we can compare output F in the network truth table of Fig. 6.8(b) to the output in the truth table of Fig. 6.2(b). Doing this, we see that both output functions are exactly the same.

EXAMPLE 6.5

Write a sum-of-products Boolean equation for the logic function in Fig. 6.9(a). Then construct a logic diagram of the NAND-NOT circuitry needed to implement the expression.

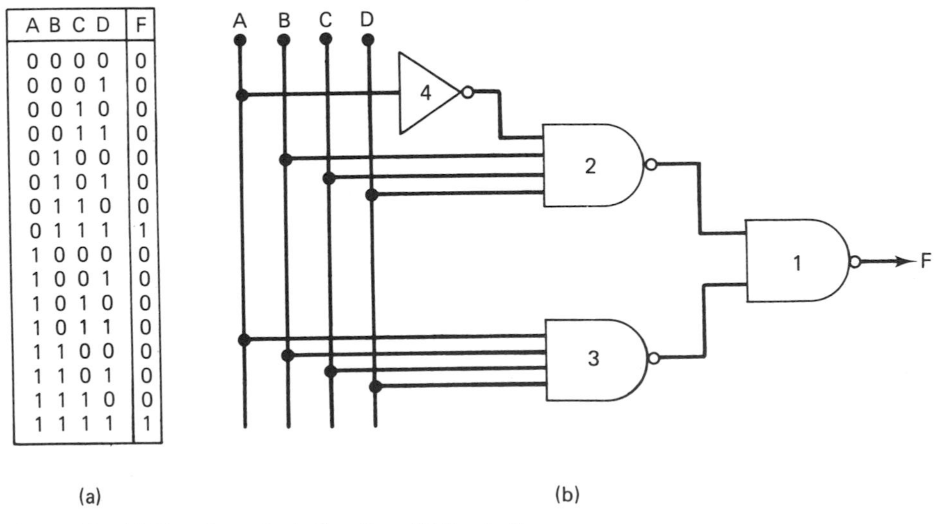

A B C D	F
0 0 0 0	0
0 0 0 1	0
0 0 1 0	0
0 0 1 1	0
0 1 0 0	0
0 1 0 1	0
0 1 1 0	0
0 1 1 1	1
1 0 0 0	0
1 0 0 1	0
1 0 1 0	0
1 0 1 1	0
1 1 0 0	0
1 1 0 1	0
1 1 1 0	0
1 1 1 1	1

Figure 6.9 (a) Four-input logic function; (b) Logic diagram.

SOLUTION:

In the truth table of Fig. 6.9(a), we observe that the output F is 1 for two input conditions. The product terms for these logic states are $\bar{A}BCD$ and $ABCD$. Writing the sum-of-products equation, we get

$$F = \bar{A}BCD + ABCD$$

To implement this expression with NAND-NOT circuitry, we need one two-input NAND gate, two four-input NANDs, and a single inverter. The two-input NAND is in the first level, as shown in the logic diagram of Fig. 6.9(b). In the second level, the two four-input NAND gates are used, and the inverter makes the $\bar{A}$ complemented input.

6.6 PRODUCT-OF-SUMS BOOLEAN EQUATION

Another approach to implementing logic circuitry is to begin with a product-of-sums Boolean equation instead of the sum-of-products expression. Here we shall show how to write sum terms and use them to form a product-of-sums equation. Besides this, the product-of-sums equation will be implemented with AND-OR-NOT circuitry.

Sum Terms

As with the product term, we can write a *sum term* for each combination of binary inputs in a truth table. When making a sum term, the OR function of each input is used. If the input logic state is binary 1, the corresponding Boolean variable is complemented.

The sum terms of the four input states in a two-variable truth table are listed in Fig. 6.10(a). In this table, the first term is written as A OR B because both inputs are binary 0. On the other hand, the B input is binary 1 in the second input condi-

A B	Sum terms
0 0	$A + B$
0 1	$A + \bar{B}$
1 0	$\bar{A} + B$
1 1	$\bar{A} + \bar{B}$

(a)

A B C	Sum terms
0 0 0	$A + B + C$
0 0 1	$A + B + \bar{C}$
0 1 0	$A + \bar{B} + C$
0 1 1	$A + \bar{B} + \bar{C}$
1 0 0	$\bar{A} + B + C$
1 0 1	$\bar{A} + B + \bar{C}$
1 1 0	$\bar{A} + \bar{B} + C$
1 1 1	$\bar{A} + \bar{B} + \bar{C}$

(b)

Figure 6.10 (a) Two-variable sum terms; (b) Three-variable sum terms.

tion. For this reason, the corresponding sum term requires B written with a bar. The remaining two terms are formed in a similar way.

The eight sum terms of a three-input logic function are shown in the table of Fig. 6.10(b).

The Product-of-Sums Equation

Using sum terms, we can write the product-of-sums Boolean equation of a logic function expressed by a truth table. For this purpose, the AND function is used to combine the sum terms for which the output logic level at F is 0. This is the opposite of the approach used to form a sum-of-products equation.

EXAMPLE 6.6

Form a product-of-sums Boolean equation for the truth table in Fig. 6.11(a).

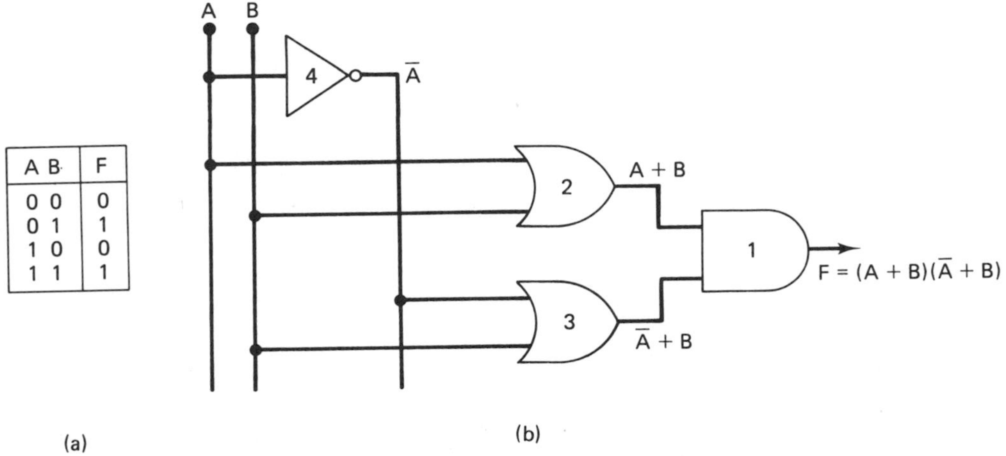

A B	F
0 0	0
0 1	1
1 0	0
1 1	1

Figure 6.11 (a) Two-input logic function; (b) AND-OR-NOT logic diagram.

SOLUTION:

The output F is binary 0 for two input conditions. These conditions have the sum terms $A + B$ and $\bar{A} + B$. Using these sum terms and the AND function, we obtain the product-of-sums equation:

$$F = (A + B)(\bar{A} + B)$$

AND-OR-NOT Logic Network

A product-of-sums Boolean equation can be implemented with AND-OR-NOT logic circuitry. To do this, the first level of logic needs a single AND gate. This gate is followed by OR gates in the second level. Inverters are used to form complemented inputs.

EXAMPLE 6.7

Construct the AND-OR-NOT logic diagram for the Boolean equation $F = (A + B)(\bar{A} + B)$. This equation was written in Example 6.6.

SOLUTION:

In this expression, there are two sum terms combined with the AND function. So the first level of gate circuits has a single two-input AND gate. This is shown in the logic diagram of Fig. 6.11(b).

The sum terms are formed with two input variables A and B. For this reason, the second level of logic in Fig. 6.11(b) contains two two-input OR gates.

One inverter is needed to provide the $\bar{A}$ complemented input.

6.7 IMPLEMENTING WITH NOR-NOT LOGIC

Another way of implementing a product-of-sums Boolean function is with NOR-NOT circuitry. For this, we again follow the same procedure. However, all gates in the first and second levels of logic are the NOR type instead of AND and OR gates. But inverters are still needed to provide complemented inputs.

For an example of NOR-NOT circuitry, let us once again take the Boolean function $F = (A + B)(\bar{A} + B)$. To implement this logic expression, two levels of NOR gates are needed, and the $\bar{A}$ complemented input must be produced by an inverter.

This equation has two sum terms. Therefore, the first level of logic contains just one two-input NOR gate. This gate is shown in the logic diagram of Fig. 6.12.

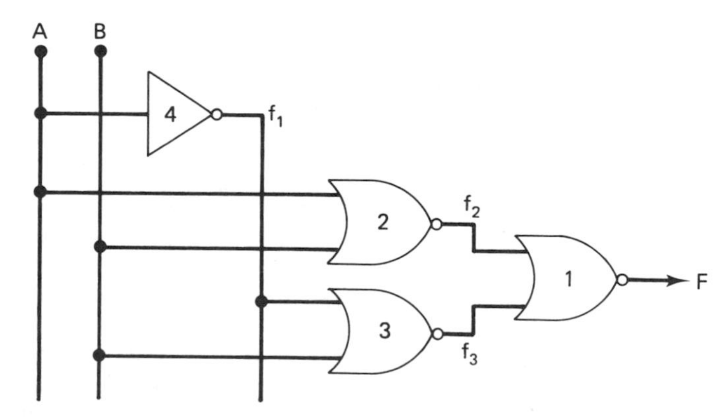

Figure 6.12 Two-input logic diagram.

Moreover, the equation uses two Boolean variables to form its sum terms. For this reason, the second level of gates in the logic diagram has two two-input NOR gates.

Besides this, the A input is inverted by a NOT gate to give the $\bar{A}$ complemented input.

EXAMPLE 6.8

Set up a product-of-sums Boolean equation and a NOR-NOT logic diagram for the truth table in Fig. 6.13(a).

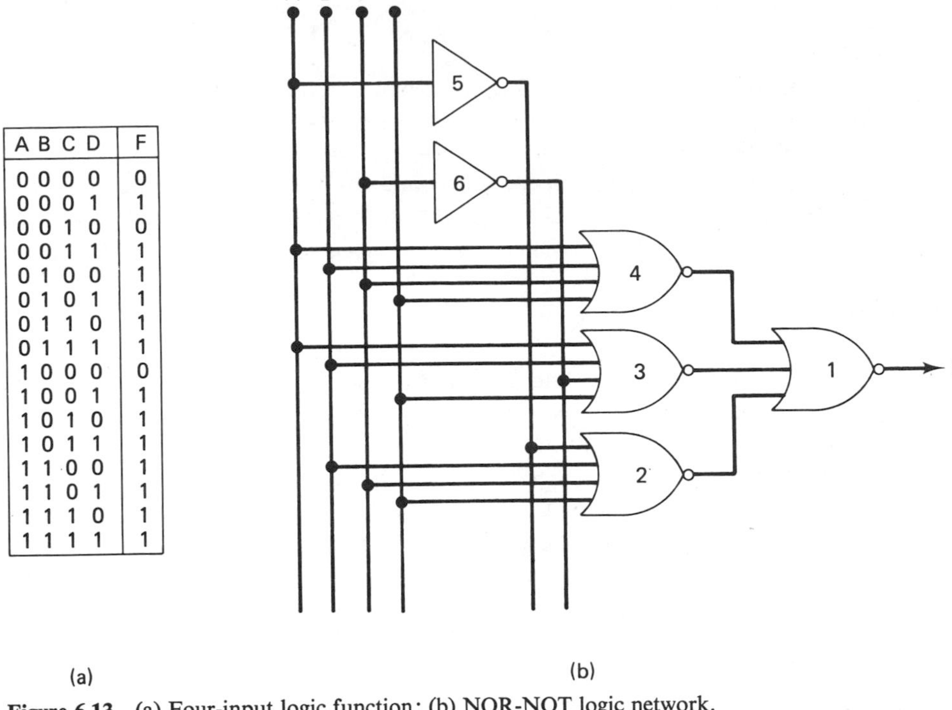

A B C D	F
0 0 0 0	0
0 0 0 1	1
0 0 1 0	0
0 0 1 1	1
0 1 0 0	1
0 1 0 1	1
0 1 1 0	1
0 1 1 1	1
1 0 0 0	0
1 0 0 1	1
1 0 1 0	1
1 0 1 1	1
1 1 0 0	1
1 1 0 1	1
1 1 1 0	1
1 1 1 1	1

Figure 6.13 (a) Four-input logic function; (b) NOR-NOT logic network.

SOLUTION:

The truth table in Fig. 6.13(a) has three input states where the output is 0. The sum terms for these cases are the following:

$$A + B + C + D$$
$$A + B + \bar{C} + D$$
$$\bar{A} + B + C + D$$

These three terms are combined with the AND function to give

$$F = (A + B + C + D)(A + B + \bar{C} + D)(\bar{A} + B + C + D)$$

For this equation, the first level in the logic diagram needs a three-input NOR gate. This is shown in the network diagram of Fig. 6.13(b).

The second logic level requires three four-input NOR gates. Besides this, the $\bar{A}$ and $\bar{C}$ complemented inputs are produced by inverters.

6.8 LOGIC NETWORKS WITH A FIRST LEVEL INPUT

In later chapters, Boolean equations will be simplified before their circuitry is implemented. This can result in a problem when the simplified expression is to be implemented with NAND-NOT or NOR-NOT circuitry. The problem is due to the fact that an input or complemented input may directly enter the first level logic gate.

For instance, we could get a simplified Boolean equation such as

$$F = A + \bar{B}C$$

This equation is consistent with the sum-of-products form and can be implemented with NAND-NOT logic. However, the first term in the expression is just the input A rather than a product term.

Using this Boolean equation and the approach for implementing NAND-NOT logic, we get the network shown in Fig. 6.14(a). But for correct operation NAND-NOT logic networks must have all inputs pass through an even number of levels of logic. In this circuit the A input bypasses the second level and goes directly to one input on the first level NAND gate.

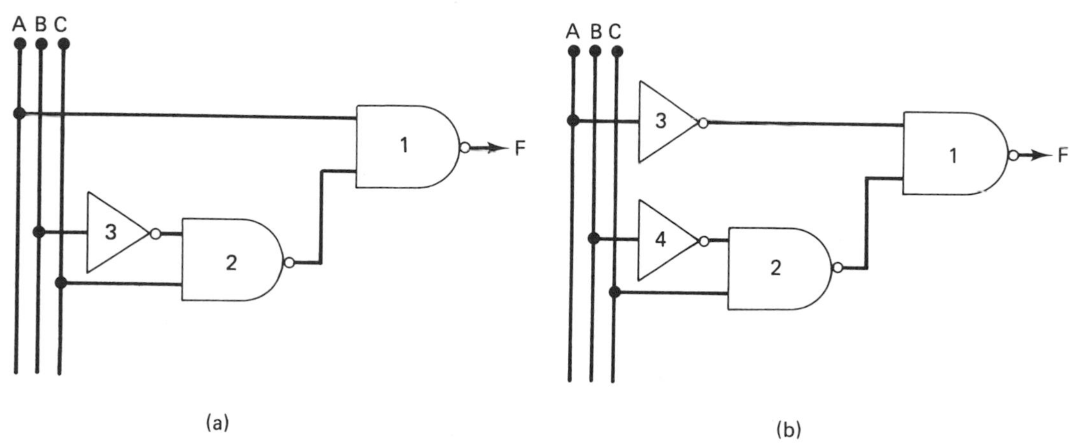

Figure 6.14 (a) Incorrect logic diagram; (b) Logic diagram corrected by inverting input A.

For this reason, the A signal passes through just 1 level or an odd number of levels of logic. The result of this input on the output is complemented by the inverting first level NAND gate. So the output of the circuit in Fig. 6.14(a) actually represents the logic function

$$F = \bar{A} + \bar{B}C$$

Here, we see that the only difference in this and the original Boolean equation is the bar over the A input. To correct this problem, we could include an inverter. So the problem of an input passing through an odd number of levels of logic is eliminated by complementing the input. The logic diagram of Fig. 6.14(b) has an inverter added for the A input.

6.9 THE AND-OR-INVERT GATE AND EXPANDER ICS

The *AND-OR-invert gate* is a special integrated circuit designed to implement sum-of-products Boolean equations. There are two types available:

1. Fixed AND-OR-invert gate
2. Expandable AND-OR-invert gate

The expandable device is normally used in conjunction with another integrated device called an *expander*.

Fixed AND-OR-Invert Gate Circuit

A fixed AND-OR-invert gate contains two levels of logic gates. This logic circuit is shown in Fig. 6.15(a). Here, we observe that the first level is a NOR gate and that the second level has two AND gates.

The first level NOR gate works like an OR gate with an inverter at the output. For this reason, the combination can be said to include the AND, OR, and invert logic functions.

This gate connection is the standard form needed to implement a sum-of-

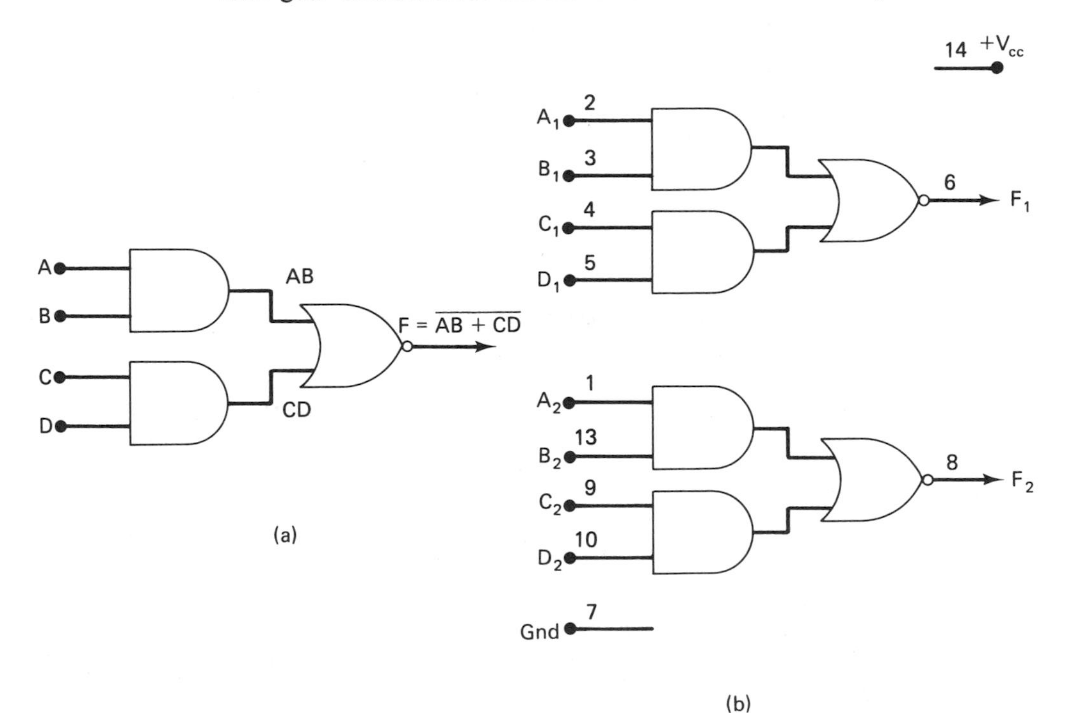

Figure 6.15 (a) Fixed AND-OR-invert gate circuit; (b) 7451 dual AND-OR-invert gate pin layout.

products expression. However, the inverter at the output due to the NOR gate gives the complement of the required logic function. This problem can be eliminated by writing the Boolean equation for the complement of the truth table output.

The 7451 integrated circuit contains AND-OR-invert networks. A pin layout for this device is shown in Fig. 6.15(b). From this diagram, we see that the 7451 contains two AND-OR-invert circuits. The output of each circuit is given by the Boolean equation

$$F = \overline{AB + CD}$$

EXAMPLE 6.9

Show how to implement the logic function F in Fig. 6.16(a) with a 7451 AND-OR-invert gate integrated circuit.

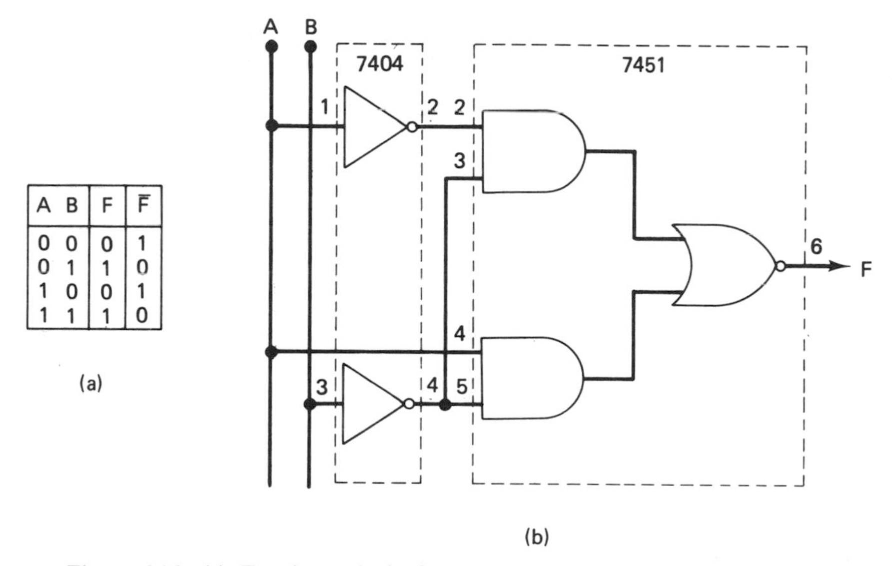

A	B	F	$\bar{F}$
0	0	0	1
0	1	1	0
1	0	0	1
1	1	1	0

Figure 6.16 (a) Two-input logic function; (b) Implemented logic function with 7451 AND-OR-invert gate.

SOLUTION:

To implement this logic function with a 7451 device, we begin by forming the complement of the output. This is indicated as $\bar{F}$ in the truth table of Fig. 6.16(a).

Now, the sum-of-products equation can be written for the $\bar{F}$ output. Using the inverted output, the equation for output F is

$$F = \bar{A}\bar{B} + A\bar{B}$$

To set up the logic circuitry one of the AND-OR-invert gates on the 7451 can be used. However, external inverter gates are needed to provide the $\bar{A}$ and $\bar{B}$ complemented inputs. These devices could be provided by a 7404 IC. A logic diagram with ICs and pin numbers indicated is shown in Fig. 6.16(b).

Expandable AND-OR-Invert Gate

In Fig. 6.17 a logic diagram of a 7453 expandable AND-OR-invert gate device is shown. This device has a gate combination that contains a NOR gate in the first level of logic and four two-input AND gates in the second level.

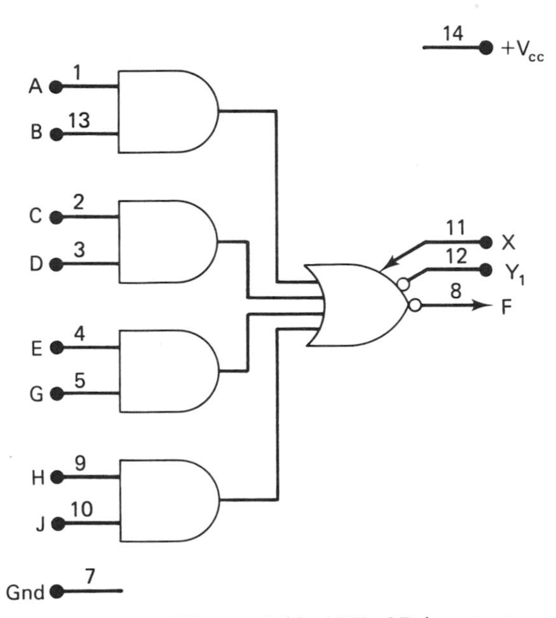

Figure 6.17 7453 expandable AND-OR-invert gate.

However, the first level gate has two extra leads marked X_1 and Y_1. These leads are known as the *expander terminals* and are used to tie additional AND gates into the second level of logic. The Boolean equation for the output of this device is

$$F = \overline{AB + CD + EG + HJ + X}$$

In this equation, the X represents the expander function.

Expander Gate

An expander gate is specially designed for connection to an expandable device such as the 7453. In this way, it provides additional inputs. For instance, the 7460 integrated circuit shown in Fig. 6.18 is a dual four-input AND expander gate.

Each gate on the device has a set of leads for expansion. These leads are labeled X_1 and Y_1 and can be connected to X_1 and Y_1 on a 7453 expandable AND-OR-invert gate.

The Boolean expression for the two devices connected together becomes

$$F = \overline{AB + CD + EG + HJ + A_1\, B_1\, C_1\, D_1}$$

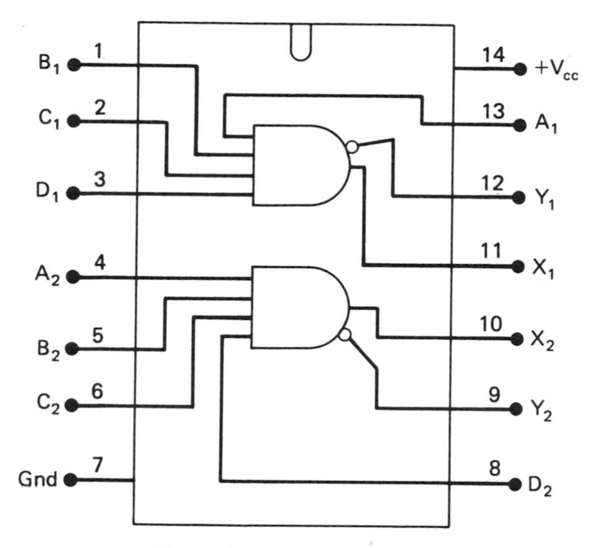

Figure 6.18 7460 expander IC.

ASSIGNMENTS

Section 6.2

1. In the block diagram of Fig. 6.1(a), we want the output F of the logic network to be at the 1 logic level for two states of inputs A and B. The first combination when the output is 1 is for both A and B equal 0. The other input state is when A equals 0 and B a logic 1. Form the truth table.

2. The inputs A, B, and C and output F of a logic network are shown in Fig. 6.19. Use these waveforms to make a truth table of the network's operation.

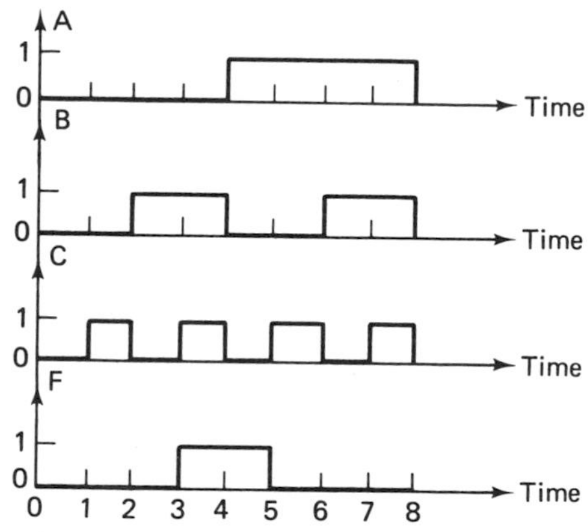

Figure 6.19 Logic network input/output waveforms.

Section 6.3

3. Make a list of the 16 product terms of a four-input truth table. Use the Boolean variables A, B, C, and D. The list should be in the form of the chart in Fig. 6.3.

4. Write a sum-of-products Boolean equation for the output F in the truth table of Fig. 6.20.

A B	F
0 0	1
0 1	0
1 0	1
1 1	0

Figure 6.20 Two-input truth table.

5. In Fig. 6.21(a), the truth table of a three-input logic function is given. Form the sum-of-products equation for output F.

6. A four-input logic function is given in the table of Fig. 6.21(b). What is the sum-of-products expression for the output?

A B C	F
0 0 0	1
0 0 1	0
0 1 0	1
0 1 1	1
1 0 0	0
1 0 1	0
1 1 0	0
1 1 1	0

(a)

A B C D	F
0 0 0 0	0
0 0 0 1	0
0 0 1 0	0
0 0 1 1	0
0 1 0 0	0
0 1 0 1	1
0 1 1 0	0
0 1 1 1	1
1 0 0 0	0
1 0 0 1	0
1 0 1 0	0
1 0 1 1	0
1 1 0 0	0
1 1 0 1	0
1 1 1 0	0
1 1 1 1	1

(b)

Figure 6.21 (a) Three-input logic function; (b) Four-input logic function.

Section 6.4

7. Make an AND-OR-NOT logic diagram for each of the Boolean equations that follow.

(a) $F = A\bar{B} + AB$ (b) $F = \bar{A}\bar{B}\bar{C}\bar{D} + \bar{A}\bar{B}CD + ABCD$

8. Set up an AND-OR-NOT logic diagram for the equation written in Problem 6. Check the output by writing the network Boolean equation from the logic diagram.

9. Write a sum-of-products equation for the truth table in Fig. 6.22. Then draw an AND-OR-NOT logic diagram for this equation. Check the operation by forming the network truth table.

A B C	F
0 0 0	1
0 0 1	0
0 1 0	0
0 1 1	0
1 0 0	1
1 0 1	0
1 1 0	0
1 1 1	0

Figure 6.22 Three-input function.

Section 6.5

10. Set up NAND-NOT logic diagrams for the equations in Problem 7.

11. Construct a NAND-NOT logic diagram of the truth table in Fig. 6.22.

12. Write a Boolean equation for the truth table formed in Problem 2. From this equation, construct a NAND-NOT logic circuit.

Section 6.6

13. Construct a chart like that in Fig. 6.10(b) of the sum terms for a four-input truth table.

14. Write a product-of-sums Boolean equation for the truth table in Fig. 6.20.

15. Draw an AND-OR-NOT logic diagram for the following equations.
(a) $F = (\bar{A} + \bar{B})(\bar{A} + B)$ (b) $F = (A + B + \bar{C} + D)(A + B + \bar{C} + \bar{D})$

16. A three-input truth table function is listed in Fig. 6.21(a). What is the product-of-sums equation for the output? Set up an AND-OR-NOT logic diagram for the equation. Check the output F by truth table analysis.

Section 6.7

17. Make NOR-NOT logic diagrams for the Boolean equations in Problem 15.

18. Write a product-of-sums equation for the logic function indicated by the table in Fig. 6.23. Construct a NOR-NOT logic diagram and check its operation by truth table analysis.

A	B	C	F
0	0	0	0
0	0	1	1
0	1	0	0
0	1	1	1
1	0	0	1
1	0	1	0
1	1	0	0
1	1	1	0

Figure 6.23 Three-input function.

Section 6.8

19. Draw a NAND-NOT logic diagram for the Boolean equation $F = \bar{A} + B\bar{C}$.

20. Construct a diagram of the NOR-NOT circuitry needed to implement the logic function $F = A(B + \bar{C})$.

Section 6.9

21. Make a Boolean equation and logic diagram to show how the truth table in Fig. 6.20 can be implemented with a 7451 AND-OR-invert gate and 7404 inverter. Check the operation at output F by truth table analysis.

Simplification of Logic Circuitry

7

7.1 INTRODUCTION

In previous chapters, we introduced the topics of Boolean variables, Boolean operators, sum-of-products Boolean equations, and product-of-sums Boolean equations. Moreover, we showed how to write the network Boolean equation from a truth table and implement it with AND-OR-NOT, NAND-NOT, or NOR-NOT logic circuitry.

However, the standard sum-of-products or product-of-sums network equations we have been using can often be simplified before getting implemented with circuitry. *Boolean algebra* is the mathematics used to simplify logic functions. With it, we can minimize the number of gates needed in a logic network.

To develop the techniques used to simplify logic circuitry, the topics that follow are included in this chapter:

1. Boolean postulates and theorems
2. Verification of Boolean postulates and theorems
3. Simplification and implementation of Boolean equations

7.2 BOOLEAN POSTULATES AND THEOREMS

Earlier, we used the truth table analysis technique to analyze the operation of existing logic networks. Another way of analyzing logic circuitry is with Boolean algebra. Boolean analysis techniques are more widely used in the design of logic networks.

The theory of Boolean algebra is represented by a set of *Boolean postulates* and a set of *Boolean theorems*. These postulates and theorems are used to reduce a complex sum-of-products or product-of-sums Boolean equation to a simpler expression.

Boolean Postulates

There are five basic postulates of Boolean algebra. A list of these postulates is shown in Fig. 7.1. Here, we see that each postulate has two parts. For instance, postulate number 1 is written in forms 1a and 1b.

P1a	$X = 1$ if $X \neq 0$
P2a	$0 \cdot 0 = 0$
P3a	$1 \cdot 1 = 1$
P4a	$1 \cdot 0 = 0$
P5a	$\overline{1} = 0$

P1b	$X = 0$ if $X \neq 1$
P2b	$1 + 1 = 1$
P3b	$0 + 0 = 0$
P4b	$0 + 1 = 1$
P5b	$\overline{0} = 1$

Figure 7.1 Boolean postulates.

Boolean postulates are obtained from the definitions of a Boolean variable and the basic Boolean operations AND, OR, and NOT. For example, earlier we said a Boolean variable such as X can take on the 0 or 1 logic states.

In this way, we get postulates 1a and 1b:

$$\text{P1a:} \quad X = 1 \text{ if } X \neq 0$$

$$\text{P1b:} \quad X = 0 \text{ if } X \neq 1$$

Postulate 1a tells that the Boolean variable X is logic 1 when it is not 0. Furthermore, 1b indicates that X must be 0 if it is not 1.

On the other hand, postulates 2a, 3a, and 4a are derived from the logical AND function:

$$\text{P2a:} \quad 0 \cdot 0 = 0$$

$$\text{P3a:} \quad 1 \cdot 1 = 1$$

$$\text{P4a:} \quad 1 \cdot 0 = 0$$

As an example, let us show how postulate 2a is obtained. This postulate says 1 AND 1 equals 1. The condition just described is one of the input states listed in the AND operation truth table of Fig. 3.8(b).

Besides this, the other two postulates, 3a and 4a, are two more logic states from the same truth table.

Similarly, postulates 2b, 3b, and 4b are formed from the OR function. These postulates are

$$\text{P2b:} \quad 1 + 1 = 1$$

$$\text{P3b:} \quad 0 + 0 = 0$$

$$\text{P4b:} \quad 0 + 1 = 1$$

They are all logic states from the OR table in Fig. 3.12(b).

The last two postulates are related to the NOT function:

$$\text{P5a:} \quad \bar{1} = 0$$

$$\text{P5b:} \quad \bar{0} = 1$$

Postulate 5a says that the complement of a logic 1 is 0 and 5b shows that the complement of 0 is 1.

Since each postulate has two parts, they are called *duals* of each other. By dual, we mean the b form is made from the a form by changing the Boolean operation AND to OR and logic 0s to 1s or 1s to 0s.

As an example, let us look at postulate 2a. By changing the AND function to OR and all three 0s to 1s, we get the dual postulate 2b.

Boolean Theorems

In Fig. 7.2, 15 Boolean theorems are listed. Here, we see that each of the theorems 1a through 15a has a dual; these duals are marked 1b through 15b, respectively. Notice that Boolean variables X, Y, and Z have been used to write

Basic theorems	T1a	$0 \cdot X = 0$
	T2a	$1 \cdot X = X$
	T3a	$X \cdot X = X$
	T4a	$X \cdot \overline{X} = 0$
	T5a	$(\overline{\overline{X}}) = \overline{X}$
Algebraic theorems	T6a	$X \cdot Y = Y \cdot X$
	T7a	$XYZ = (XY)Z$
	T8a	$XY + XZ = X(Y + Z)$
Absorption theorems	T9a	$X + XY = X$
	T10a	$X + \overline{X}Y = X + Y$
	T11a	$XY + X\overline{Y} = X$
	T12a	$ZX + Z\overline{X}Y = ZX + ZY$
	T13a	$XY + \overline{X}Z + YZ = XY + \overline{X}Z$
De Morgan's theorem	T14a	$\overline{X \cdot Y \cdot Z} = \overline{X} + \overline{Y} + \overline{Z}$
Dual theorem	T15a	$XY + \overline{X}Z = (X + Z)(\overline{X} + Y)$

Basic theorems	T1b	$1 + X = 1$
	T2b	$0 + X = X$
	T3b	$X + X = X$
	T4b	$X + \overline{X} = 1$
	T5b	$(\overline{\overline{X}}) = X$
Algebraic theorems	T6b	$X + Y = Y + X$
	T7b	$X + Y + Z = (X + Y) + Z$
	T8b	$(X + Y)(X + Z) = X + YZ$
Absorption theorems	T9b	$X(X + Y) = X$
	T10b	$X(\overline{X} + Y) = XY$
	T11b	$(X + Y)(X + \overline{Y}) = X$
	T12b	$(Z + X)(Z + \overline{X} + Y) = (Z + X)(Z + Y)$
	T13b	$(X + Y)(\overline{X} + Z)(Y + Z) = (X + Y)(\overline{X} + Z)$
De Morgan's theorem	T14b	$\overline{X + Y + Z} = \overline{X} \cdot \overline{Y} \cdot \overline{Z}$
Dual theorem	T15b	$(X + Y)(\overline{X} + Z) = XZ + \overline{X}Y$

Figure 7.2 Boolean theorems.

these theorems instead of the variables A, B, and C we used for Boolean equations.

In this diagram, we have grouped the theorems into five categories:

1. Basic theorems
2. Algebraic theorems
3. Absorption theorems
4. De Morgan's theorem
5. Dual theorem

Here, we shall show how some of the *basic theorems* can be found from the definitions of the AND, OR, and NOT logic functions.

Basic theorem 1a states that the AND operation of X with logic 0 always gives logic 0:

$$\text{T1a:} \quad 0 \cdot X = 0$$

To show how this theorem is obtained, we can use the logic symbol in Fig. 7.3.

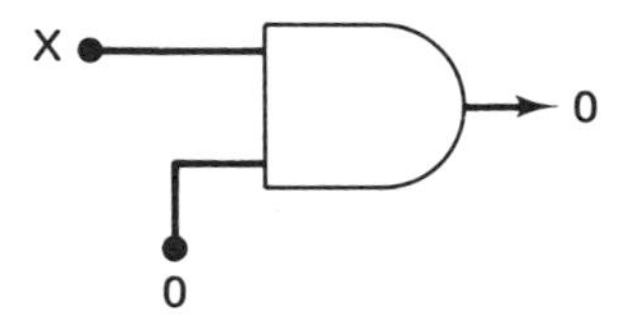

Figure 7.3 Basic theorem 1a.

On this AND gate, one input is fixed at logic 0, and the other is Boolean variable X. The X input can take on the value 0 or 1 and gives the two input conditions 00 or 01, respectively. However, for both cases the output F of the AND gate is logic 0.

Looking at this result, we can say that the output of an AND gate is 0 when one input is held at logic 0. This description is consistent with theorem 1a.

Using a similar approach, we can verify the remaining basic theorems.

EXAMPLE 7.1

Show how to obtain Boolean identity 3a.

SOLUTION:

Theorem 3a indicates that the output of an AND gate for which both inputs are made the same is equal to the input:

$$\text{T3a:} \quad X \cdot X = X$$

The gate connection in Fig. 7.4 illustrates this condition. On this gate, both inputs are wired together and made the variable X.

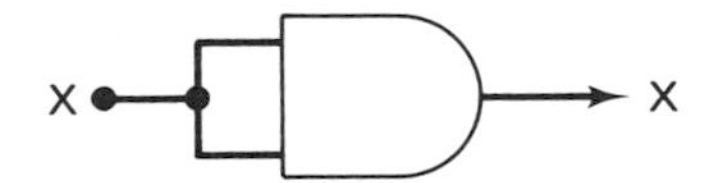

Figure 7.4 Basic theorem 3a.

When X is set to logic 0, both inputs to the AND gate are 0, and the output is also logic 0. However, for X equal to 1 both inputs are logic 1, and the output goes to the 1 level.

In this way, we find that the output is the same as the input or just X.

7.3 VERIFICATION OF BOOLEAN THEOREMS

In the last section, the basic Boolean theorems were described and verified by using logic symbols. Now we shall consider the remaining theorems.

Each of these identities states the equivalence of two Boolean expressions. Both sides of these theorems can be illustrated with a logic diagram, and their equivalence can be verified by truth table analysis.

Algebraic Theorems

The Boolean equations 6a, 7a, and 8a are called *algebraic theorems* because they represent the commutative, associative, and distributive properties of algebra:

T6a: $X \cdot Y = Y \cdot X$ commutative law

T7a: $XYZ = (XY)Z$ associative law

T8a: $XY + XZ = X(Y + Z)$ distributive law

The *commutative law* 6a shows that the order in which the two variables X and Y are ANDed does not matter. That is, the AND of X and Y is the same as Y AND X.

On the other hand, the *associative law* 7a indicates that variables can be grouped together without changing the result. For example, this theorem says that the three-input AND gate in Fig. 7.5(a) is the same as the logic diagram in Fig. 7.5(b).

Theorem 8a is the *distributive law*. This theorem shows that the algebraic operation of factoring can be performed on Boolean functions. Here, we see that variable X is common to each term on the left-hand side of the equation and that it can be factored out. After factoring, the result is X ANDed with the OR of variables Y and Z.

The duals of these theorems are as follows:

T6b: $X + Y = Y + X$

T7b: $X + Y + Z = (X + Y) + Z$

T8b: $(X + Y)(X + Z) = X + YZ$

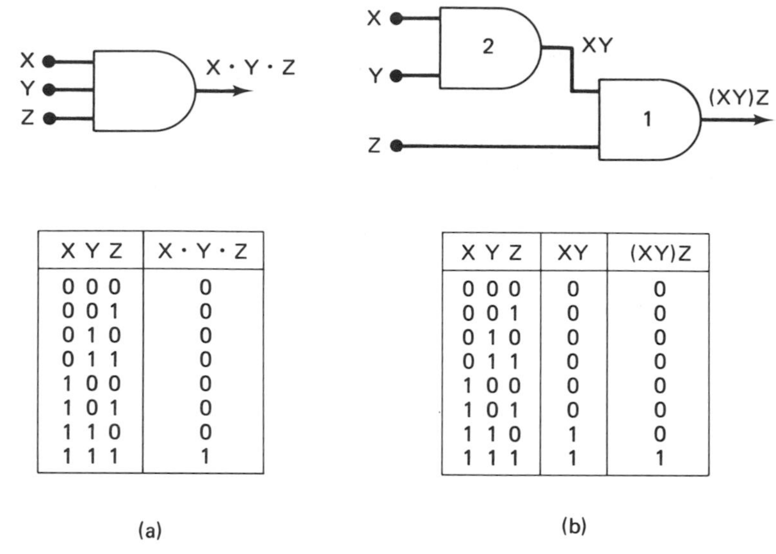

X Y Z	X · Y · Z
0 0 0	0
0 0 1	0
0 1 0	0
0 1 1	0
1 0 0	0
1 0 1	0
1 1 0	0
1 1 1	1

X Y Z	XY	(XY)Z
0 0 0	0	0
0 0 1	0	0
0 1 0	0	0
0 1 1	0	0
1 0 0	0	0
1 0 1	0	0
1 1 0	1	0
1 1 1	1	1

(a) (b)

Figure 7.5 (a) Three-input AND gate; (b) Grouped two-input AND gates.

Each of these theorems can be verified by truth table analysis.

As an example, let us verify identity 7a. The left-hand side of this equation is just a three-input AND function. In Fig. 7.5(a), the truth table for a three-input AND gate is shown.

The other side of this equation says that two two-input AND gates connected together as in Fig. 7.5(b) work the same as a three-input AND gate. The truth table of this connection is also listed in Fig. 7.5(b). From this table, we find that the X and Y inputs are first ANDed together. This gives two output states that are logic 1, as shown under the XY column. After this, the XY output is ANDed with Z to give the output function. Doing this, we find that the output is again logic 1 when all inputs are 1.

In this way, we can see that the associative law of Boolean algebra is valid.

Absorption Theorems

The *absorption theorems* show how some standard Boolean variable forms can be reduced to a simpler form. There are five absorption theorems listed in the table of Fig. 7.2:

$$\text{T9a:} \quad X + XY = X$$

$$\text{T10a:} \quad X + \bar{X}Y = X + Y$$

$$\text{T11a:} \quad XY + X\bar{Y} = X$$

$$\text{T12a:} \quad ZX + Z\bar{X}Y = ZX + ZY$$

$$\text{T13a:} \quad XY + \bar{X}Z + YZ = XY + \bar{X}Z$$

Using these equations, a complex variable combination with the form on the left-hand side can be located in a Boolean expression and replaced by the simplified form on the right.

The duals for the absorption theorems are 9b through 13b:

T9b: $X(X + Y) = X$

T10b: $X(\bar{X} + Y) = XY$

T11b: $(X + Y)(X + \bar{Y}) = X$

T12b: $(Z + X)(Z + \bar{X} + Y) = (Z + X)(Z + Y)$

T13b: $(X + Y)(\bar{X} + Z)(Y + Z) = (X + Y)(\bar{X} + Z)$

EXAMPLE 7.2

Verify absorption equation 9a by the truth table analysis technique.

SOLUTION:

To verify this equation, we need a two-input truth table. This table is made with inputs X and Y in Fig. 7.6.

X Y	XY	X + XY
0 0	0	0
0 1	0	0
1 0	0	1
1 1	1	1

Figure 7.6 Verification of identity 9a.

First, the AND of inputs X and Y is formed. This function is 1 when both X and Y are 1 together, as shown under the XY column of the table.

Now, we must OR input X with the XY column. This gives a logic 1 if either X or XY is 1. The result is shown in the last column of the truth table.

Comparing the $X + XY$ column to the X input, we see that they are the same. This verifies absorption theorem 9a.

De Morgan's Theorem

Two of the most important Boolean identities are known as *De Morgan's theorems*. These equations are 14a and 14b:

T14a: $\overline{X \cdot Y \cdot Z} = \bar{X} + \bar{Y} + \bar{Z}$

T14b: $\overline{X + Y + Z} = \bar{X} \cdot \bar{Y} \cdot \bar{Z}$

De Morgan's theorems are important because they show how the NAND and NOR gates are related to AND-OR-NOT logic. Besides this, they are needed for the analysis of both NAND-NOT and NOR-NOT logic circuitry.

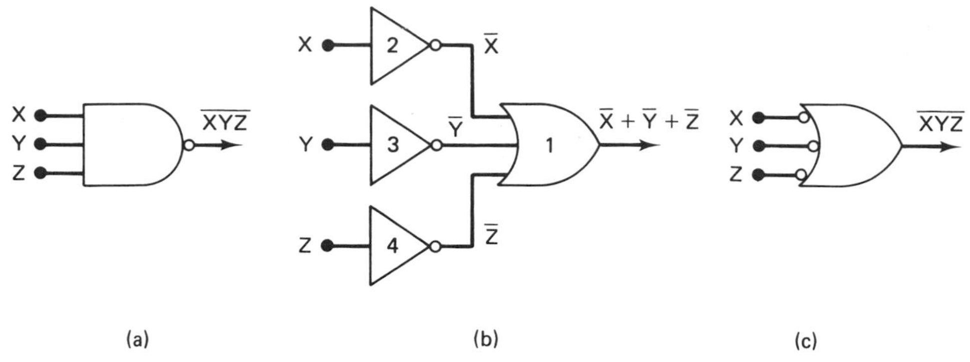

Figure 7.7 (a) NAND symbol; (b) OR-NOT equivalent; (c) Alternative NAND gate symbol.

For instance, the first of De Morgan's theorems says that a NAND gate like that in Fig. 7.7(a) is equivalent to an OR gate with inverters at its inputs. This OR-NOT gate combination is shown in Fig. 7.7(b).

Another way of indicating a NAND gate is with the symbol of Fig. 7.7(c). Here the NAND function is represented as an OR gate with a dot at each input to show the inversions.

Similarly, the second of De Morgan's theorems shows that a NOR gate works the same as an AND gate with inverted inputs. This logic relationship is illustrated in Figs. 7.8(a) and (b). In Fig. 7.8(c), an alternative NOR gate symbol is shown.

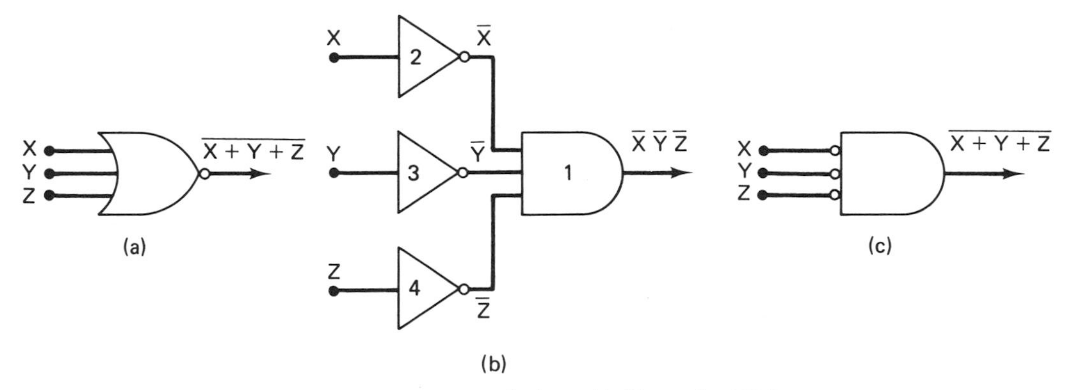

Figure 7.8 (a) NOR symbol; (b) AND-NOT equivalent; (c) Alternative NOR gate symbol.

EXAMPLE 7.3

Use truth table analysis to verify theorem 14a.

SOLUTION:

The left-hand side of this equation is a three-input NAND gate function, and its truth table is shown in Fig. 7.9(a).

X Y Z	$\overline{XYZ}$
0 0 0	1
0 0 1	1
0 1 0	1
0 1 1	1
1 0 0	1
1 0 1	1
1 1 0	1
1 1 1	0

(a)

X Y Z	$\bar{X}$	$\bar{Y}$	$\bar{Z}$	$\bar{X}+\bar{Y}+\bar{Z}$
0 0 0	1	1	1	1
0 0 1	1	1	0	1
0 1 0	1	0	1	1
0 1 1	1	0	0	1
1 0 0	0	1	1	1
1 0 1	0	1	0	1
1 1 0	0	0	1	1
1 1 1	0	0	0	0

(b)

Figure 7.9 (a) Left-hand side of 14a; (b) Right-hand side of 14a.

For the right-hand side, we must invert each input. This gives the $\bar{X}$, $\bar{Y}$, and $\bar{Z}$ columns in the truth table of Fig. 7.9(b).

After this, the OR function of these complemented inputs is formed. This gives a result identical to the NAND gate output in the table of Fig. 7.9(a).

Dual Theorems

The last two theorems are the *dual theorems* 15a and 15b:

$$\text{T15a:}\quad XY + \bar{X}Z = (X + Z)(\bar{X} + Y)$$

$$\text{T15b:}\quad (X + Y)(\bar{X} + Z) = XZ + \bar{X}Y$$

These theorems are useful when converting Boolean equations between the sum-of-products and product-of-sums form.

In theorem 15a, we replace the two AND operations by OR functions and the single OR operation with an AND. Moreover, the Z variable is grouped with X instead of $\bar{X}$, and Y is grouped with $\bar{X}$. The logic diagrams for this equation are shown in Figs. 7.10(a) and (b).

7.4 SIMPLIFICATION OF BOOLEAN EQUATIONS

The Boolean equations we have set up to this point have not been simplified before they were implemented with circuitry. In this section, we use the postulates and theorems of Boolean algebra to obtain a simplified expression. By simplifying the equation, we can make a logic network with fewer gates. For this reason, the logic function can be implemented with fewer integrated circuits and less expensive circuitry.

To simplify a sum-of-products or product-of-sums equation, we must apply in an orderly way the postulates listed in Fig. 7.1 and the theorems of Fig. 7.2 to the expression.

As an example, let us simplify the Boolean equation

$$F = \bar{A}B + AB$$

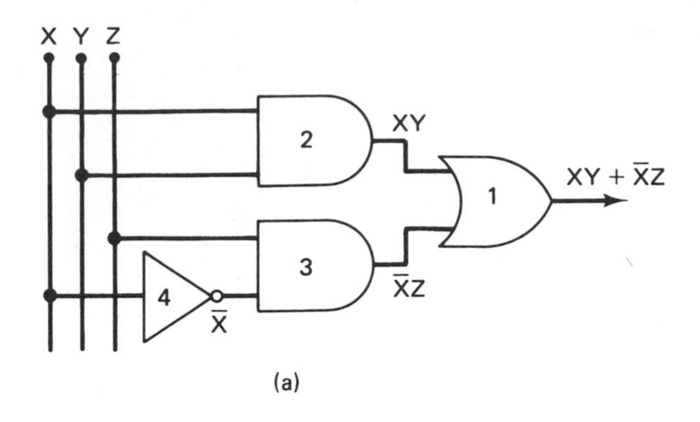

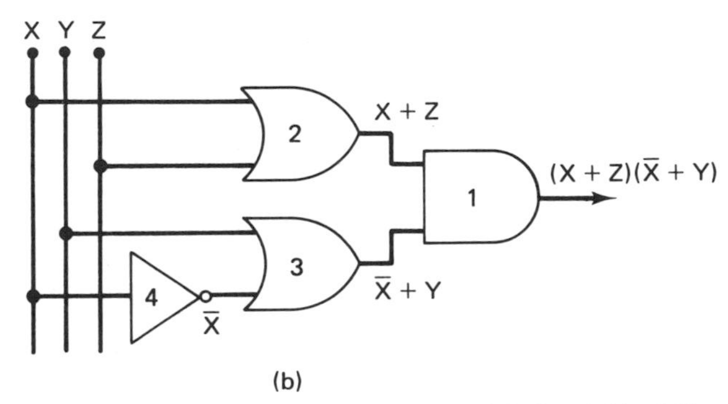

Figure 7.10 (a) Left-hand side of 15a; (b) Right-hand side of 15a.

This expression has the same form as identity 8a. Here the variable common to both terms is B, and it can be factored out. In this way, we get

$$F = B(\bar{A} + A)$$

This equation can be further simplified. Using identity 6b, we can change the position of the two A terms. The result is

$$F = B(A + \bar{A})$$

Now, the sum part of the expression is consistent with the form of theorem 4b. Replacing this term by 1, we obtain

$$F = B(1)$$

In this form, the two terms can be interchanged by using theorem 6a, and the final simplification is done with law 2a:

$$F = 1 \cdot B$$

$$F = B$$

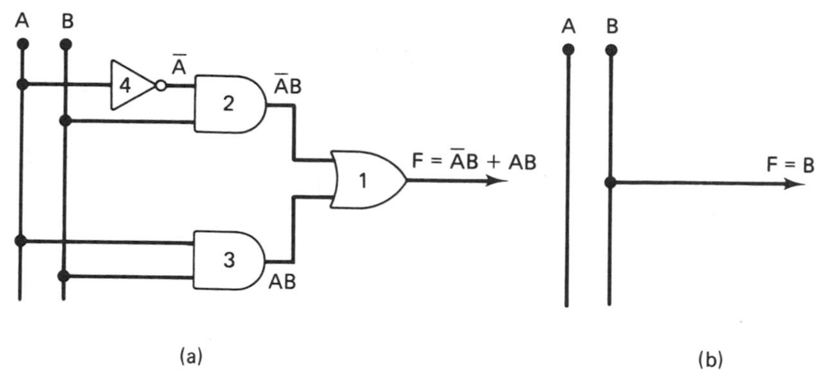

Figure 7.11 (a) Original logic diagram; (b) Simplified logic diagram.

In Fig. 7.11(a), the AND-OR-NOT logic diagram for the original equation is drawn. Here we see that four logic gates are needed to implement the function. However, through Boolean algebra we find that the expression simplifies to just B. So the logic function can be implemented without any logic gates. Instead, the B input is directly sent to the output, as shown in Fig. 7.11(b).

EXAMPLE 7.4

Use Boolean algebra to simplify the expression

$$F = A\bar{B}CD + ABCD$$

Show how both the original and simplified logic functions can be implemented with logic circuitry.

SOLUTION:

With identity 6a, we can group the common ACD variables together:

$$F = ACD\bar{B} + ACDB$$

Next, the ACD term is factored according to identity 8a:

$$F = ACD(\bar{B} + B)$$

Changing the order of the B terms by law 6b, we get

$$F = ACD(B + \bar{B})$$

Now the sum term can be eliminated by theorem 4b:

$$F = ACD(1)$$

Again, the order can be changed by identity 6a and reduced to find the final form with 2a:

$$F = 1 \cdot ACD$$

$$F = ACD$$

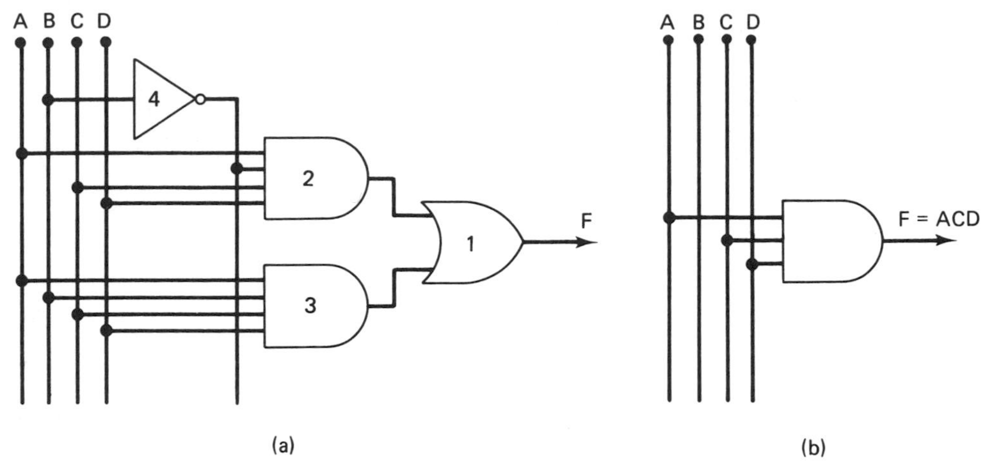

Figure 7.12 (a) Original logic diagram; (b) Simplified logic diagram.

A logic diagram of the original network is shown in Fig. 7.12(a) and the simplified circuit in Fig. 7.12(b). From these diagrams, we see that a four-gate network is reduced to a single logic gate.

EXAMPLE 7.5

Simplify the equation $F = \bar{A}BC + AB\bar{C} + ABC$.

SOLUTION:

First, we shall treat the last two terms of the expression with theorem 8a:

$$F = \bar{A}BC + AB(\bar{C} + C)$$

Now identity 4b can be used to reduce the sum term $\bar{C} + C$:

$$F = \bar{A}BC + AB$$

This form is the same as absorption theorem 12a. This gives

$$F = AB + BC$$

We could reduce this equation one step further. However, the present form is consistent with the form of a sum-of-products equation.

EXAMPLE 7.6

Convert the simplified equation of Example 7.5 to a product-of-sums form and implement with NOR-NOT logic.

SOLUTION:

We begin with the simplified sum-of-products equation

$$F = AB + BC$$

Using theorem 8a, the common B variable is factored. This gives

$$F = B(A + C)$$

This is the simplified product-of-sums form equation and can be directly implemented with NOR-NOT logic. The logic diagram obtained is shown in Fig. 7.13. In this diagram, we see that the B input is inverted. This occurs because it passes through one level of logic gates.

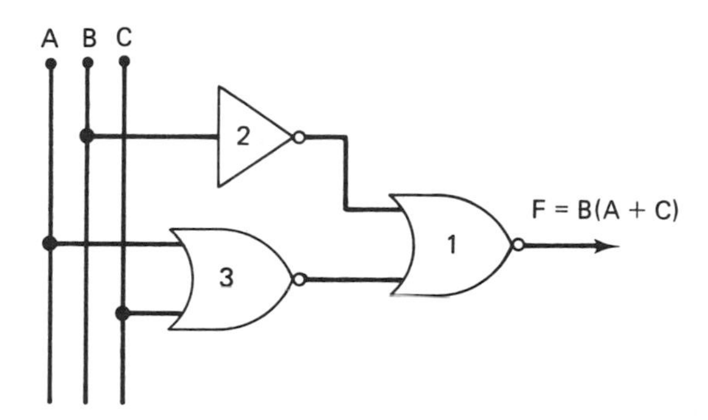

Figure 7.13 NOR-NOT logic diagram.

EXAMPLE 7.7

Simplify the Boolean equation

$$F = (A + B + C + D)(A + B + \bar{C} + D)(\bar{A} + B + C + D)$$

SOLUTION:

Our simplification begins by applying identity 11b to the first two sum terms. In this way, we get

$$F = (A + B + D)(\bar{A} + B + C + D)$$

Next, theorem 8b is used and results in

$$F = (B + D) + A(\bar{A} + C)$$

In this expression, the second part is the form of identity 10b. So we get the final form as

$$F = B + D + AC$$

7.5 LOGIC NETWORK DESIGN EXAMPLES

In the last section, we showed how to simplify Boolean equations and implement the simplified expression with gate circuitry. However, in practical design problems the Boolean expression is normally not available. Instead, a four-step procedure can be followed to design a logic network:

1. Form a truth table.
2. Write a Boolean equation.
3. Simplify the expression.
4. Implement the logic circuitry.

Here, this procedure will be illustrated with examples.

EXAMPLE 7.8

Write a sum-of-products Boolean equation for the truth table in Fig. 7.14(a). Simplify the expression and implement with NAND-NOT logic circuitry.

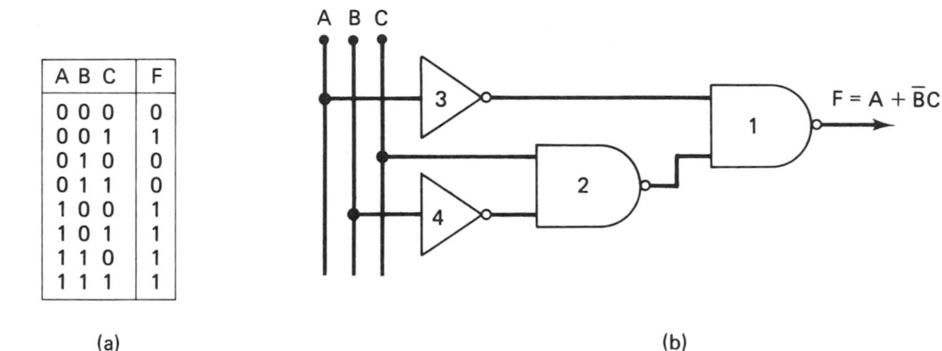

A B C	F
0 0 0	0
0 0 1	1
0 1 0	0
0 1 1	0
1 0 0	1
1 0 1	1
1 1 0	1
1 1 1	1

Figure 7.14 (a) Three-input logic function; (b) NAND-NOT logic network.

SOLUTION:

Using the sum terms for the five input states in which the output is 1, we get the sum-of-products equation

$$F = \bar{A}\bar{B}C + A\bar{B}\bar{C} + A\bar{B}C + AB\bar{C} + ABC$$

The simplification begins with the last two terms. Applying absorption identity 11a, we get

$$F = \bar{A}\bar{B}C + A\bar{B}\bar{C} + A\bar{B}C + AB$$

This same identity can be used on the second and third terms. This gives

$$F = \bar{A}\bar{B}C + A\bar{B} + AB$$

Now we can use identity 11a once again on the last two terms:

$$F = \bar{A}\bar{B}C + A$$

The function that remains has the form of identity 10a. With this theorem, we get the expression in its final simplified form:

$$F = A + \bar{B}C$$

To implement this equation with NAND-NOT logic, we need four gates connected as shown in Fig. 7.14(b). The first level of logic has a two-input NAND gate and the second level another two-input NAND gate.

Both the $\bar{A}$ and $\bar{B}$ inputs are provided by inverters. The A input must be inverted because this signal would enter the first level of logic.

EXAMPLE 7.9

Form a truth table for the waveforms in Fig. 7.15(a). Using this truth table, write a sum-of-products Boolean equation and simplify. Construct an appropriate logic diagram.

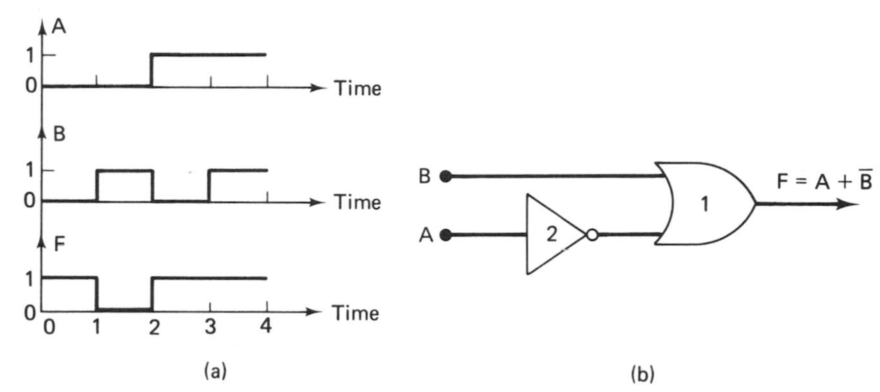

Figure 7.15 (a) Timing diagram; (b) NOR-NOT logic diagram.

SOLUTION:

Looking at the waveforms, we see that the A and B inputs are 0 in the time interval from 0 to 1. During this time, the output is logic 1.

In the interval 1 to 2, the A input is 0, and B equals 1. For this input state, the output F switches to logic 0.

The next interval corresponds to A equals 1 and B is 0. Here the output goes back to logic 1.

The last inputs are both 1, and the output remains at logic 1.

Formalizing these input and output conditions, we get

A B	F
0 0	1
0 1	0
1 0	1
1 1	1

Using the 1 output states, we get the Boolean equation

$$F = \bar{A}\bar{B} + A\bar{B} + AB$$

Simplication using absorption theorem 11a yields

$$F = \bar{A}\bar{B} + A$$

After this reduction, identity 10a can be applied. This results in

$$F = A + \bar{B}$$

A quicker way of doing this problem is to directly write the product-of-sums equation. This gives the simplified expression $F = A + \bar{B}$ without reductions.

To implement this expression, OR-NOT logic circuitry can be used, and the logic diagram is shown in Fig. 7.15(b).

ASSIGNMENTS

Section 7.2

1. Which postulate makes the statement 0 OR 1 is equal to 1?

2. Describe how the dual of a postulate can be formed.

3. Draw a logic symbol for basic theorem 2b. Explain why the output will always be X.

4. For basic identity 3b, draw a logic symbol and show why the output is equal to X.

5. Explain why the AND gate indicated by theorem 4a will always have an output equal to 0.

Section 7.3

6. Make a logic diagram for both sides of the 7b form of the associative law. Verify the theorem by making a truth table for each side of the expression.

7. Verify theorem 8a by truth table analysis.

8. Draw a logic diagram for both sides of theorem 8b.

9. Use the truth table analysis technique to verify absorption identity 10b.

10. Make a logic diagram for each side of theorem 13a.

11. Verify theorem 12a by truth table analysis.

12. Show how the left-hand side of theorem 9a can be reduced to the simplified form on the right by using the basic and algebraic identities.

13. With the basic and algebraic theorems, show how to reduce identity 10b to its simplified form.

14. Verify the 14b form of De Morgan's theorem by truth table analysis.

15. Draw a logic symbol with inverted inputs like that in Fig. 7.7(c) to give AND gate operation.

16. Verify dual theorem 15b by truth table analysis.

Section 7.4

17. Use Boolean algebra to simplify the equations that follow.

(a) $F = A\bar{B} + AB$ (b) $F = \bar{A}\bar{B} + \bar{A}B$

18. Simplify each of the sum-of-products equations that follow.

(a) $F = ABC\bar{D} + ABCD$ (b) $F = \bar{A}B\bar{C}D + AB\bar{C}D$

19. Simplify the following equations and make an AND-OR-NOT logic diagram for the simplified expression.

(a) $F = A\bar{B}C + A\bar{B}\bar{C} + ABC$ (b) $F = \bar{A}\bar{B}\bar{C} + A\bar{B}\bar{C} + AB\bar{C}$

20. Simplify the following product-of-sums equations and make an AND-OR-NOT logic diagram for the simplified function.

(a) $F = (A + B)(A + \bar{B})(\bar{A} + \bar{B})$ (b) $F = (\bar{A} + \bar{B} + \bar{C})(A + \bar{B} + \bar{C})$

Section 7.5

21. For the truth table in Fig. 7.16(a), write a sum-of-products Boolean equation, simplify the expression, and implement with NAND-NOT circuitry.

(a)

A B C	F
0 0 0	0
0 0 1	0
0 1 0	0
0 1 1	0
1 0 0	0
1 0 1	1
1 1 0	1
1 1 1	1

(b)

A B C	F
0 0 0	0
0 0 1	0
0 1 0	0
0 1 1	1
1 0 0	1
1 0 1	1
1 1 0	1
1 1 1	1

Figure 7.16 (a) Three-input logic function; (b) Three-input truth table.

22. Write a product-of-sums equation for the logic function in the table of Fig. 7.16(b). Simplify the expression and construct a NOR-NOT logic diagram.

23. Use the waveforms in Fig. 7.17 to form a truth table. From this table, write a sum-of-products equation. Simplify the equation and implement with NAND-NOT circuitry.

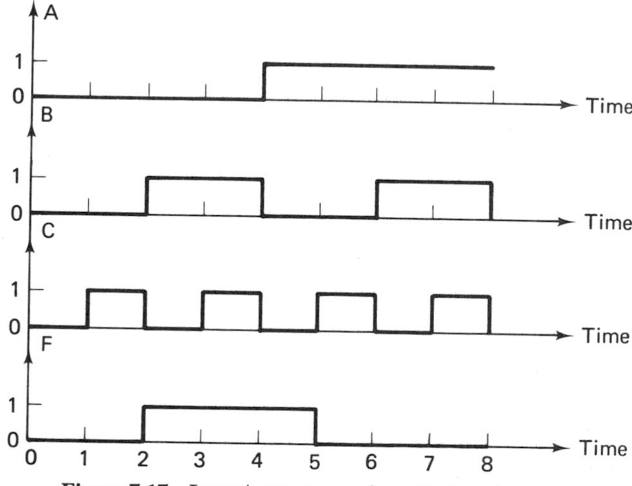

Figure 7.17 Input/output waveform timing diagram.

The Exclusive-OR Gate, Equality Gate, and Comparator Circuitry

8

8.1 INTRODUCTION

In this chapter we introduce a first group of closely related standard combinational logic networks. They include the following circuits:

1. Exclusive-OR gate
2. Equality gate
3. Equality comparator
4. Inequality comparator
5. Magnitude comparator

8.2 THE EXCLUSIVE-OR GATE

The *exclusive-OR gate* is a widely used logic gate combination. It can be constructed using AND-OR-NOT logic, NAND-NOT logic, or NOR-NOT circuitry. However, it is also available as a standard integrated circuit.

This gate is the building block for many MSI combinational logic devices. For instance, the comparator, adder, subtractor, and parity checker/generator circuits all depend on the exclusive-OR function for their operation.

Truth Table, Boolean Expression, and Logic Symbol

The truth table of exclusive-OR circuit operation is shown in Fig. 8.1(a). Here we see the circuit has two inputs A and B with a single output F.

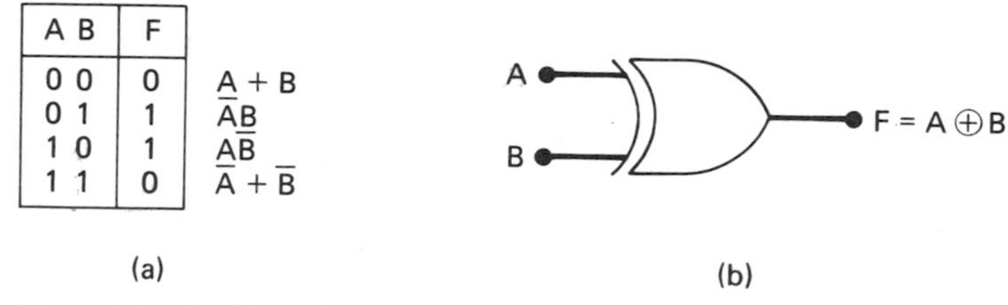

A B	F	
0 0	0	$A + B$
0 1	1	$\bar{A}B$
1 0	1	$A\bar{B}$
1 1	0	$\bar{A} + \bar{B}$

Figure 8.1 (a) Exclusive-OR truth table; (b) Logic symbol and Boolean expression.

From the truth table, the operation of the exclusive-OR is found. It says that the output is 1 if either input A or B is at the 1 logic level. This differs from the standard OR gate (inclusive-OR) in that the output is 0 for both inputs equal to 1.

Looking at the output in the truth table once again, we see that it is 1 when the input logic levels are unequal. So the exclusive-OR gate is sometimes called an *inequality gate*.

Using the truth table, we can write both a sum-of-products and product-of-sums Boolean equation for exclusive-OR operation. The product terms are marked next to the 1 level outputs and sum terms by the 0 outputs. With the product terms, we write the sum-of-products equation for exclusive-OR operation. This gives

$$F = \bar{A}B + A\bar{B}$$

To get the product-of-sums expression, the sum terms for the 0 outputs are used. Therefore, we get

$$F = (\bar{A} + \bar{B})(A + B)$$

Since the exclusive-OR function is so common to combinational logic circuitry, a new Boolean operator is defined to indicate the logic function. This symbol is $\oplus$, and the exclusive-OR of inputs A and B can be rewritten as

$$F = A \oplus B$$

Besides this, a special logic symbol is used to illustrate an exclusive-OR gate in a logic diagram. This symbol is shown in Fig. 8.1(b). In this diagram, the inputs are marked A and B. The output F is labeled with the Boolean expression of the device.

EXAMPLE 8.1

What is the output logic level of the exclusive-OR gate in Fig. 8.2?

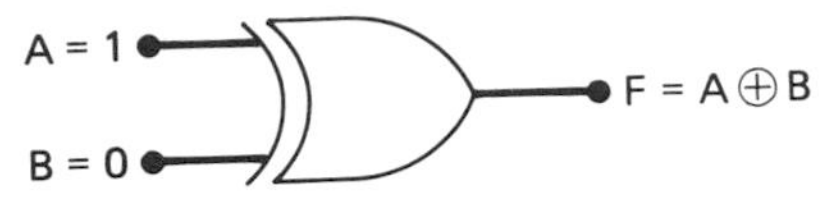

Figure 8.2 Exclusive-OR gate with inputs.

SOLUTION:

The inputs of the gate are $A = 1$ and $B = 0$. Using the truth table in Fig. 8.1(a), the exclusive-OR gate output is found to be 1.

8.3 EXCLUSIVE-OR GATE CIRCUITRY

The exclusive-OR logic function can be constructed with several different combinational logic networks. In fact, from the sum-of-products equation $F = \bar{A}B + A\bar{B}$ we can set up the AND-OR-NOT logic diagram of Fig. 8.3(a) and the NAND-NOT connection of Fig. 8.3(b).

However, using the product-of-sums expression $F = (\bar{A} + \bar{B})(A + B)$, we can make a different AND-OR-NOT gate connection and its equivalent NOR-NOT logic combination.

The AND-OR-NOT exclusive-OR gate circuit in Fig. 8.3(a) has a single OR gate in the first logic level and two AND gates in the second level. Each of these logic gates has just two inputs. Furthermore, both inputs A and B are complemented with inverters.

To show that the operation of the circuit in Fig. 8.3(a) is consistent with the exclusive-OR function, we shall now write the Boolean equation for its output. The output f_1 of the upper inverter is $f_1 = \bar{A}$, and that of the lower inverter is $f_2 = \bar{B}$. The upper second level logic AND gate has inputs f_1 and B. So its output is given by the expression $f_3 = f_1B$ or $\bar{A}B$. For the other AND gate the inputs are f_2 and A. This gives output f_4 as $f_4 = Af_2 = A\bar{B}$. The last step is to find the output F of the OR gate in the first logic level. To do this, we write $F = f_3 + f_4$, and substituting gives $F = \bar{A}B + A\bar{B}$. In this way, we see that the logic diagram does correctly produce exclusive-OR gate operation.

The NAND-NOT exclusive-OR gate connection in Fig. 8.3(b) is a lot like the AND-OR-NOT diagram. But here all AND and OR gates are replaced with NAND gates. The advantage of this approach is that fewer integrated circuits are needed to construct the actual circuitry.

EXAMPLE 8.2

Use the truth table analysis approach to show that the NAND-NOT logic connection repeated in Fig. 8.4(a) operates like an exclusive-OR gate.

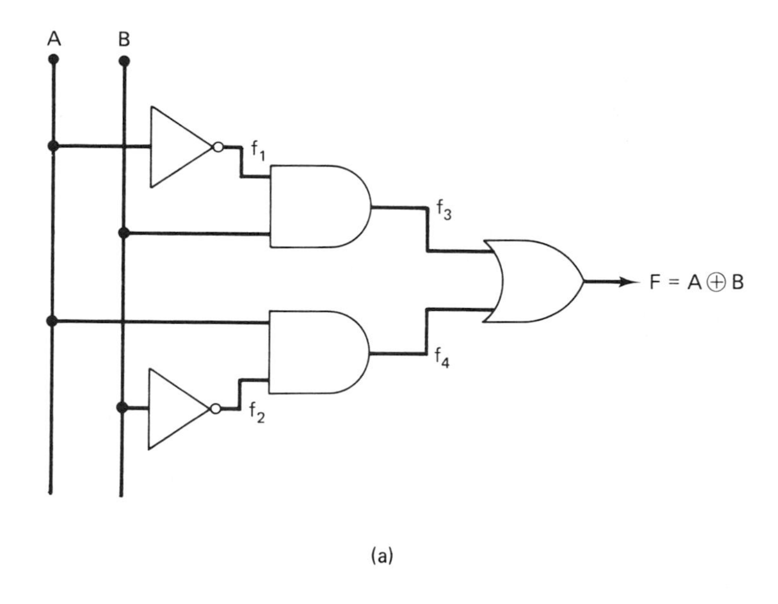

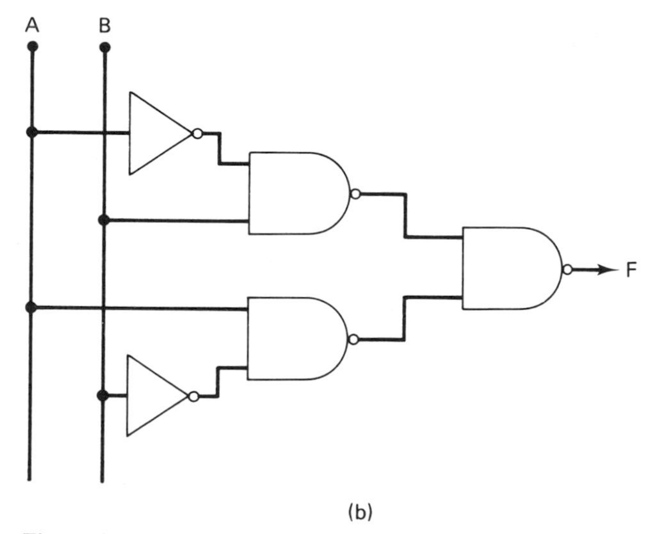

Figure 8.3 (a) AND-OR-NOT exclusive-OR gate; (b) NAND-NOT exclusive-OR gate.

SOLUTION:

The output f_1 of the inverter, gate 1, is 1 when input A is 0. This is shown in the truth table of Fig. 8.4(b). Now the output f_2 of the other inverter is found to be 1 for input $B = 0$. This gives the fourth column of the truth table.

The first NAND circuit, gate 3, has an output due to input B and inverter output f_1. This output is 0 if both B and f_1 are equal to 1, as shown in the fifth

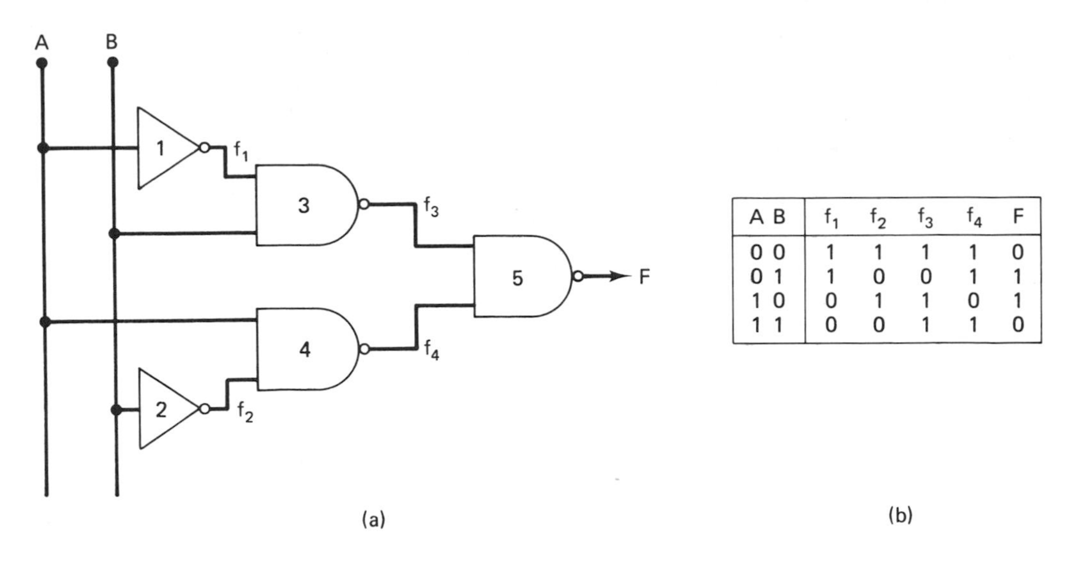

A B	f_1	f_2	f_3	f_4	F
0 0	1	1	1	1	0
0 1	1	0	0	1	1
1 0	0	1	1	0	1
1 1	0	0	1	1	0

Figure 8.4 (a) NAND-NOT logic diagram; (b) Truth table.

column of the truth table. The output of the other NAND gate in the second level of logic is determined by A and f_2. So output f_4 is 0 when both inputs A and f_2 are in the 1 state.

The output of the exclusive-OR circuit comes from the NAND gate labeled 5. Its inputs are f_3 and f_4. This makes output $F = 0$ for both f_3 and f_4 equal to 1. This is shown in the last column of the truth table and indicates correct exclusive-OR operation.

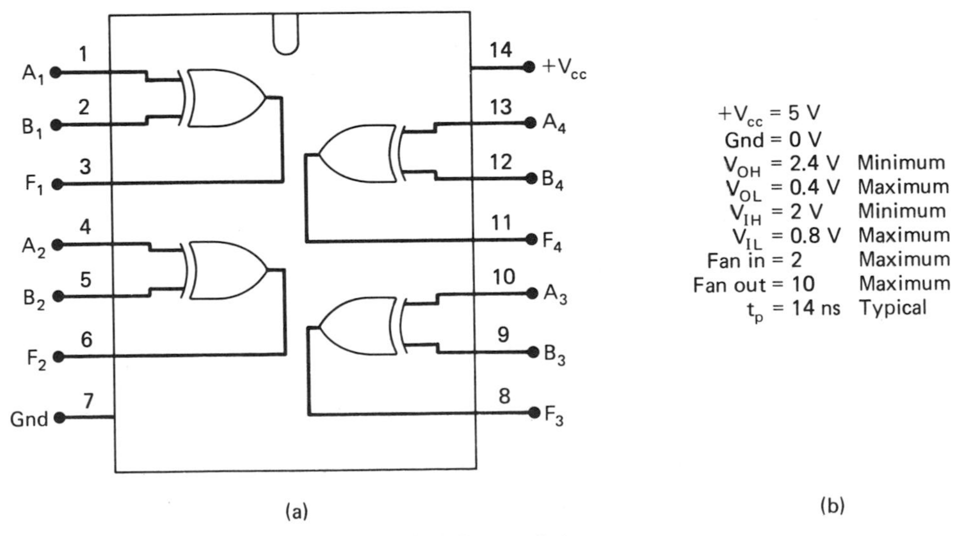

Figure 8.5 (a) 7486 pin layout; (b) Electrical characteristics.

7486 Integrated Exclusive-OR Gate

The 7486 integrated circuit is a TTL exclusive-OR device. It contains four two-input exclusive-OR gates powered from a single set of power supply terminals. A pin layout of the quad two-input exclusive-OR gate IC is shown in Fig. 8.5(a).

Some of the electrical characteristics of the 7486 are listed in Fig. 8.5(b). Here we see that its power supply is +5 V to ground and that its logic levels are consistent with those of the other basic logic gates.

Besides this, we find that the fan in is 2 and the fan out, 10. So the output of each gate in the device can be tied to 10 TTL loads.

Finally, the propagation delay of the device is 14 ns. This says that the output will switch logic levels about 14 ns after the input signals are set.

8.4 THE EQUALITY GATE

Another important combinational logic network is the *equality gate.* It operates like an exclusive-OR gate with an inverter at the output. This connection is shown in the logic diagram of Fig. 8.6. For this reason, the equality gate is sometimes called an *exclusive-NOR gate.*

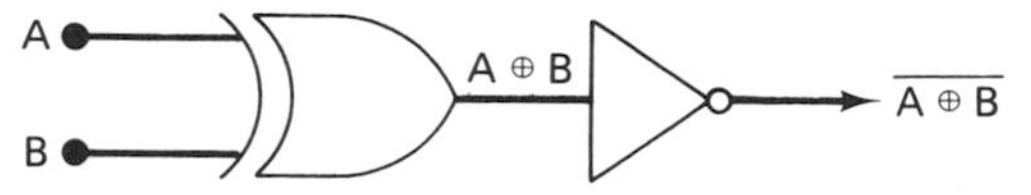

Figure 8.6 Equality gate from exclusive-OR gate.

Truth Table, Boolean Expression, and Logic Symbol

In Fig. 8.7(a), the truth table of the equality function is shown. The output F to inputs A and B is the complement of that for the exclusive-OR function. Looking at this table, we see that the output of the equality gate is 1 if both inputs are at the same logic level.

From the output in the truth table, we can write Boolean equations for equality gate operation. Using the 1 output levels, we get the sum-of-products

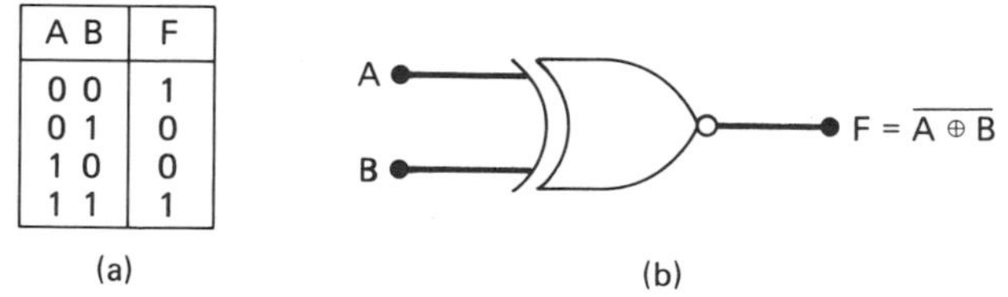

A B	F
0 0	1
0 1	0
1 0	0
1 1	1

Figure 8.7 (a) Equality gate truth table; (b) Logic symbol and Boolean expression.

equation

$$F = \bar{A}\bar{B} + AB$$

and the product-of-sums equation is found from the 0 outputs to be

$$F = (A + \bar{B})(\bar{A} + B)$$

Another way of expressing equality operation is as the complement of the exclusive-OR equation. This gives

$$F = \overline{A \oplus B}$$

The logic symbol used to indicate an equality gate is the same as for the exclusive-OR gate except its output has a dot. In Fig. 8.7(b), this logic symbol is shown with inputs A and B and output F. The output Boolean expression is also indicated.

EXAMPLE 8.3

If the input logic levels of an equality gate are $A = 1$ and $B = 1$, what is the output F?

SOLUTION:

Looking at the equality gate truth table in Fig. 8.7(a), we see that output F equals 1 when both inputs are logic 1.

Circuitry

From the sum-of-products equation $F = \bar{A}\bar{B} + AB$, we can set up two different combinational logic networks for the equality gate. Using AND-OR-NOT logic, we get the logic diagram of Fig. 8.8(a). Replacing the AND and OR gates with NAND gates, we make the NAND-NOT logic connection in Fig. 8.8(b).

EXAMPLE 8.4

Write the Boolean expression for output F in the equality gate logic diagram of Fig. 8.8(a).

SOLUTION:

The outputs of the inverters are given by the expressions

$$f_1 = \bar{A}$$
$$f_2 = \bar{B}$$

Then the outputs of the second level logic AND gates are

$$f_3 = f_1 f_2 = \bar{A}\bar{B}$$
$$f_4 = AB$$

Finally, the output is found to be

$$F = f_3 + f_4 = \bar{A}\bar{B} + AB$$

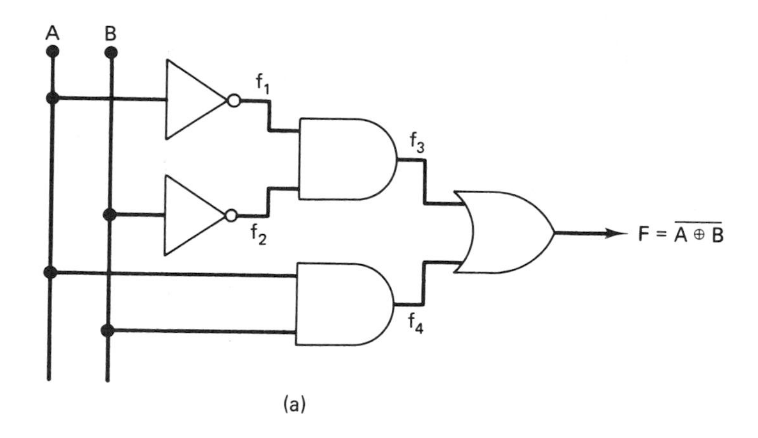

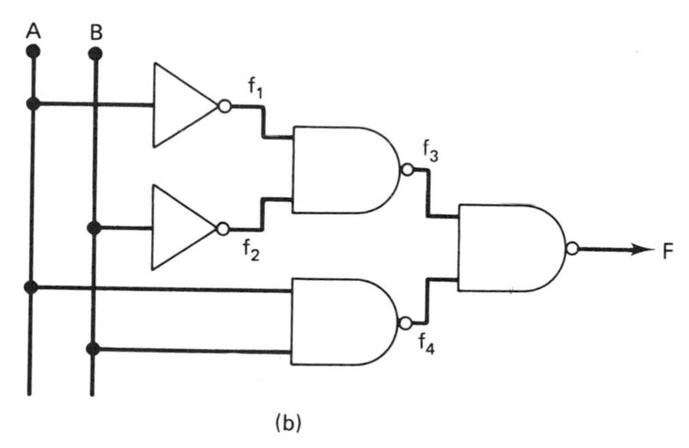

Figure 8.8 (a) AND-OR-NOT equality gate; (b) NAND-NOT equality gate.

8.5 THE COMPARATOR CIRCUIT

In many digital electronic systems, binary codes are compared, and the processing that follows depends on the result of this comparison. The type of circuit used to compare multibit binary words is called a *comparator*.

As an example, let us look at the comparator block of Fig. 8.9. Inputs to the comparator are two 2-bit binary words A and B:

$$A = A_1A_0$$
$$B = B_1B_0$$

The output is labeled F and signals if the two words are equal or unequal. For the two words to be equal, their corresponding bits must be the same. That is,

$$A_0 = B_0$$

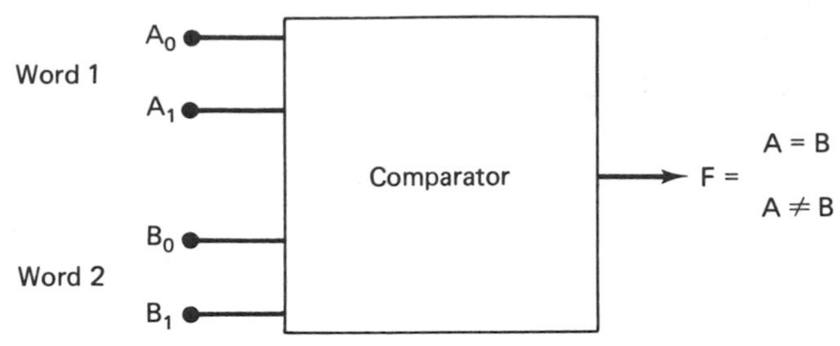

Figure 8.9 Two-bit comparator block.

and

$$A_1 = B_1$$

The comparator circuit is a combinational logic network and is built using the exclusive-OR or equality gate.

8.6 INEQUALITY AND EQUALITY COMPARATORS

The two fundamental comparator circuits are the *equality comparator* and *inequality comparator*. In this section, we shall look into the circuitry and operation of each of these circuits.

Inequality Comparator

An inequality comparator is used to indicate if two incoming words are unequal. Therefore, its output is logic 1 if $A \neq B$ and 0 when $A = B$.

The exclusive-OR gate is a circuit that indicates 1 when its inputs are of unlike logic levels. So it can be used to make an inequality comparator. The logic diagram of Fig. 8.10 shows how two exclusive-OR gates and an OR gate can be connected to make a 2-bit inequality comparator circuit.

In this circuit exclusive-OR gates are used to compare the separate bits of the two words. For instance, the upper gate compares the least significant bits A_0 and B_0. If they are unequal, the output of the gate becomes 1. The most significant bits

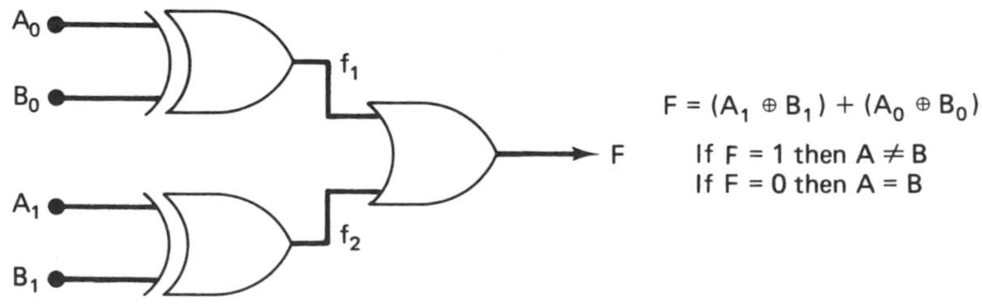

Figure 8.10 Two-bit inequality comparator circuit.

A_1 and B_1 are compared by the lower exclusive-OR gate. The output F of the comparator is from the OR gate, and it is 1 if either or both inputs are 1. This condition occurs if A and B are not equal.

EXAMPLE 8.5

Write a Boolean equation for the output of the inequality comparator in Fig. 8.10.

SOLUTION:

The output of the upper exclusive-OR gate is given by the expression

$$f_1 = A_0 \oplus B_0$$

and that of the lower gate is

$$f_2 = A_1 \oplus B_1$$

Combining these two outputs with the OR function, we get

$$F = f_1 + f_2 = (A_0 \oplus B_0) + (A_1 \oplus B_1)$$

This says that the output will be 1 if $A_0 \neq B_0$ or $A_1 \neq B_1$.

EXAMPLE 8.6

The inequality comparator circuit of Fig. 8.10 is repeated in Fig. 8.11(a). Check its operation with truth table analysis.

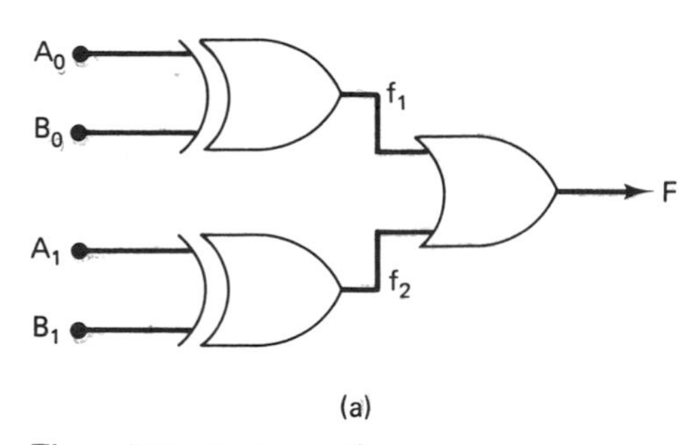

(a)

A_1	A_0	B_1	B_0	f_1	f_2	F
0	0	0	0	0	0	0
0	0	0	1	1	0	1
0	0	1	0	0	1	1
0	0	1	1	1	1	1
0	1	0	0	1	0	1
0	1	0	1	0	0	0
0	1	1	0	1	1	1
0	1	1	1	0	1	1
1	0	0	0	0	1	1
1	0	0	1	1	1	1
1	0	1	0	0	0	0
1	0	1	1	1	0	1
1	1	0	0	1	1	1
1	1	0	1	0	1	1
1	1	1	0	1	0	1
1	1	1	1	0	0	0

(b)

Figure 8.11 (a) Inequality comparator; (b) Truth table.

SOLUTION:

The inequality comparator can be treated as a four-input logic network. The inputs are A_1, A_0, B_1, and B_0, and there are three gate outputs f_1, f_2, and F. This is tabulated in the truth table of Fig. 8.11(b). In this table, all possible combinations of the four inputs are entered.

The f_1 output is 1 if its inputs A_0 and B_0 are unequal. This result is recorded in the f_1 column of the truth table. On the other hand, output f_2 is 1 when A_1 and B_1 are unequal. Then the F output in the table is 1 if f_1 or f_2 is 1.

From the truth table, we see that the output is 1 for all but four sets of binary words. These words are the following:

$$A_1A_0 = 00 \qquad B_1B_0 = 00$$
$$A_1A_0 = 01 \qquad B_1B_0 = 01$$
$$A_1A_0 = 10 \qquad B_1B_0 = 10$$
$$A_1A_0 = 11 \qquad B_1B_0 = 11$$

So the circuit does correctly indicate unequal words.

Equality Comparator

When comparing two binary words, the equality comparator will signal when they are equal. For this reason, the output is 1 when $A = B$ and 0 if $A \neq B$.

The operation of the equality comparator is the exact opposite of the inequality comparator. Therefore, its circuit can be made by inverting the output of the inequality comparator circuit. However, we can do this by just replacing the OR gate at the output with a NOR gate as shown in Fig. 8.12(a).

A second way of making an equality comparator circuit is to use equality

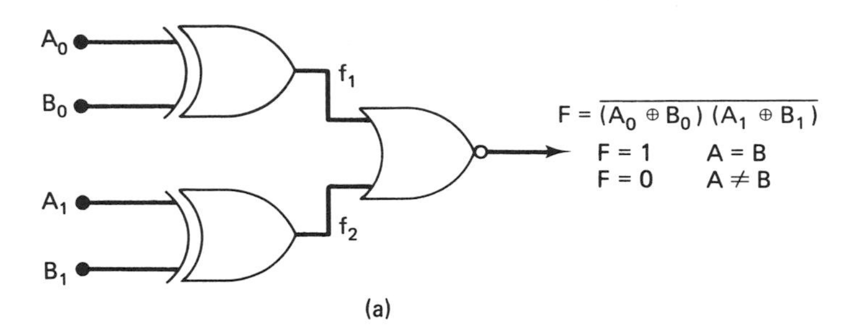

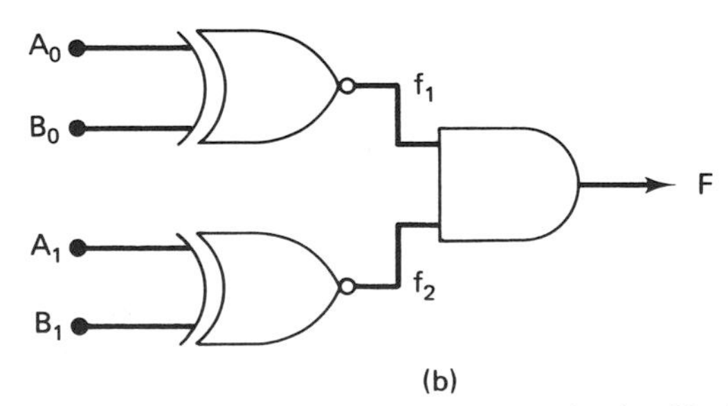

Figure 8.12 (a) Equality comparator circuit; (b) Alternative equality comparator.

logic gates to test the words and an AND gate at the output. This logic connection is shown in Fig. 8.12(b).

EXAMPLE 8.7

Find the output logic level in the equality comparator circuit of Fig. 8.12(a) if the input words are $A_1A_0 = 00$ and $B_1B_0 = 00$.

SOLUTION:

The inputs to the upper exclusive-OR gate are $A_0 = 0$ and $B_0 = 0$. This makes its output $f_1 = 0$. For the other exclusive-OR, the inputs are $A_1 = 0$ and $B_1 = 0$. So its output f_2 is also 0.

Now, the NOR gate inputs are $f_1 = 0$ and $f_2 = 0$. This gives a 1 level at the output of the comparator and is consistent with equality comparator operation.

8.7 MAGNITUDE COMPARATOR

The *magnitude comparator* is another type of comparator circuit. It is like the inequality and equality comparators in that it compares multibit words; however, the magnitude comparator can determine if one word is greater, equal to, or less than the other.

In Fig. 8.13(a), a 2-bit magnitude comparator block is shown. Again, the inputs are two words A and B:

$$A = A_1A_0$$
$$B = B_1B_0$$

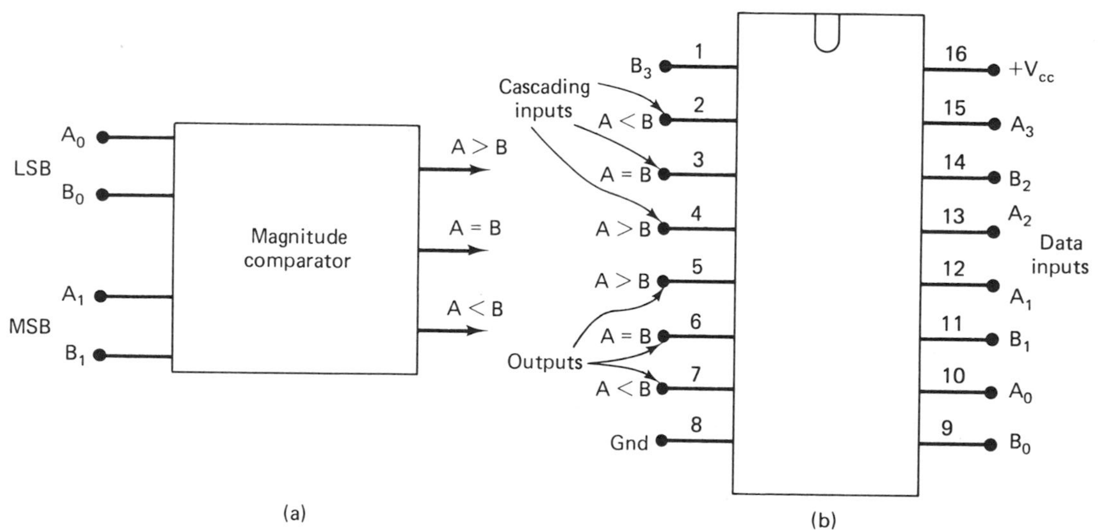

Figure 8.13 (a) Magnitude comparator block; (b) 7485 magnitude comparator pin layout.

But the circuit has three output terminals: one for word A greater than B, a second for A equal to B, and the last for A less than B:

$$F_1 = A > B$$
$$F_2 = A = B$$
$$F_3 = A < B$$

To find if word A is greater, equal to, or less than B, the circuit compares their equivalent decimal values. After this, the corresponding output lead goes to logic 1.

EXAMPLE 8.8

If the input data words to a 2-bit magnitude comparator are $A = 01$ and $B = 10$, what will be the outputs F_1, F_2, and F_3?

SOLUTION:

The decimal equivalents of the input words are

$$A = 01 = 1$$
$$B = 10 = 2$$

Therefore, word A is less than B, and the outputs become

$$F_1 = A > B = 0$$
$$F_2 = A = B = 0$$
$$F_3 = A < B = 1$$

The 7485 Integrated 4-Bit Magnitude Comparator

Multibit magnitude comparators are available as a standard MSI circuit. The 7485 is a TTL 4-bit magnitude comparator device.

A pin layout of the 7485 magnitude comparator is shown in Fig. 8.13(b). From this diagram, we see that the device has three sets of input leads. Two sets are for the *data input words* A and B:

$$A = A_3A_2A_1A_0$$
$$B = B_3B_2B_1B_0$$

The remaining are *control inputs* called the *cascading inputs*. There is one cascading input for each relationship between A and B:

$$C_1 = A > B$$
$$C_2 = A = B$$
$$C_3 = A < B$$

The cascading inputs are provided so that 7485 devices can be interconnected to make larger input word capacity. For example, two 7485 ICs can be tied together to make an 8-bit magnitude comparator.

The *output leads* produce the result of the comparison of input words A and B. These three leads are

$$F_1 = A > B$$
$$F_2 = A = B$$
$$F_3 = A < B$$

and the output corresponding to the correct relation between A and B goes to the 1 logic level.

To make the 7485 operate like a 4-bit comparator, the cascading inputs must be correctly set. As shown in Fig. 8.14(a), the less than and greater than inputs are grounded, but the $A = B$ input is returned to the + end of the V_{cc} supply. By just using the $A = B$ output of this circuit, we can get equality comparator operation. However, to obtain inequality comparator operation the $A = B$ output must be inverted.

Some of the important electrical characteristics of the 7485 comparator are listed in Fig. 8.14(b).

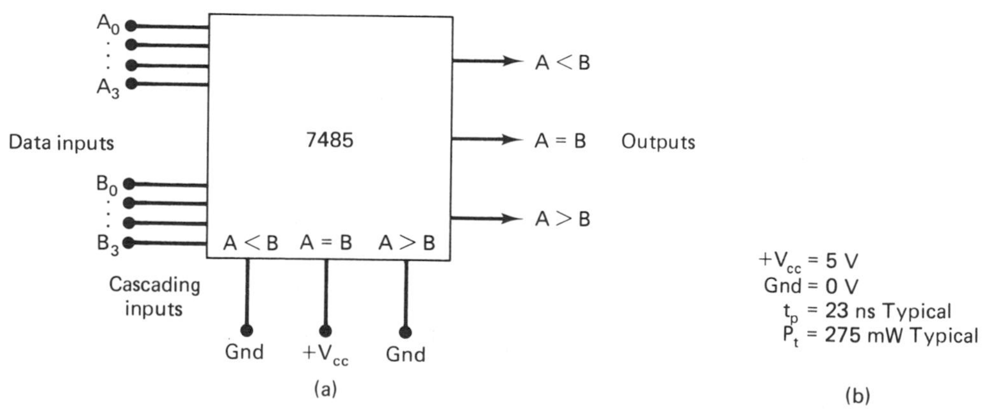

Figure 8.14 (a) Connection as a four-bit magnitude comparator; (b) Electrical characteristics.

The block diagram of Fig. 8.15 shows how two 7485 devices are cascaded to make an 8-bit comparator. Here we see that the comparator for the lower bits, A_0 through A_3 and B_0 through B_3, has its cascading inputs set for 4-bit operation. On the other hand, the cascading inputs of the upper bit comparator are activated by the outputs of the lower comparator. The outputs of the 8-bit circuit are those of the upper comparator.

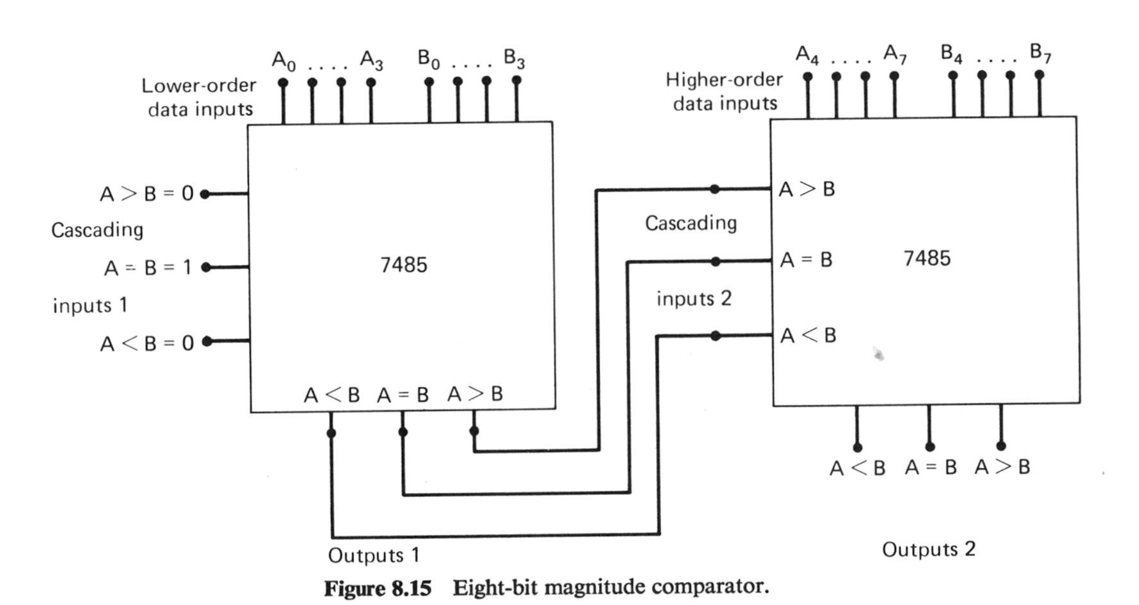

Figure 8.15 Eight-bit magnitude comparator.

ASSIGNMENTS

Section 8.2

1. Describe the operation of an exclusive-OR gate in terms of its output logic level in the 1 state.

2. What is the output of an exclusive-OR gate when its inputs are $A = 1$ and $B = 1$?

3. The inputs of an exclusive-OR gate are called B and C. Write the sum-of-products equation for the gate output, and then rewrite the equation using the exclusive-OR Boolean operator.

Section 8.3

4. Write Boolean equations for outputs f_1, f_2, f_3, f_4, and F in the logic diagram of Fig. 8.16(a). Is output F consistent with exclusive-OR operation?

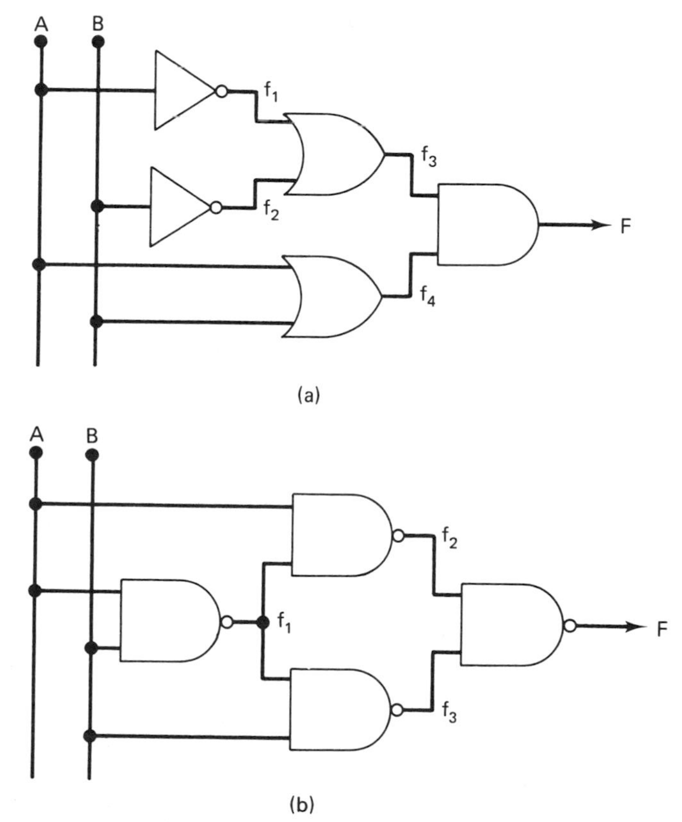

Figure 8.16 (a) AND-OR-NOT exclusive-OR gate; (b) NAND-NOT exclusive-OR gate.

5. The logic diagram in Fig. 8.16(b) is a variation on the NAND-NOT logic diagram of Fig. 8.4(a). Check its operation by truth table analysis. Is the output correct for exclusive-OR operation?

6. Use the product-of-sums Boolean equation to set up a NOR-NOT logic diagram for an exclusive-OR gate.

7. The waveforms in Fig. 5.17 are the input signals to an exclusive-OR gate. Draw the output waveform.

Section 8.4

8. Describe the operation of an equality gate in terms of a 1 level output.

9. If the inputs of an equality gate are $A = 1$ and $B = 0$, what is the output?

10. Perform a truth table analysis of the NAND-NOT equality gate of Fig. 8.8(b). Is output F consistent with equality gate operation?

11. The inputs to the equality gate circuit in Fig. 8.8(a) are $A = 0$ and $B = 0$. What are the logic levels at outputs f_1, f_2, f_3, f_4, and F?

12. Construct the output waveform of an equality gate if its inputs are the waveforms shown in Fig. 5.17.

Section 8.5

13. Draw a block diagram of a 4-bit comparator.

Section 8.6

14. Describe the operation of an inequality comparator in terms of its output logic level in the 0 state.

15. What are outputs f_1, f_2, and F in the inequality comparator of Fig. 8.11(a) for inputs $A = 01$ and $B = 11$?

16. Use truth table analysis to find output F in the equality comparator of Fig. 8.12(b). Is the output operation correct?

17. Write a Boolean equation for outputs f_1, f_2, and F in the comparator circuit of Fig. 8.12(b).

Section 8.7

18. If the inputs to a 4-bit magnitude comparator are $A = 1101$ and $B = 1100$, what will be outputs F_1, F_2, and F_3?

19. To what logic levels must the cascading inputs of a 7485 comparator be set to give 4-bit operation?

20. Draw a block diagram to show how 7485 magnitude comparators must be connected to make a 12-bit word comparator.

21. If the input words to an 8-bit comparator are $A = 01001111$ and $B = 01011111$, indicate the logic levels at the outputs of both the lower and upper 4-bit comparators in Fig. 8-15.

22. The A and B waveforms at the inputs of a 7485 comparator are shown in Fig. 8-17. This comparator is connected for 2-bit operation. Draw the waveforms of the $A > B$, $A = B$, and $A < B$ outputs.

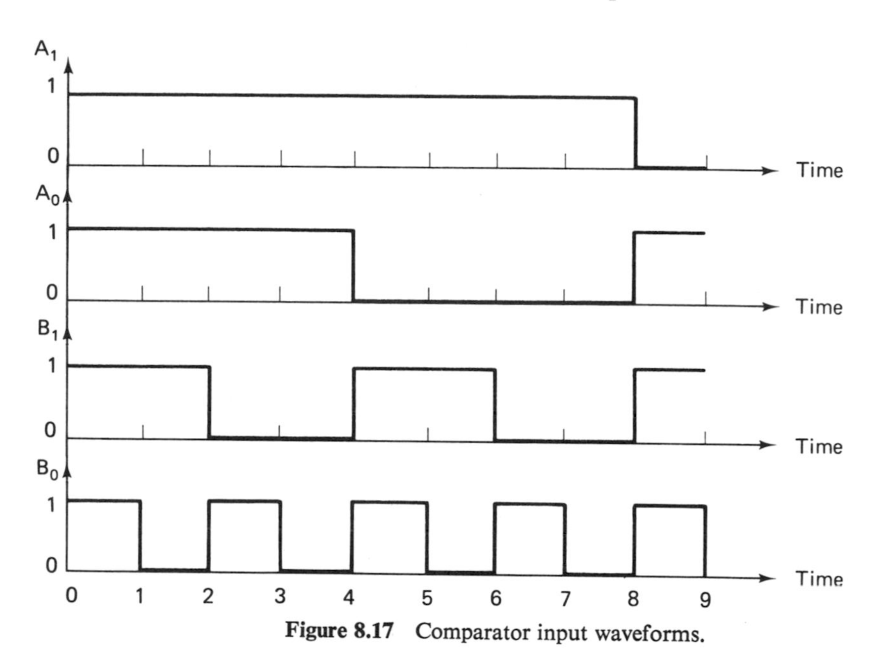

Figure 8.17 Comparator input waveforms.

Digital Coding 9

9.1 INTRODUCTION

In electronic equipment, decimal numbers and other information are represented with binary words or numbers. For each number or piece of information, an equivalent binary combination is defined, and the complete group of these combinations is called a *code*. For instance, 4-bit binary bytes can be used to form a numerical code. In this case, a different binary combination is used for each number from 0 through 9.

This chapter is an introduction to digital coding. For this purpose, the topics that follow are covered:

1. Standard codes
2. Weighted and unweighted codes
3. Parity, parity checkers, and parity generators

9.2 STANDARD NUMERIC CODES

In modern digital equipment, many different codes are used to represent numerical information. Some of the more popular codes are listed here:

1. BCD code
2. XS-3 code

3. Gray code
4. XS-3 Gray code

In certain applications, the use of one code or the other simplifies and reduces the circuitry needed to process information. By limiting the amount of switching circuitry, the reliability of the digital system is improved.

Weighted and Unweighted Codes

Codes are generally grouped into one of two categories: *weighted codes* and *unweighted codes*. For a code to be weighted, each of its binary combinations must be numerically equivalent to the corresponding decimal number. That is, each bit in the code is given a weight, and the sum of the weights for the bits that are logic 1 equals the decimal number.

This idea is the same as that used for the weights of binary, octal, and hexadecimal numbers in Chapter 2. However, for a code the weights of all bits are not necessarily related to the same base number.

When a code is unweighted, the separate bits do not have a numerical weight. For this reason, there is no direct relation between a coded number and its equivalent decimal value. Instead, binary words are arbitrarily selected to indicate numbers 0 through 9.

9.3 BINARY-CODED DECIMAL CODE

One of the most widely used codes in digital circuitry is the *binary-coded decimal*. This code is also called the *BCD code* and *8421 code*.

To form the BCD code, the first 10 4-bit binary numbers are selected to represent decimal numbers 0 through 9, respectively. The BCD code is shown in the table of Fig. 9.1(a). Looking at this table, we see that the decimal number 0 is

Decimal	BCD
0	0000
1	0001
2	0010
3	0011
4	0100
5	0101
6	0110
7	0111
8	1000
9	1001

(a)

Bits	A_3	A_2	A_1	A_0
Weights	2^3	2^2	2^1	2^0
	8	4	2	1

(b)

Figure 9.1 (a) BCD code; (b) BCD code weights.

written as 0000 in BCD code. Moreover, the number 9 is given by 1001. Waveforms of the BCD code are shown in Fig. 9.2.

The BCD code is a weighted code. In Fig. 9.1(b), the four bits A_3, A_2, A_1, and A_0 are shown with their weights. A_0 is the least significant bit and has a weight of 2^0 or 1. The remaining bits A_1, A_2, and A_3 have weights of $2^1 = 1$, $2^2 = 4$, and $2^3 = 8$, respectively.

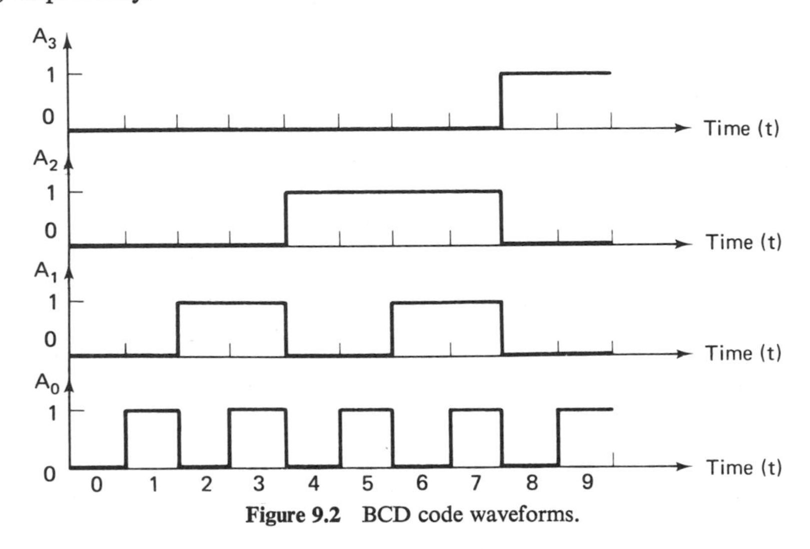

Figure 9.2 BCD code waveforms.

To show that the BCD code is weighted, let us check the code for decimal number 7:

$$\begin{aligned} 7 &= 0111 \\ &= 0(2^3) + 1(2^2) + 1(2^1) + 1(2^0) \\ &= 4 + 2 + 1 \\ 7 &= 7 \end{aligned}$$

Using the same approach, the other nine BCD combinations can be shown to be correctly weighted.

EXAMPLE 9.1

Write the decimal number 84 in BCD code.

SOLUTION:

To write this two-digit number in BCD code, the value in each digit must be replaced with its 4-bit equivalent from the BCD code. Using the table of Fig. 9.1(a), we get

$$8 = 1000$$

$$4 = 0100$$

Combining these two codes, we find that

$$84_{10} = 10000100_{BCD}$$

EXAMPLE 9.2

Find the decimal value for the BCD-coded number 00010010.

SOLUTION:

Starting from the LSB, the umber is divided into two groups, each with 4 bits. These groups represent BCD digits and are to be replaced with their equivalent decimal numbers:

$$00010010 = 0001 \ \ 0010$$

$$00010010 = 12$$

EXAMPLE 9.3

Rewrite the binary number 11000000_2 using BCD code.

SOLUTION:

To express this binary number in BCD code, we must first find its decimal value. This value is obtained as follows:

$$\begin{aligned} 11000000 &= 1(2^7) + 1(2^6) \\ &= 1(128) + 1(64) \\ 11000000_2 &= 192_{10} \end{aligned}$$

Next, each decimal digit is replaced with its equivalent 4-bit BCD code:

$$192_{10} = 000110010010_{BCD}$$

9.4 EXCESS-3 CODE

Another code widely used to represent numerical data in digital equipment is the *excess-3 code*. The name for this code is abbreviated as *XS-3*.

A list of the XS-3 code for numbers 0 through 9 is shown in Fig. 9.3(a). From this table, we see that the XS-3 code is formed from the BCD code by shifting the code by three numbers.

For instance, the BCD code starts with decimal number 0 equal to 0000. However, the XS-3 value for number 0 is 0011. This binary combination is used for 3 in the BCD code. After this, the code for XS-3 number 1 is the same as that of BCD number 4 and so on through number 9. Waveforms for the XS-3 code are shown in Fig. 9.3(b).

The XS-3 code is unweighted. For this reason, no bit weights are defined.

EXAMPLE 9.4

Express the number 42_{10} in XS-3 code.

SOLUTION:

To form the XS-3 equivalent of this decimal number, we must represent each digit as a 4-bit XS-3 code:

$$4 = 0111$$

$$2 = 0101$$

Decimal	XS-3
0	0011
1	0100
2	0101
3	0110
4	0111
5	1000
6	1001
7	1010
8	1011
9	1100

(a)

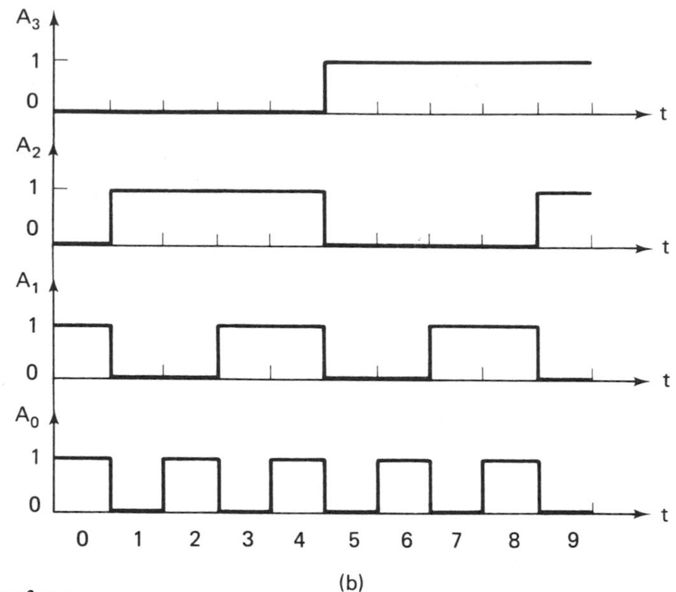

Figure 9.3 (a) Excess-3 code; (b) Waveforms.

This results in

$$42_{10} = 01110101_{\text{XS-3}}$$

EXAMPLE 9.5

The number 10010110 is expressed in XS-3 code. What is the decimal value?

SOLUTION:

When converting an XS-3-coded number to decimal form, the bits must be separated into groups of four starting from the LSB. After this, each group is replaced by its equivalent value from Fig. 9.3(a):

$$10010110 = 1001\ 0110$$

$$10010110_{\text{XS-3}} = 63_{10}$$

9.5 GRAY CODE AND EXCESS-3 GRAY CODE

Two other codes important in digital electronics are the *Gray code* and *excess-3 Gray code*. The Gray code for numbers 0 through 9 is shown in Fig. 9.4. It is another unweighted code.

From this table, we see that the binary words in the Gray code are selected so just 1 bit changes logic levels when going from number to number. In this way, the amount of switching is minimized and the reliability of the switching circuitry is improved. This code is mainly used in mechanical switching systems.

A variation of the Gray code is shown in Fig. 9.5. It is called the excess-3 Gray

Decimal	Gray
0	0000
1	0001
2	0011
3	0010
4	0110
5	0111
6	0101
7	0100
8	1100
9	1101

Figure 9.4 Gray code.

Decimal	Excess-3 Gray
0	0010
1	0110
2	0111
3	0101
4	0100
5	1100
6	1101
7	1111
8	1110
9	1010

Figure 9.5 Excess-3 Gray code.

code (XS-3 Gray code) and is the original Gray code shifted by three binary combinations. This code exhibits the same properties as the Gray code.

EXAMPLE 9.6

Express the number 937 with the Gray code.

SOLUTION:

Using 4-bit words for numbers 9, 3, and 7 from the table in Fig. 9.4, we get the Gray code number that follows:

$$937_{10} = 110100100100_{\text{Gray}}$$

EXAMPLE 9.7

Convert the Gray code number 011001011100 to decimal form.

SOLUTION:

Dividing the bits of the number into three groups of four and replacing them by the values from the table in Fig. 9.4, we get the decimal number that follows:

$$011001011100_{\text{Gray}} = 468_{10}$$

EXAMPLE 9.8

Write the decimal number from Example 9.6 in excess-3 Gray code.

SOLUTION:

Replacing each number with a 4-bit word from Fig. 9.5, we find the XS-3 Gray code number:

$$937_{10} = 101001011111_{\text{XS-3 Gray}}$$

9.6 PARITY AND THE PARITY BIT

In digital electronics, coded information is transferred between circuits and systems. This exchange of data must be done without error.

However, a problem found in some equipment is noise or transients that cause errors in the transfer of data. For instance, noise at an input of a circuit can cause a 0 logic level to be changed to a 1. In a similar way, a logic 1 can be changed to a 0.

To improve reliability of information transfer, codes with *parity* are often used. In this way, circuits can be used to test input data for correct transmission.

Odd and Even Parity

Coded information can have one of two types of parity. They are called *odd parity* and *even parity*. For a code to have odd parity, each word must have an odd number of bits at the 1 logic level. On the other hand, even parity means that each word in the code has an even number of bits that are 1.

Parity Bit

A code without parity such as the BCD, XS-3, Gray, or XS-3 Gray codes can be made to have parity by adding another bit. This fifth bit is called the *parity bit*, and its logic level can be set to give even or odd parity.

For example, in Fig. 9.6 the BCD code is shown with a parity bit set for even parity. Here we see that the parity bit P is made 1 when there is an odd number of bits at logic 1 and left 0 when an even number of bits are already 1. From this table, we see that the word 0000 is considered to have even parity.

P	A	B	C	D
0	0	0	0	0
1	0	0	0	1
1	0	0	1	0
0	0	0	1	1
1	0	1	0	0
0	0	1	0	1
0	0	1	1	0
1	0	1	1	1
1	1	0	0	0
0	1	0	0	1

Figure 9.6 BCD code with even parity.

EXAMPLE 9.9

Make a table of the Gray code with parity bit P set for odd parity.

SOLUTION:

The table in Fig. 9.7 shows the Gray code with parity bit. To make odd parity, P is set to 1 when the $ABCD$ combination has 0 or 2 bits that are at logic 1.

P	A	B	C	D
1	0	0	0	0
0	0	0	0	1
1	0	0	1	1
0	0	0	1	0
1	0	1	1	0
0	0	1	1	1
1	0	1	0	1
0	0	1	0	0
1	1	1	0	0
0	1	1	0	1

Figure 9.7 Gray code with odd parity.

9.7 THE PARITY CHECKER

When a code with parity is used to represent digital information to be transferred between systems, reliability is improved by checking the data for correct parity after it has been transferred. To do this, a circuit called a *parity checker* is used.

In Fig. 9.8, a block diagram of a parity checker circuit is shown. This parity checker has four data inputs A, B, C, and D. Besides this, it has two outputs marked P_o and P_e for odd parity and even parity, respectively.

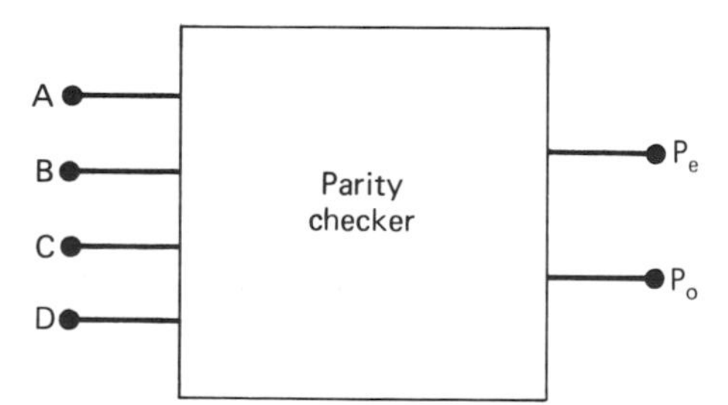

Figure 9.8 Parity checker block.

If the 4-bit input code $ABCD$ has odd parity, the P_o output switches to logic 1 and output P_e remains at logic 0. On the other hand, for an input with even parity the P_e output is logic 1 and P_o stays at logic 0.

The P_o or P_e output can be sent to logic circuitry to indicate that the input information has correct parity. If the parity is wrong, a signal can be set up to initiate resending the data.

Odd/Even Parity Checker Circuit

An *odd/even parity checker circuit* can be used to indicate if the different words in a code have even or odd parity.

In Fig. 9.9(a), a simplified 4-bit parity checker circuit is shown. Here, we see that the circuit is formed with three exclusive-OR gates and an inverter to provide the P_e output. A truth table of this circuit's operation for a BCD code input is shown in Fig. 9.9(b).

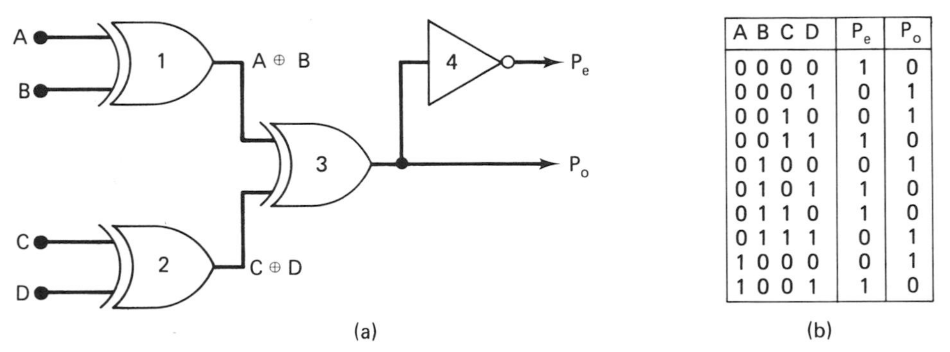

A B C D	P_e	P_o
0 0 0 0	1	0
0 0 0 1	0	1
0 0 1 0	0	1
0 0 1 1	1	0
0 1 0 0	0	1
0 1 0 1	1	0
0 1 1 0	1	0
0 1 1 1	0	1
1 0 0 0	0	1
1 0 0 1	1	0

Figure 9.9 (a) Four-bit odd/even parity checker; (b) Truth table.

EXAMPLE 9.10

Check the operation of the odd/even parity checker in Fig. 9.9(a) for input $ABCD = 0010$.

SOLUTION:

The input to gate 1 is $AB = 00$. For an exclusive-OR gate, the output is at logic 0:

$$A \oplus B = 0$$

At the same time, the input to gate 2 is $CD = 10$, and the output is at logic 1:

$$C \oplus D = 1$$

These two outputs are passed on to inputs of gate 3 to form the odd parity output P_o. On gate 3 one input is logic 0, and the other is 1. This makes the output equal to 1:

$$P_o = 1$$

Therefore, the input data have odd parity.

The P_o output is inverted by gate 4 to make even parity output P_e. When P_o equals 1, P_e is at logic 0:

$$P_e = 0$$

Both parity outputs are consistent with those indicated in the table of Fig. 9.9(b).

9.8 THE PARITY GENERATOR

To produce parity for a code without parity such as the BCD or XS-3 code, a *parity generator circuit* is used.

A general parity generator block is shown in Fig. 9.10. From this diagram, we see that a 4-bit code is entered at the input and that a 5-bit code with parity is provided at the output. Of these five outputs, four are inputs *A*, *B*, *C*, and *D*. The fifth output is parity bit *P*.

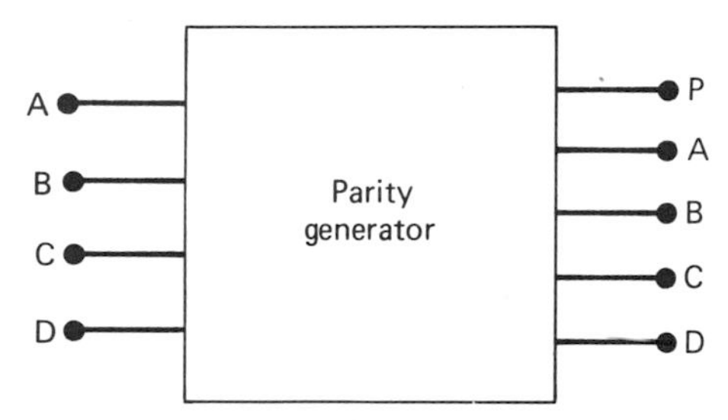

Figure 9.10 Parity generator block.

Even Parity Generator

A parity generator circuit can be made from a parity checker like that shown in Fig. 9.9(a). For example, the *even parity generator circuit* of Fig. 9.11(a) is just an odd parity checker circuit in which the odd parity output is used as the parity bit.

The truth table of an even parity generator is shown in Fig. 9.11(b). Here, we notice that the input is the BCD code; however, the output is the BCD code re-

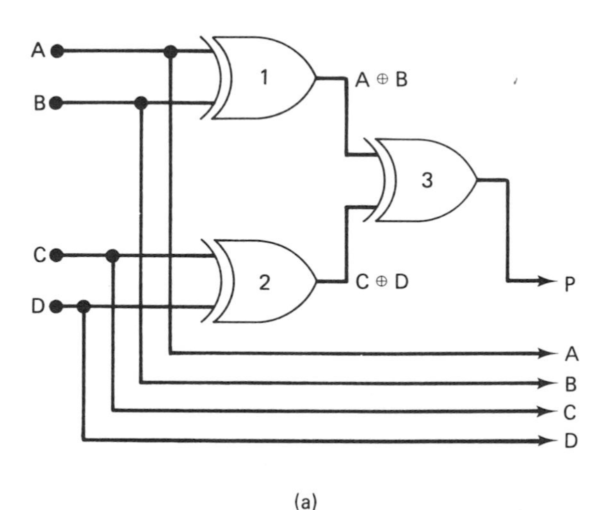

Input	Output
A B C D	P A B C D
0 0 0 0	0 0 0 0 0
0 0 0 1	1 0 0 0 1
0 0 1 0	1 0 0 1 0
0 0 1 1	0 0 0 1 1
0 1 0 0	1 0 1 0 0
0 1 0 1	0 0 1 0 1
0 1 1 0	0 0 1 1 0
0 1 1 1	1 0 1 1 1
1 0 0 0	1 1 0 0 0
1 0 0 1	0 1 0 0 1

(b)

Figure 9.11 (a) Even parity generator circuit; (b) Truth table.

peated with parity bit P. Looking at the output states, we find that each combination has an even number of 1s or even parity.

EXAMPLE 9.11

Form a network truth table for the odd parity generator circuit in Fig. 9.12(a). Use the XS-3 code as the input.

SOLUTION:

The output of gate 1 is $f_1 = A \oplus B$, and this output is 1 if input A or B is at 1. However, if both A and B are 0 or 1, the output remains at 0. The output f_1 to the XS-3 code is listed in the table of Fig. 9.12(b).

Next, the output of gate 2, $f_2 = C \oplus D$, is to be found. It is logic 1 if input C or D is at logic 1 but not if both C and D are 1 or 0. This output is indicated under the f_2 column of the truth table.

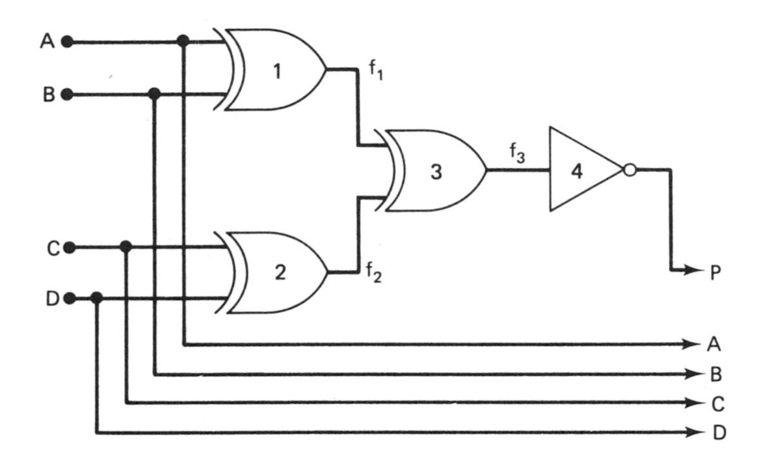

(a)

Input		f₁	f₂	f₃	Output
A B C D		f_1	f_2	f_3	P A B C D
0 0 1 1		0	0	0	1 0 0 1 1
0 1 0 0		1	0	1	0 0 1 0 0
0 1 0 1		1	1	0	1 0 1 0 1
0 1 1 0		1	1	0	1 0 1 1 0
0 1 1 1		1	0	1	0 0 1 1 1
1 0 0 0		1	0	1	0 1 0 0 0
1 0 0 1		1	1	0	1 1 0 0 1
1 0 1 0		1	1	0	1 1 0 1 0
1 0 1 1		1	0	1	0 1 0 1 1
1 1 0 0		0	0	0	1 1 1 0 0

(b)

Figure 9.12 (a) Odd parity generator; (b) Network truth table.

After this, gate 3 produces the exclusive-OR of inputs $A \oplus B$ and $C \oplus D$. This output is marked f_3 in the table and is logic 1 if one of the two inputs is 1.

The parity bit P is the output of gate 4. In the truth table, P equals the complement of f_3.

This P output is combined with inputs A, B, C, and D to give the 5-bit XS-3 code with odd parity. In the output, the parity bit is in the MSB location. Looking at the output in the table, we see that the code has an odd number of 1s in each combination and has odd parity.

9.9 INTEGRATED PARITY GENERATOR/CHECKER DEVICE

Parity generator/checker devices are available as standard MSI circuits. The block diagram of a 74180 parity generator/checker is shown in Fig. 9.13(a) and its pin layout in Fig. 9.13(b).

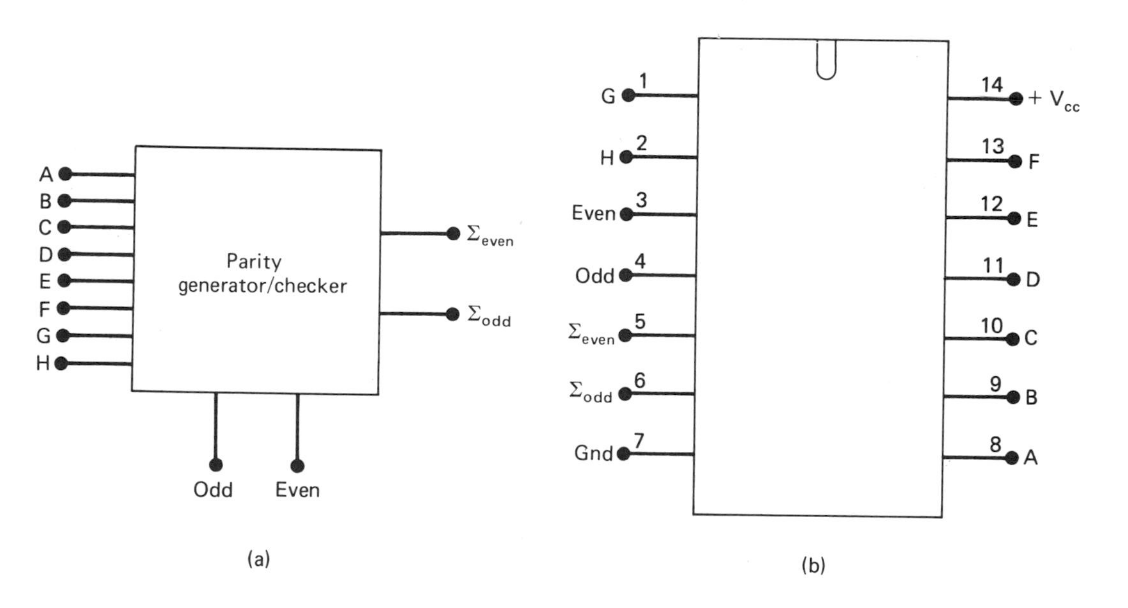

Odd	Even	Operation
0	0	Inhibit
0	1	Indicates parity by logic 1
1	0	Indicates parity by logic 0
1	1	Inhibit

(c)

Figure 9.13 (a) 74180 parity checker/generator block; (b) 74180 pin layout; (c) Modes of operation.

In this block diagram, we see that the device has eight input leads. These leads are marked *A* through *H*. Moreover, two outputs are produced. One is called $\sum_{odd}$ and is used to indicate odd parity. The other is the even parity output $\sum_{even}$. Here, the Greek symbol sigma means sum, and the outputs are used to indicate if the sum of the 1 states in the 8-bit input code is even or odd.

Besides this, two control leads are supplied. These leads are labeled *odd* and *even*. By changing the logic levels at these leads, the way in which the parity generator/checker works is changed. For instance, making both controls 0 or 1 inhibits the device from operating, and no output is produced when a code appears at the inputs.

However, by setting the even control to logic 1 and odd to 0 the IC is set to indicate odd or even parity with a logic 1 at the corresponding output. On the other hand, by making the even control 0 and odd equal to 1 the parity checker signals the correct parity by a logic 0 at the corresponding output. These modes of operation are summarized in the table of Fig. 9.13(c).

Parity Checker Operation

The 74180 IC can be used to make an odd or even parity checker circuit. Furthermore, it can be set to indicate the parity of the code at the input with a logic 0 or logic 1 level.

In Fig. 9.14, a 74180 is connected for operation as an 8-bit even parity checker. Here the odd control lead is wired to ground and the even control to $+V_{cc}$. These connections make the control signals odd = 0 and even = 1.

As shown in the table of Fig. 9.13(c), the parity checker will indicate even parity by a logic 1 at the $\sum_{even}$ output. When an 8-bit code with odd parity is entered into the circuit, the $\sum_{even}$ output switches to logic 0.

EXAMPLE 9.12

The input code to the parity checker of Fig. 9.14 is 01011111. What is the logic level at the $\sum_{even}$ output?

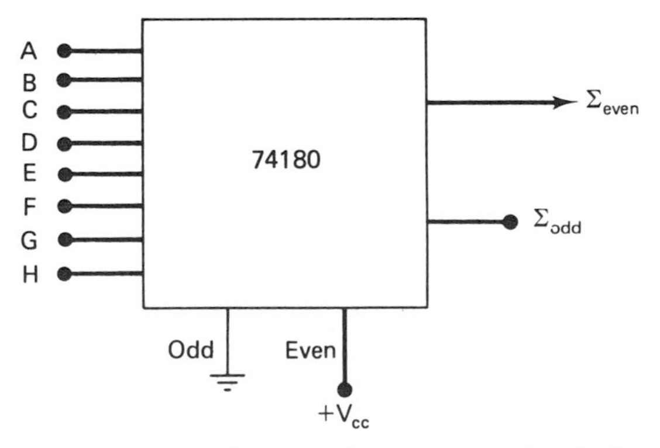

Figure 9.14 74180 connected as an even parity checker.

SOLUTION:

The circuit is set up to indicate even parity by a logic 1 at the $\sum_{\text{even}}$ output. This input code has 6 bits that are logic 1. This number of 1 states is consistent with even parity. For this reason, the $\sum_{\text{even}}$ output is logic 1:

$$\sum_{\text{even}} = 1$$

Parity Generator Operation

A 74180 device can also be connected to work as an odd or even parity generator. In Fig. 9.15(a), a 4-bit even parity generator circuit is shown. Here the *A*, *B*, *C*, and *D* inputs are used, and all others are wired to ground to make logic 0.

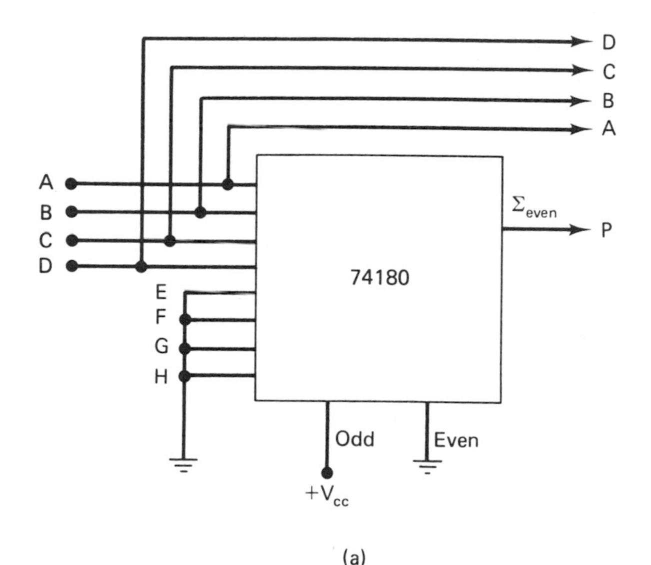

A B C D	P(Σ_{even})	A B C D
0 0 1 0	1	0 0 1 0
0 1 1 0	0	0 1 1 0
0 1 1 1	1	0 1 1 1
0 1 0 1	0	0 1 0 1
0 1 0 0	1	0 1 0 0
1 1 0 0	0	1 1 0 0
1 1 0 1	1	1 1 0 1
1 1 1 1	0	1 1 1 1
1 1 1 0	1	1 1 1 0
1 0 1 0	0	1 0 1 0

(b)

Figure 9.15 (a) 74180 connected as an even parity generator; (b) Excess-3 Gray code with even parity.

In this circuit, the control leads odd and even are set to 1 and 0, respectively. This connection makes the parity generator produce a 0 at the output corresponding to the parity of the input code. In this way, the $\sum_{\text{even}}$ output generates the logic level needed for an even parity bit. The $\sum_{\text{even}}$ output is used as the parity bit along with the 4-bit code at the input to make a 5-bit code with even parity at the output.

To illustrate the operation of this circuit, let us find its output when the excess-3 Gray code is applied to the input. As a first input, let us take the code 0101 that represents number 3. This input has 2 bits at logic 1. For this reason, the $\sum_{\text{even}}$ output goes to logic 0 and signals even parity. In this way, the 5-bit code with parity is 00101. The parity bit is again used in the most significant bit location.

For a second example, we shall take the XS-3 Gray number for 4. In this case, the input is $ABCD = 0100$. This code has odd parity, and the $\sum_{\text{even}}$ output switches to logic 1. Combining the inputs with the parity bit, we get 10100. Again, the 5-bit code at the output has even parity. A complete listing of the outputs for the excess-3 Gray code is shown in Fig. 9.15(b).

ASSIGNMENTS

Section 9.2

1. Give the names of three widely used numeric codes.

2. Define a weighted code and an unweighted code.

Section 9.3

3. What are two other names for the binary-coded decimal code?

4. Is the binary-coded decimal code a weighted or unweighted code?

5. Find the decimal value for the binary-coded decimal numbers that follow.
(a) 1001 (b) 00111001 (c) 100001110000

6. Write the decimal numbers that follow in BCD code.
(a) 29 (b) 99 (c) 106

7. Convert the binary numbers that follow to decimal form and then express in BCD code.
(a) 11010110_2 (b) 1000011_2

Section 9.4

8. Is the XS-3 code weighted?

9. The numbers that follow are written in excess-3 code. Find their decimal value.
(a) 010000110110 (b) 11001100

10. Express the decimal numbers that follow in excess-3 code.
(a) 29 (b) 67

11. Convert the binary numbers that follow to decimal form and express in XS-3 code.
(a) 1001001_2 (b) 1100110001_2

Section 9.5

12. The number 11011101 is written in Gray code. What is the decimal value?

13. Find the decimal value for the excess-3 Gray code number 01010100.

14. Write the decimal number 205 in both Gray code and XS-3 Gray code.

15. Draw waveforms like those shown in Fig. 9.3(b) for the Gray code.

16. Draw waveforms like those shown in Fig. 9.3(b) for the XS-3 Gray code.

Section 9.6

17. Describe a code with even parity and a code with odd parity.

18. What is the term used to refer to an extra bit added to a code to give even or odd parity?

19. Make a table of the XS-3 code with a parity bit (P) set for odd parity.

Section 9.7

20. What type of circuit is used to find if a code has even or odd parity?

21. Make a truth table of the operation of the circuit in Fig. 9.16 for an XS-3 code input. Does the output indicate odd or even parity with a logic 1 level?

22. The waveforms of Fig. 9.3(b) are applied to the input of the parity checker circuit in Fig. 9.16. Draw the output waveform.

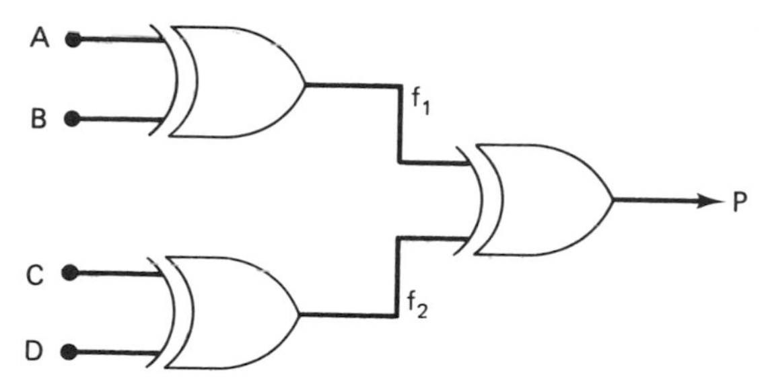

Figure 9.16 Parity checker.

Section 9.8

23. What is the name of the circuit used to make a code without parity have even or odd parity?

24. Use the waveforms drawn in Problem 15 as the input to the circuit in Fig. 9.17. Construct the output waveforms.

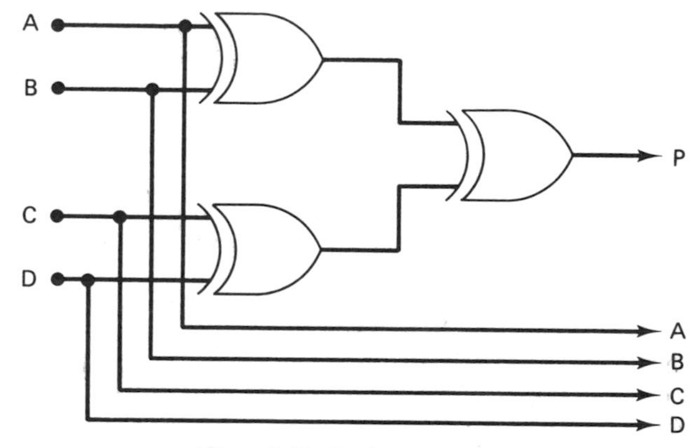

Figure 9.17 Parity generator.

Section 9.9

25. If the circuit in Fig. 9.14 has its controls reset to odd $= 1$ and even $= 0$ with a new input of 11101111, what are the $\sum_{odd}$ and $\sum_{even}$ outputs?

26. Draw the connection of the 74180 as a 4-bit odd parity checker that indicates the correct parity with a logic 1 at the output.

27. Show the hookup of the 74180 that is needed to make a parity generator with a 9-bit odd parity output.

28. Assume that the input to the parity checker in Problem 26 is the Gray code; draw the output waveform.

Decoder Circuitry 10

10.1 INTRODUCTION

In many digital electronic systems, information is processed using different codes in some sections of circuitry. For this reason, the separate sections of the system are interconnected with circuitry to change between codes. This type of circuit is known as a *decoder*.

In this chapter, we shall study gate decoder circuits and MSI decoder devices. The topics included are as follow:

1. Types of decoding circuitry
2. Gate decoder matrix
3. Integrated BCD to decimal decoder
4. Other integrated decoders

10.2 TYPES OF DECODING CIRCUITRY

A decoder circuit is used to change codes between input and output. Figure 10.1 shows a block diagram of a general decoder circuit. In this diagram, we see that the circuit has a 4-bit data input marked as *ABCD*. Besides this, most decoders have 10 outputs. These outputs usually represent decimal numbers 0 through 9.

The input to a decoder circuit can be one of the many codes introduced in

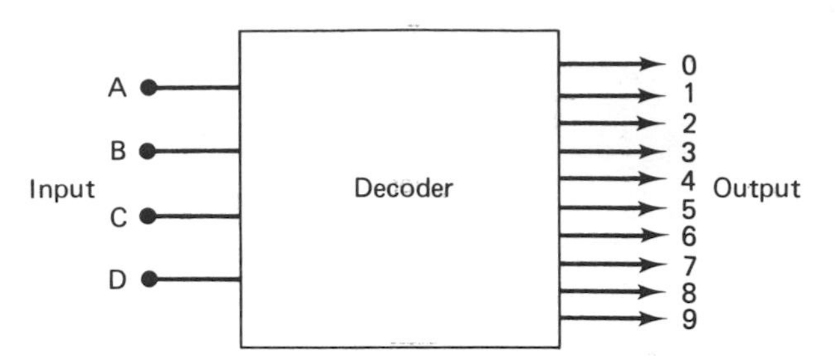

Figure 10.1 Decoder block diagram.

Chapter 9. For instance, the BCD code can be applied at the input, and the output produces a decimal code. In this case, the circuit is called a *BCD to decimal decoder*.

Some of the other widely used decoder circuits are the *excess-3 to decimal decoder*, *Gray to decimal decoder*, and *excess-3 Gray to decimal decoder*.

10.3 GATE DECODER MATRIX

The basic circuit used to make a decoder is called a *matrix*. A matrix is an array of logic gates, and it can be constructed from integrated AND gates or NAND gates. In this section, we shall show how different types of decoder circuits can be constructed with logic gates.

BCD to Decimal Decoder

The AND gate matrix of a BCD to decimal decoder is shown in Fig. 10.2(a). This circuit consists of 4 inverter gates and 10 4-input AND gates.

The BCD to decimal decoder has four inputs A, B, C, and D. Four inverters are used to produce the complemented inputs $\bar{A}$, $\bar{B}$, $\bar{C}$, and $\bar{D}$. These signals are produced by gates 1 through 4 in the logic diagram.

The logic gate matrix has one AND gate for each of the 10 decimal outputs. Looking at Fig. 10.2(a), we see that the outputs of gate 5 through gate 14 are marked 0 through 9, respectively. These gates are wired to the inputs and complemented inputs needed to give the correct decimal output corresponding to the input BCD code.

For instance, gate 5 has as its inputs $\bar{A}$, $\bar{B}$, $\bar{C}$, and $\bar{D}$. Its output is given by the Boolean equation

$$f_0 = \bar{A}\bar{B}\bar{C}\bar{D}$$

When the BCD code at the input is for decimal number 0, the input code is

$$ABCD = 0000$$

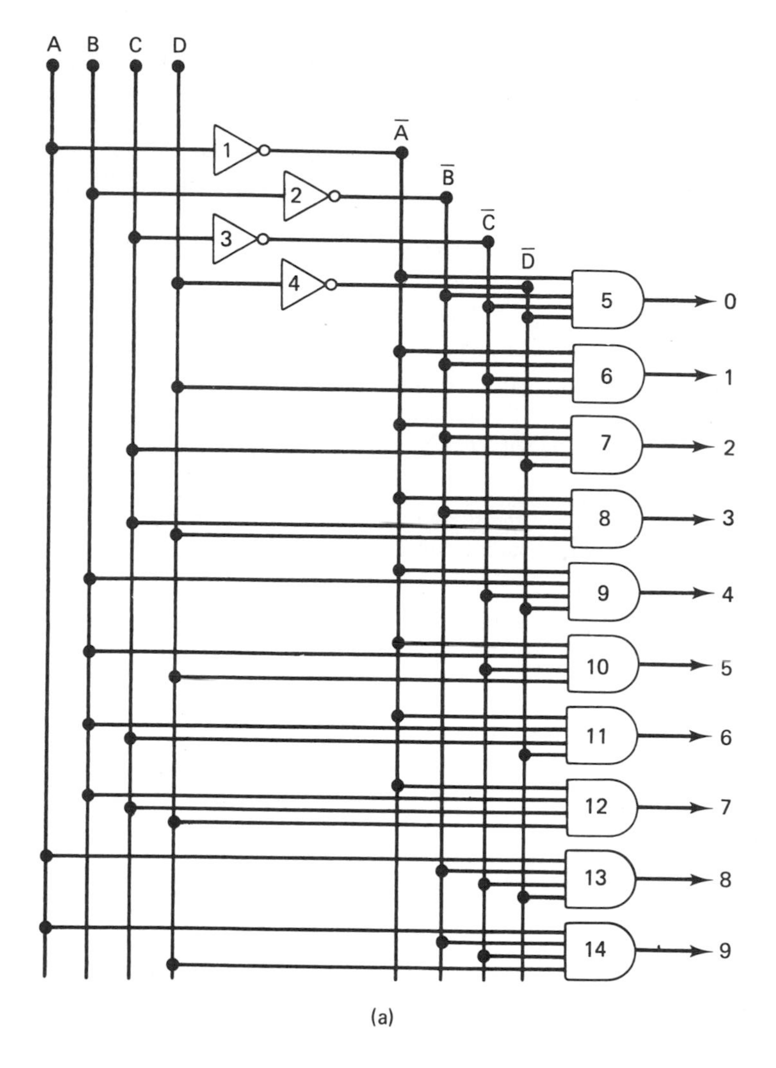

(a)

A B C D	$\overline{A}$ $\overline{B}$ $\overline{C}$ $\overline{D}$	0 1 2 3 4 5 6 7 8 9
0 0 0 0	1 1 1 1	1 0 0 0 0 0 0 0 0 0
0 0 0 1	1 1 1 0	0 1 0 0 0 0 0 0 0 0
0 0 1 0	1 1 0 1	0 0 1 0 0 0 0 0 0 0
0 0 1 1	1 1 0 0	0 0 0 1 0 0 0 0 0 0
0 1 0 0	1 0 1 1	0 0 0 0 1 0 0 0 0 0
0 1 0 1	1 0 1 0	0 0 0 0 0 1 0 0 0 0
0 1 1 0	1 0 0 1	0 0 0 0 0 0 1 0 0 0
0 1 1 1	1 0 0 0	0 0 0 0 0 0 0 1 0 0
1 0 0 0	0 1 1 1	0 0 0 0 0 0 0 0 1 0
1 0 0 1	0 1 1 0	0 0 0 0 0 0 0 0 0 1

(b)

Figure 10.2 (a) BCD to decimal decoder; (b) Truth table.

Now the complemented inputs are all logic 1, as shown in the truth table of Fig. 10.2(b). Therefore, all inputs on AND gate 5 are logic 1. This makes the output switch to logic 1 and signal a decimal number 0.

At the same time, at least one input on all other AND gates in the matrix is logic 0. For this reason, the outputs for decimal numbers 1 through 9 remain at the 0 logic level.

In Fig. 10.2(b), a complete truth table of operation for a BCD to decimal decoder is given. From this table, we see that each input condition causes just one output to switch to logic 1. This is the output corresponding to the decimal equivalent number of the BCD code.

A decoder circuit can be made that sinks the decimal output to logic 0 instead of sourcing to logic 1. This is done by replacing the AND gates in the logic gate matrix of Fig. 10.2(a) with NAND gates.

EXAMPLE 10.1

The input code to the BCD to decimal decoder in Fig. 10.2(a) is $ABCD = 0111$. Find which output switches to logic 1.

SOLUTION:

When the input is $ABCD = 0111$, the $\bar{A}$, B, C, and D signals are all logic 1:

$$\bar{A}BCD = 1111$$

Looking at the AND gate matrix, we see that gate 12 has inputs $\bar{A}$, B, C, and D. All inputs on this gate are logic 1, and the decimal 7 output is at the 1 logic level.

EXAMPLE 10.2

Make a logic drawing of a decoder circuit that converts the excess-3 Gray code to the decimal code.

SOLUTION:

A truth table of the excess-3 Gray code is shown in Fig. 10.3(a). From this table, we see that all input variables appear in both the logic 0 and 1 states. For this reason, inverters are needed to provide the $\bar{A}$, $\bar{B}$, $\bar{C}$, and $\bar{D}$ complemented inputs. These gates are numbered 1 through 4 in the logic diagram of Fig. 10.3(b). Along with this, we needed one 4-input AND gate for each of the 10 decimal outputs.

The excess-3 Gray code for decimal 0 in the truth table is 0010. Writing a Boolean equation for this input, we get

$$f_0 = \bar{A}\bar{B}C\bar{D}$$

To implement this output equation, the four inputs on AND gate 5 in Fig. 10.3(b) are wired to the $\bar{A}$, $\bar{B}$, C, and $\bar{D}$ inputs.

The next input down the table is $ABCD = 0110$ for decimal number 1. This gives the Boolean equation

$$f_1 = \bar{A}BC\bar{D}$$

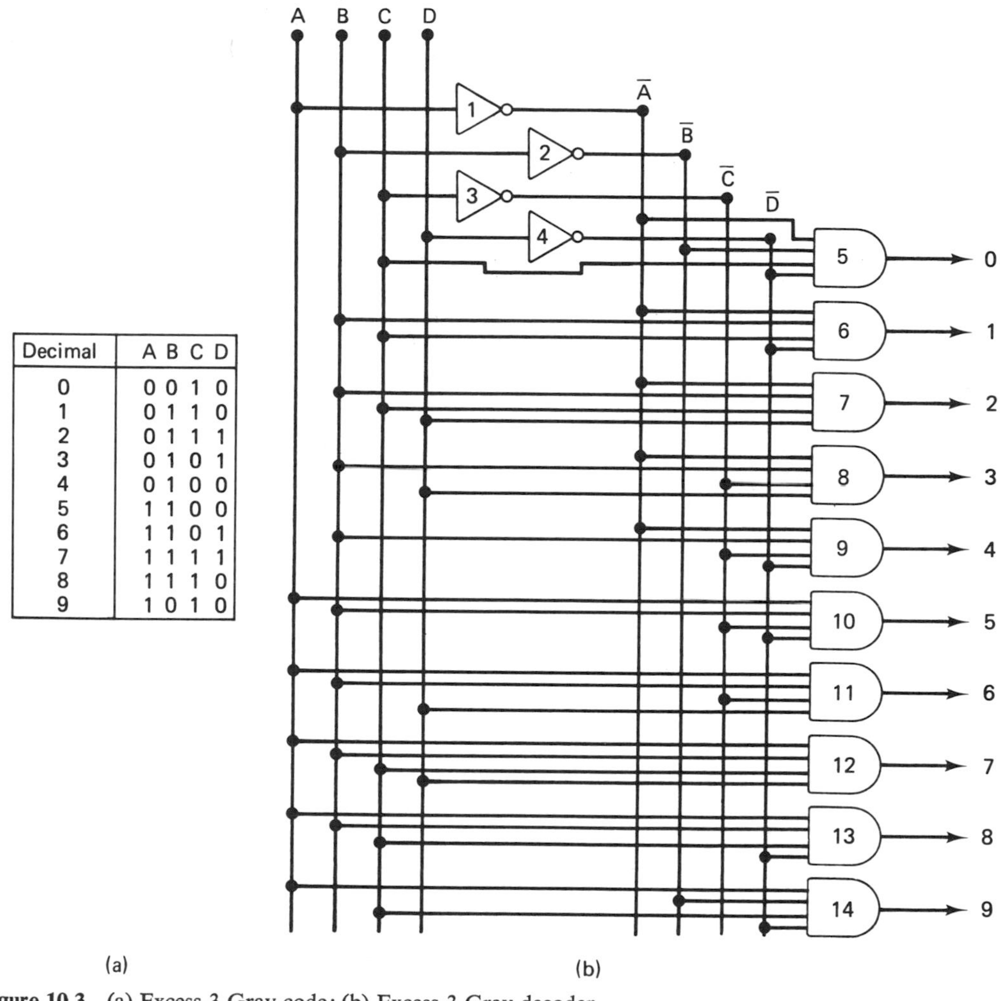

Decimal	A B C D
0	0 0 1 0
1	0 1 1 0
2	0 1 1 1
3	0 1 0 1
4	0 1 0 0
5	1 1 0 0
6	1 1 0 1
7	1 1 1 1
8	1 1 1 0
9	1 0 1 0

Figure 10.3 (a) Excess-3 Gray code; (b) Excess-3 Gray decoder.

For this reason, the inputs on AND gate 6 are tied to the $\bar{A}$, B, C, and $\bar{D}$ inputs. The input connections on the remaining AND gates 7 through 14 are made in the same way.

10.4 INTEGRATED BCD TO DECIMAL DECODER

Decoder circuits are available as standard MSI circuits. In integrated decoder circuits, the inputs to the decoder matrix are buffered, as shown in the block diagram of Fig. 10.4. This buffering limits the loading effect on the circuits that supply signals to the A, B, C, and D inputs.

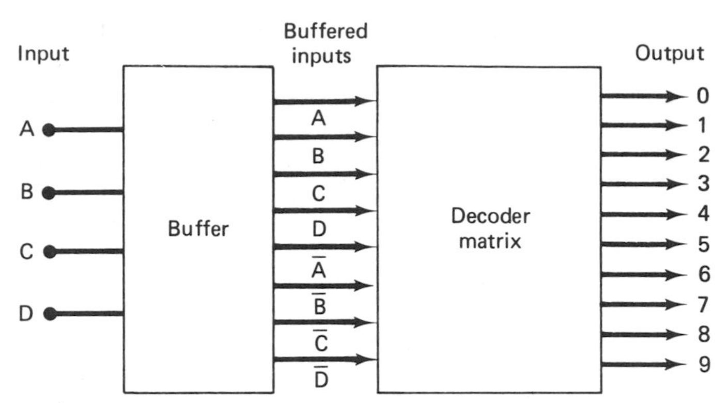

Figure 10.4 Integrated decoder block diagram.

7442 BCD to Decimal Decoder

One of the most widely used integrated decoder devices is the 7442 BCD to decimal decoder. A logic diagram of the circuitry inside a 7442 decoder is shown in Fig. 10.5(a) and its pin layout in Fig. 10.5(b).

At the input of this circuit, eight inverter gates are used to make an inverting buffer. The gates numbered 1 through 4 invert the A, B, C, and D input signals to provide complemented inputs $\bar{A}$, $\bar{B}$, $\bar{C}$, and $\bar{D}$, respectively. After this, the complemented inputs are inverted once more by gates 5 through 8 to give the A, B, C, and D inputs for the decoder matrix.

When buffered in this way, the A, B, C, and D inputs of the integrated circuit act like one TTL load to circuits driving the inputs. If a direct connection were made from the A input to the decoder matrix, this input would go directly to five NAND gates and one inverter. This represents a fan out of 6 instead of 1 to a circuit supplying the A input.

The decoder matrix section has 10 4-input NAND gates. These gates are marked 9 through 18 and produce the decimal 0 through 9 outputs, respectively.

Since a NAND gate decoder array is used, the decimal output corresponding to the value of the BCD input switches to logic 0. At the same time, all other outputs stay at logic 1.

Most integrated decoders sink the output to ground. Each output on a 7442 can sink a maximum of 16 mA.

Operation of the 7442 Decoder

A complete truth table of operation for the 7442 BCD to decimal decoder is shown in Fig. 10.5(c). Here we see that D is used to indicate the most significant bit of the input and A the LSB in the table. So the input is indicated as $DCBA$ instead of $ABCD$ as in earlier circuitry. This notation is used for most integrated decoder devices.

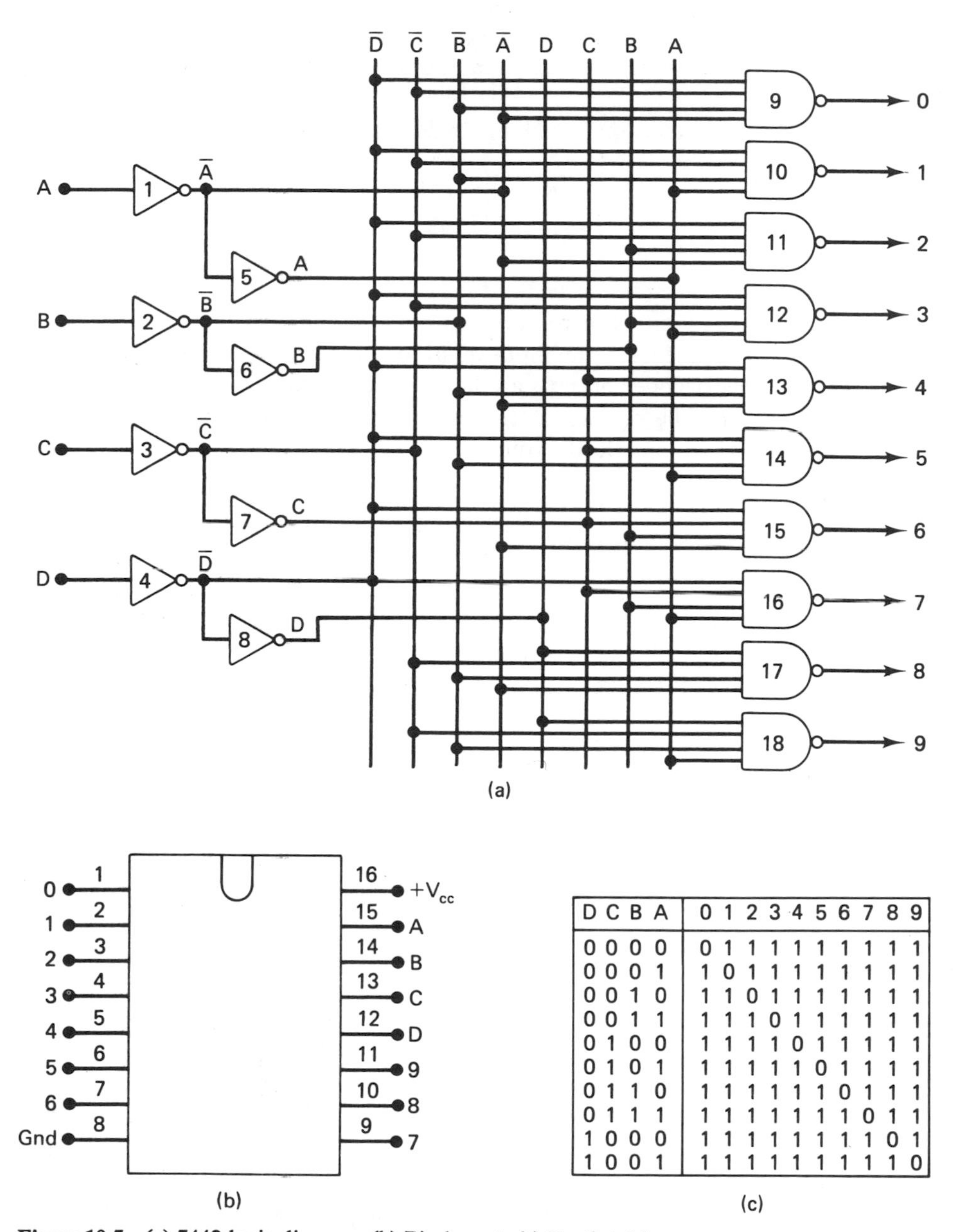

D	C	B	A	0	1	2	3	4	5	6	7	8	9
0	0	0	0	0	1	1	1	1	1	1	1	1	1
0	0	0	1	1	0	1	1	1	1	1	1	1	1
0	0	1	0	1	1	0	1	1	1	1	1	1	1
0	0	1	1	1	1	1	0	1	1	1	1	1	1
0	1	0	0	1	1	1	1	0	1	1	1	1	1
0	1	0	1	1	1	1	1	1	0	1	1	1	1
0	1	1	0	1	1	1	1	1	1	0	1	1	1
0	1	1	1	1	1	1	1	1	1	1	0	1	1
1	0	0	0	1	1	1	1	1	1	1	1	0	1
1	0	0	1	1	1	1	1	1	1	1	1	1	0

Figure 10.5 (a) 7442 logic diagram; (b) Pin layout; (c) Truth table.

Let us check the operation of the decoder circuit in Fig. 10.5(a) for input $DCBA = 0000$. When all inputs are at the 0 logic level, complemented inputs $\bar{D}$, $\bar{C}$, $\bar{B}$, and $\bar{A}$ at outputs of gates 1 through 4 are all logic 1.

These complemented inputs are inverted by gates 5 through 8. Their outputs give the D, C, B, and A input signals to the decoder matrix. Each of these signals is at logic 0.

Now the eight inputs to the NAND gate decoder matrix are known to be

$$\bar{D}\bar{C}\bar{B}\bar{A} = 1111$$

$$DCBA = 0000$$

Looking at the NAND gates, we see that gate 9 has as its inputs $\bar{D}$, $\bar{C}$, $\bar{B}$, and $\bar{A}$. All of these inputs are logic 1; therefore, its output switches to logic 0.

At the same time, at least one input on all other NAND gates is logic 0. For this reason, all other decoder outputs are held at the 1 logic level.

Here we see that the decoder indicates that the input is the BCD code for decimal number 0 by making the 0 output switch to logic 0. This result is consistent with the truth table in Fig. 10.5(c).

7445 BCD to Decimal Decoder

In some applications, the output circuitry of a BCD to decimal decoder must drive a high current load. For this purpose, devices such as the 7445 BCD to decimal decoder are available. The 7445 IC is identical in pin layout and logic operation to the 7442; however, its outputs are buffered to sink up to 80 mA each instead of 16 mA.

EXAMPLE 10.3

Figure 10.6 shows the BCD input waveform of a 7442 decoder. Draw the waveforms that appear at the 0, 1, and 2 outputs.

SOLUTION:

From the 7442 truth table in Fig. 10.5(c), we find that the 0 output is logic 1 unless all input signals are logic 0. From the input waveforms in Fig. 10.6, we see that all inputs are 0 during the first time interval. Therefore, the 0 output is at the 0 logic level in the first time interval and 1 thereafter. This waveshape is shown in Fig. 10.6

The decimal 1 output goes to logic 0 when the input is 0001. This condition occurs during the second interval of time in the input waveforms. Therefore, the decimal 1 output switches to logic 0 for the second time interval in the output waveform of Fig. 10.6.

For the 2 output to be logic 0, the input must be $DCBA$ equals 0010. This happens in the third time interval of the input waveforms. So the decimal 2 output switches to logic 0 during the third time interval of the output waveforms.

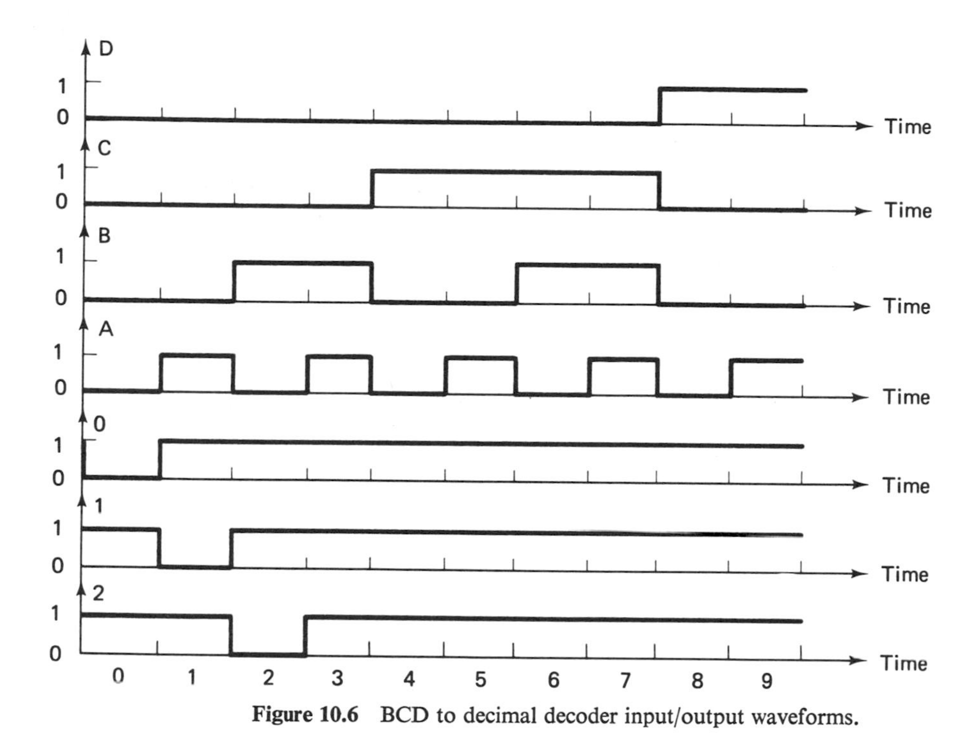

Figure 10.6 BCD to decimal decoder input/output waveforms.

10.5 OTHER INTEGRATED DECODERS

Many other decoder devices are available as standard MSI TTL integrated circuits. Two such devices are the 7443 and 7444 decoders.

7443 Decoder

The 7443 device is an excess-3 to decimal decoder. Its pin layout is shown in Fig. 10.7(a). Here, we see that the pin layout is the same as for the 7442 BCD to decimal decoder IC.

A truth table of 7443 decoder operation is shown in Fig. 10.7(b). Looking at this table, we find that the *DCBA* input columns represent the excess-3 code. At the outputs, the lead with the decimal number corresponding to the coded input switches to a logic 0. On the other hand, all other outputs stay at logic 1. Each output of the 7443 can sink a load that needs up to 16 mA.

Figure 10.7(c) shows the logic circuitry found within the 7443 device. From this logic diagram, we see that the circuitry of this decoder is similar to that used to make a 7442 decoder. However, the interconnection of the NAND gates to the *DCBA* inputs and $\bar{D}\bar{C}\bar{B}\bar{A}$ complemented inputs are changed to produce a decimal output when the excess-3 code is applied at the input.

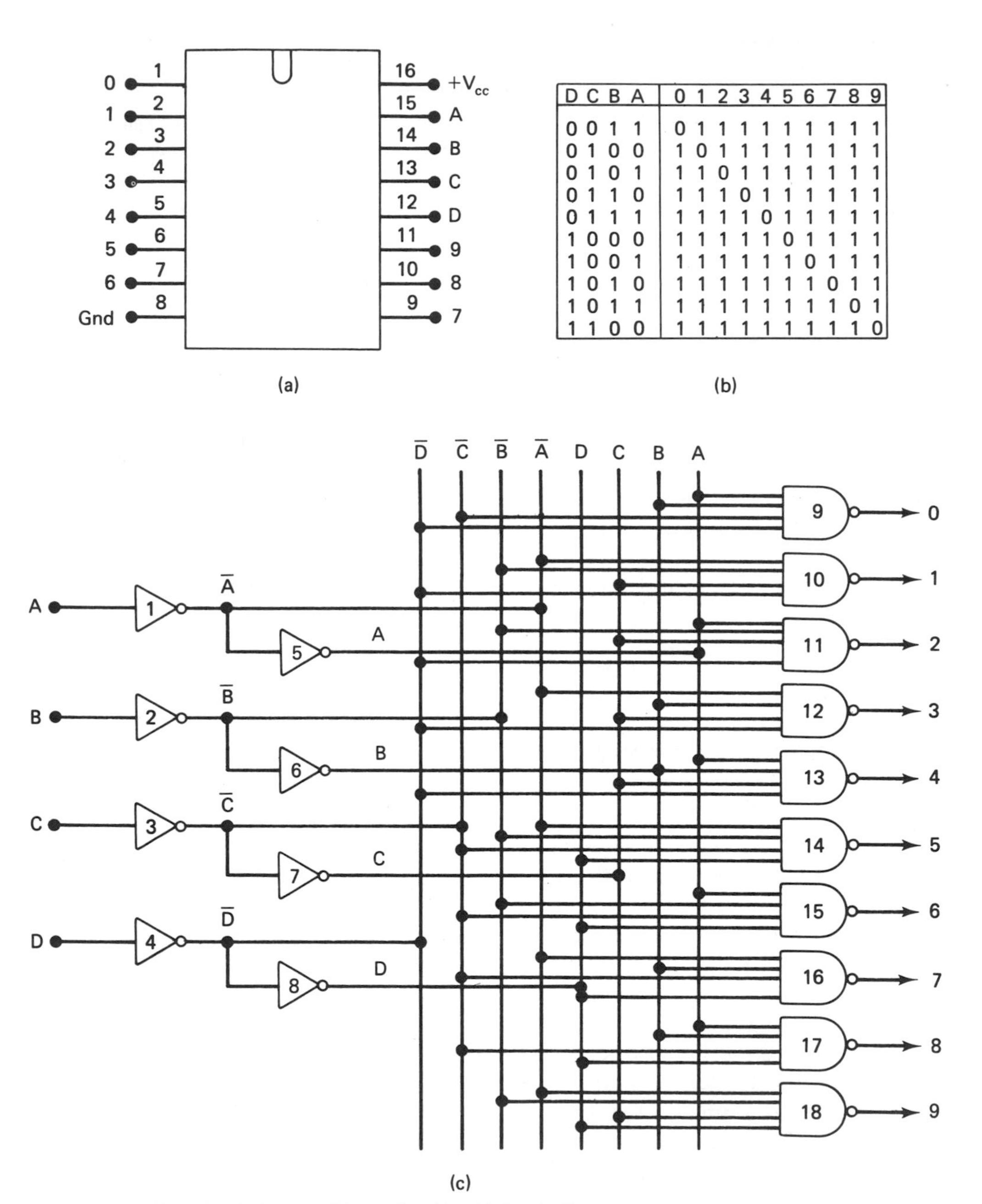

D	C	B	A	0	1	2	3	4	5	6	7	8	9
0	0	1	1	0	1	1	1	1	1	1	1	1	1
0	1	0	0	1	0	1	1	1	1	1	1	1	1
0	1	0	1	1	1	0	1	1	1	1	1	1	1
0	1	1	0	1	1	1	0	1	1	1	1	1	1
0	1	1	1	1	1	1	1	0	1	1	1	1	1
1	0	0	0	1	1	1	1	1	0	1	1	1	1
1	0	0	1	1	1	1	1	1	1	0	1	1	1
1	0	1	0	1	1	1	1	1	1	1	0	1	1
1	0	1	1	1	1	1	1	1	1	1	1	0	1
1	1	0	0	1	1	1	1	1	1	1	1	1	0

Figure 10.7 (a) 7443 pin layout; (b) Truth table; (c) Logic diagram.

7444 Decoder

Another integrated decoder circuit is the 7444 excess-3 Gray decoder. This device takes an excess-3 Gray code at the *DCBA* input and converts it to a decimal output.

The pin layout of the 7444 device is identical to that shown for the 7443 in Fig. 10.7(a). Besides this, the logic circuitry inside the 7444 is again like that of the 7442 and 7443 devices, but the NAND gate input wiring is changed to correspond to an excess-3 Gray code.

ASSIGNMENTS

Section 10.2

1. What is the function of a decoder circuit?
2. Give the names of three widely used decoder circuits.
3. What is the most common output code supplied by a decoder circuit?

Section 10.3

4. Write a Boolean equation for each of the decimal outputs in the Gray to decimal decoder circuit of Fig. 10.8.
5. Perform a truth table analysis of the decoder circuit in Fig. 10.8. Use the Gray code as the input.
6. Draw a logic diagram like that in Fig. 10.8 for an excess-3 to decimal decoder circuit.
7. Make a drawing of a BCD to decimal decoder that will sink the decimal outputs to ground.
8. What is the output of each gate in the XS-3 Gray decoder in Fig. 10.3(b) for input $ABCD = 1111$?

Section 10.4

9. What kind of decoder circuit is in the 7442 device?
10. For what reason are the inputs of the 7442 device buffered?
11. Is *D* the MSB or LSB of the input on a 7442 decoder IC?
12. Does the 7442 source the decoded output to $+V_{cc}$ or sink it to ground?
13. What is the output current rating of the 7442 decoder?
14. How is the 7445 device different from the 7442 decoder?
15. Complete Example 10.3 for the waveforms of decoder outputs 3 through 9.

Section 10.5

16. What type of decoder is the 7443 device? How much current can it sink at an output?

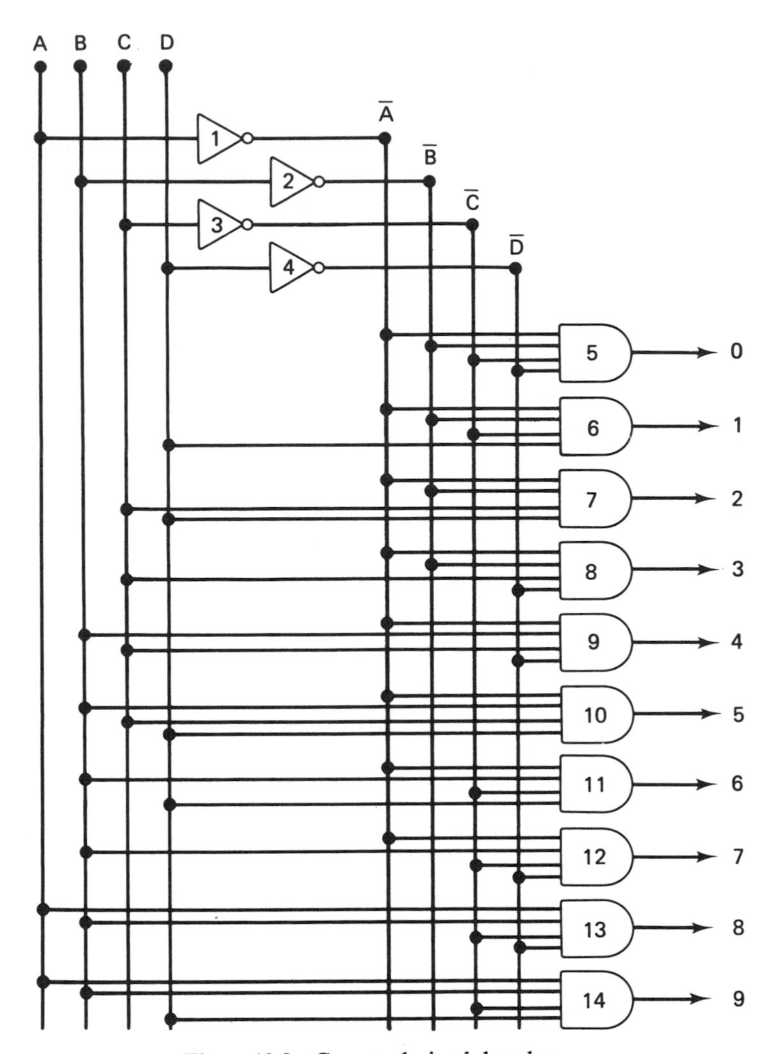

Figure 10.8 Gray to decimal decoder.

17. What kind of decoder is the 7444?

18. Construct a logic diagram like that in Fig. 10.7(c) for the 7444 decoder device.

19. Make a drawing of the input and output waveforms of a 7443 device.

Readouts and Driver Circuitry

11

11.1 INTRODUCTION

The output unit in many digital systems is a *decimal readout* or *display*. For example, calculators are made with a multidigit numeric digital display. Numeric display devices are divided into two categories. These groups are the *indicator tube* and the *seven-segment display*. In this chapter, we shall study the different types of numeric display devices available and the circuitry used to make them operate. For this purpose, the topics that follow are covered in this chapter:

1. The numeric indicator tube display
2. BCD to decimal decoder/nixie driver
3. The seven-segment display
4. Seven-segment decoder/drivers

11.2 THE NUMERIC INDICATOR TUBE DISPLAY

The indicator tube is one of the earlier numeric display devices. It is also known as a *nixie tube*. In Fig. 11.1(a), a nixie tube display device is shown. Here, we see that the device has a separate indicator for each number and that they are stacked back through the device.

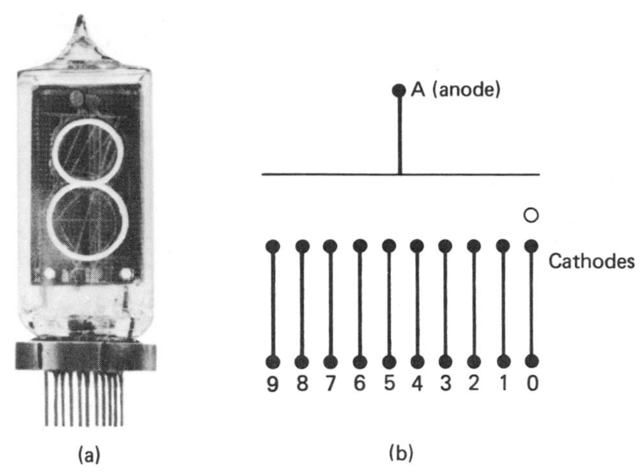

Figure 11.1 (a) Nixie tube (Burroughs Corporation); (b) Symbol.

Figure 11.1(b) shows the logic symbol for a nixie tube. This device is a *gas-filled cold cathode tube*. By cold cathode, we mean that the tube does not have to be heated with a *filament*. From the symbol, it is seen that the different numeric elements have a common lead called the *anode*. On the other hand, separate *cathode* leads are provided for each number. Cathodes are labeled 0 through 9 and the anode A.

In most uses, the anode is supplied with a high positive dc voltage. This voltage value is normally about +170 V. To make a number light, ground must be applied to the corresponding cathode.

The advantage of the numeric indicator tube is that it displays numbers with their true shape. However, the need for high voltage is a disadvantage in some digital equipment.

EXAMPLE 11.1

Which cathode on the nixie tube in Fig. 11.1(b) must be grounded to make the number 5 light.

SOLUTION:

To light the number 5, the cathode marked 5 must be supplied with ground.

11.3 THE BCD TO DECIMAL DECODER/NIXIE DRIVER

Numeric information in digital equipment is often expressed using the BCD code. For this reason, a nixie tube display is driven with a BCD to decimal decoder circuit. As shown in Fig. 11.2, the decoder takes a 4-bit BCD input and supplies a decimal output to the cathodes of the display. It is used to supply ground to the cathode corresponding to the decimal value of the BCD input.

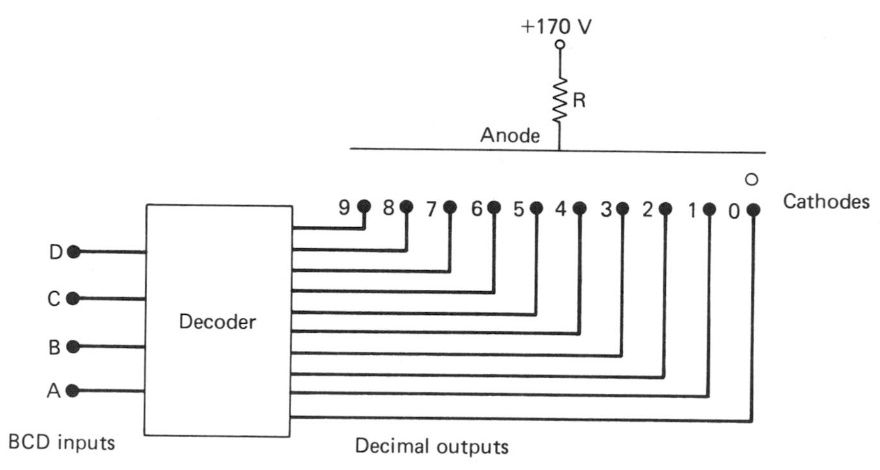

Figure 11.2 Driving the nixie tube with a BCD to decimal decoder.

For instance, the BCD input could be the code 0101 for decimal number 5. This input makes the decoder output marked 5 switch to logic 0. In this way, cathode 5 of the nixie tube is returned to ground. Now the anode for all numbers is supplied with +170 V, but just the cathode for number 5 is grounded. Therefore, the number 5 lights on the readout. The anode is returned to the +170-V supply through a current-limiting resistor *R*.

74141 BCD to Decimal Decoder/Nixie Driver

The 74141 is an integrated BCD to decimal decoder that is made to drive nixie tube displays. Its output circuitry is buffered to withstand the high voltage needed to operate a nixie tube. This device can be used as the decoder in the circuit of Fig. 11.2.

A logic diagram of the circuitry inside a 74141 decoder is shown in Fig. 11.3. Here, we see that the output circuitry is protected with *zener diodes* by holding the voltage at each output to a safe value. Pin numbering for the leads on the 74141 is marked in the diagram of Fig. 11.3.

EXAMPLE 11.2

The decoder device in Fig. 11.2 is a 74141 BCD to decimal decoder. If the BCD input is 0011, what are the logic outputs of the decoder? What number is displayed on the readout?

SOLUTION:

This BCD input corresponds to the number 3. Therefore, all outputs of the decoder are at logic 1 except the lead for number 3, which switches to logic 0. This 0 level at the 3 output of the decoder supplies ground to the cathode marked 3 on the nixie tube and makes number 3 light.

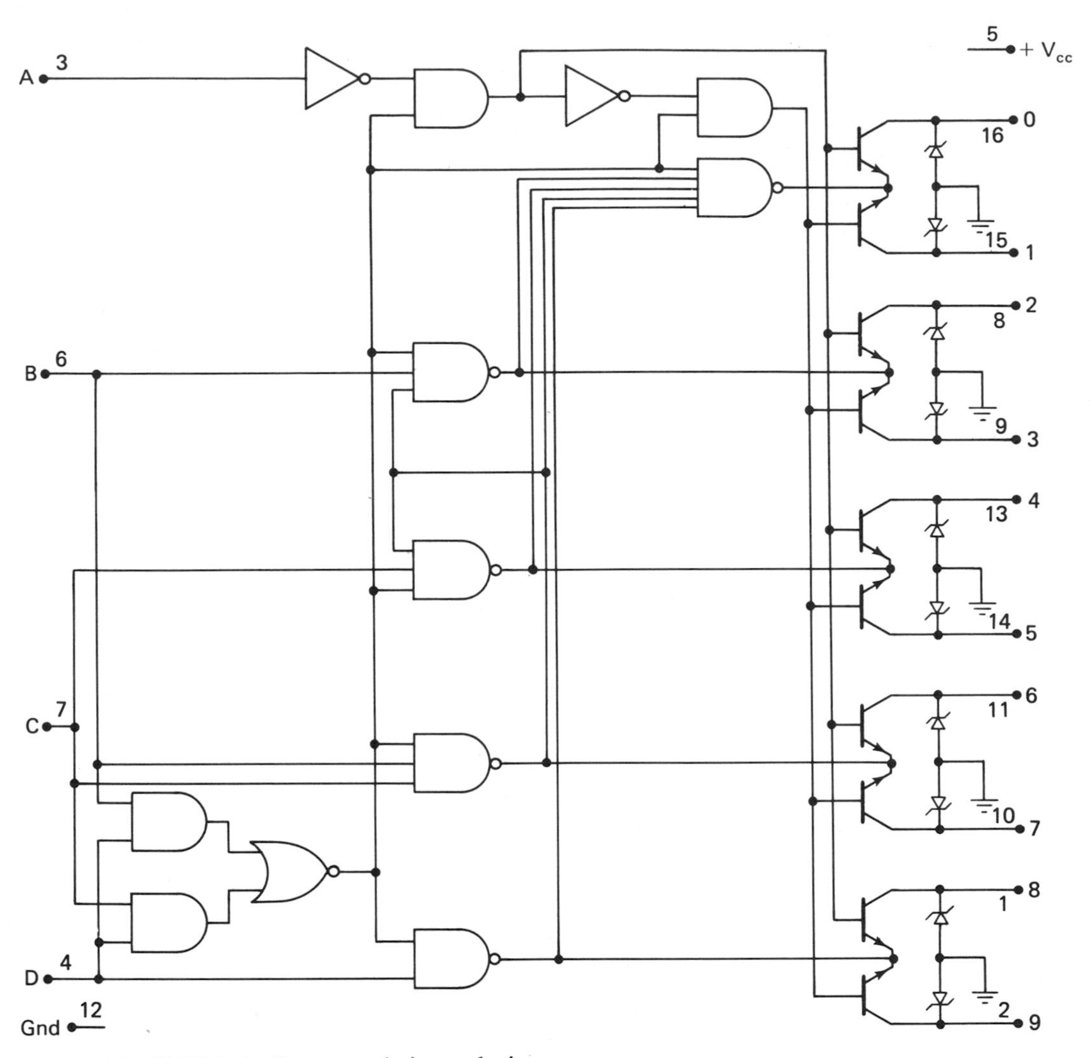

Figure 11.3 74141 logic diagram and pin numbering.

11.4 THE SEVEN-SEGMENT DISPLAY

The most widely used readout in modern digital equipment is the seven-segment display. A seven-segment display is shown in Fig. 11.4(a). This device is a *light-emitting diode* or *LED-type* seven-segment readout. Two other popular types of seven-segment readouts are the *gas discharge display* and the *liquid crystal display*.

In Fig. 11.4(b), we see that the seven segments are layed out in a single plane. Here, it is found that the segments are labeled *a* through *g*. Besides this, a decimal point *DP* is supplied.

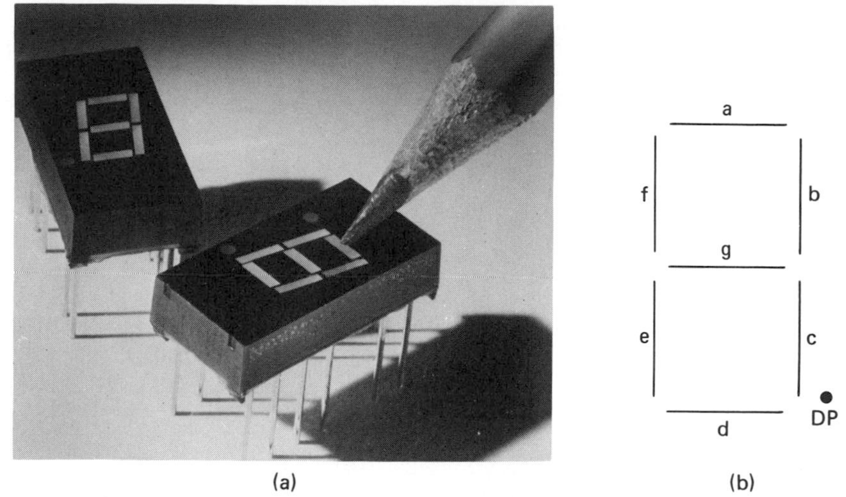

(a) (b)

Figure 11.4 (a) Seven-segment LED readout (Monsanto Company); (b) Seven-segment layout.

To form numbers 0 through 9, different combinations of segments are lit. For instance, to make the number 0 all segments except *g* are turned ON.

EXAMPLE 11.3

Which segments on the seven-segment display of Fig. 11.4(b) must be lit to display the number 1?

SOLUTION:

To make the number 1 appear on the display, the *b* and *c* segments must light.

Light-Emitting Diode Display

A single-digit LED seven-segment display is shown in Fig. 11.4(a). It uses an array of seven light-emitting diodes to form numbers 0 through 9 and another diode to provide the decimal point.

Figure 11.5(a) shows the connection of diodes within an LED display. Here we see that the anode leads on all diodes are tied together and labeled with *A*. On the other hand, the cathodes *a* through *g* are independent. For this reason, the device is called a *common anode LED readout.*

The pin layout of a typical seven-segment display device is shown in Fig. 11.5(b). In this diagram, it is seen that the common anode connection is divided between three power supply pins marked A_1, A_2, and A_3.

To operate the LED display, all anode terminals must be supplied by a positive voltage indicated by $+V_{cc}$. For a number to light, the correct cathode leads must be supplied with ground. When grounded the diodes go ON and produce lighted segments.

For correct operation, current flow through each diode must be limited to about 20 mA. This is done by connecting a fixed resistor in series with each diode

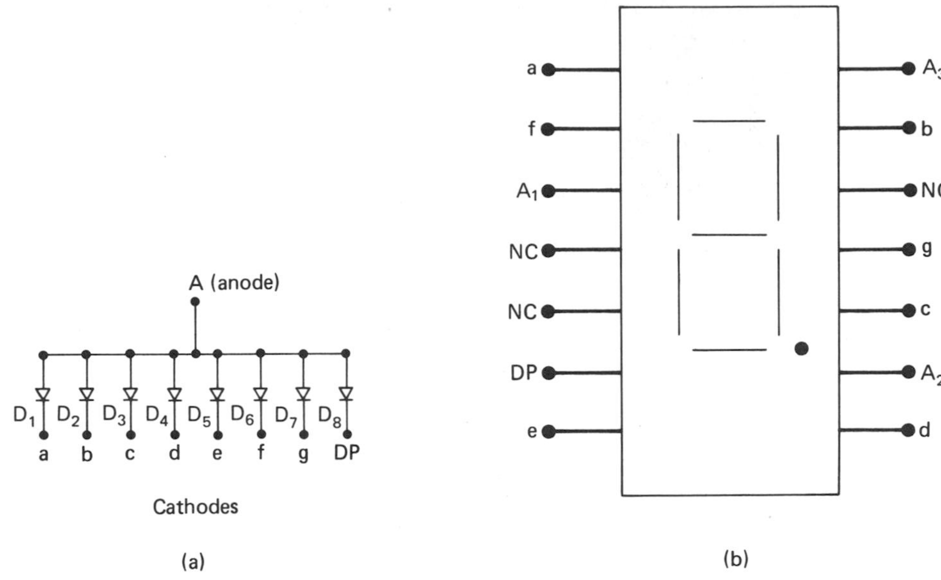

Figure 11.5 (a) LED diode array; (b) Pin layout.

lead. If a +5-V dc supply is used for $+V_{cc}$, the current-limiting resistor value is 250 ohms (250 Ω).

LED displays are available in different sizes and colors. Most LEDs are red; however, green- and amber-colored devices are also in use. A common size display has .3-in. high numbers. Furthermore, they are available with a *common cathode diode connection* as well as the common anode configuration shown in Fig. 11.5(a). Their main advantages are rugged construction and the ability to work off a low voltage supply. However, they have a disadvantage in that they cannot be viewed in direct sunlight.

EXAMPLE 11.4

Which diodes in the array of Fig. 11.5(a) are ON when number 7 is displayed?

SOLUTION:

The number seven is formed by lighting segments *a*, *b*, and *c*. So diodes D_1, D_2, and D_3 are ON.

Multidigit Displays

For most applications, a readout with several digits is needed. *Multidigit displays* are available in LED, gas discharge, and liquid crystal types.

Figure 11.6(a) shows a three-digit light-emitting diode display and Fig. 11.6(b) a gas discharge device. The LED display is set up in a DIP layout and can work off a +5-V dc supply. On the other hand, the gas discharge display is a high-voltage device and works more like a nixie tube.

Liquid crystal or LCD displays are also low-voltage devices like the LED

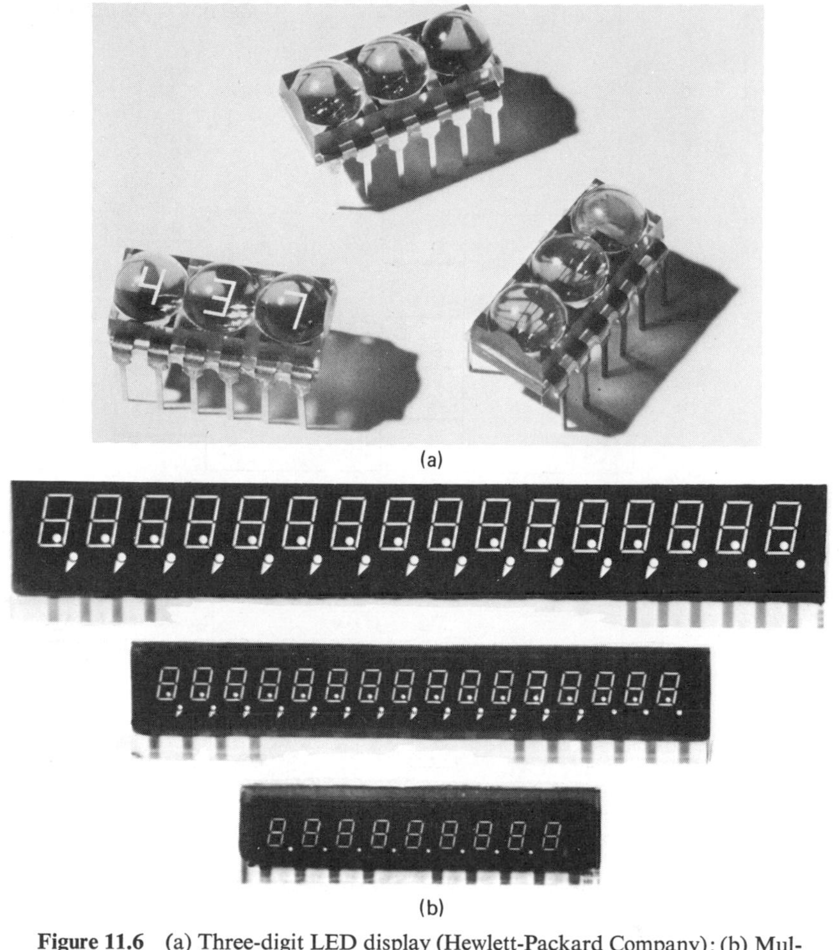

(a)

(b)

Figure 11.6 (a) Three-digit LED display (Hewlett-Packard Company); (b) Multidigit gas discharge display (Burroughs Corporation).

display. However, they have the added advantage of drawing very low current. In fact, an LCD display draws as low as 20 nanoamperes (20 nA) per segment. This is 1 million times lower than the current supplied to an equivalent LED display. For this reason, they are used in miniature battery-operated equipment such as wristwatches.

11.5 THE SEVEN-SEGMENT DECODER

Seven-segment decoder circuits are used to interface BCD calculating circuitry with a seven-segment display. In Fig. 11.7, the connection of the decoder and display is shown. Here, we see that the BCD code outputs *D*, *C*, *B*, and *A* of the calculating unit are sent to inputs on the decoder. At the other side of the decoder, segment outputs *a* through *g* are wired to the display.

A general seven-segment decoder block is shown in Fig. 11.8(a). The input and

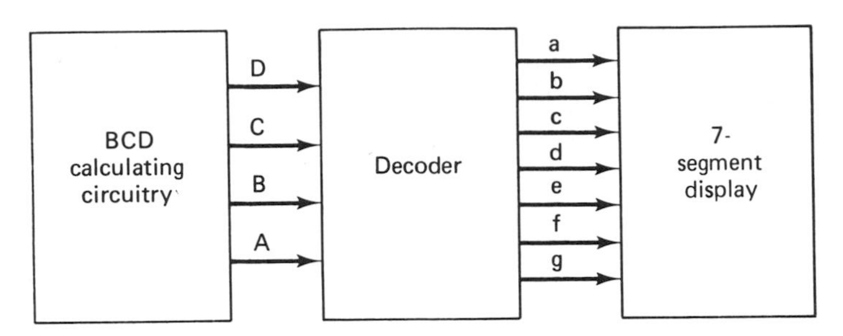

Figure 11.7 Interfacing calculating circuitry with an LED seven-segment display.

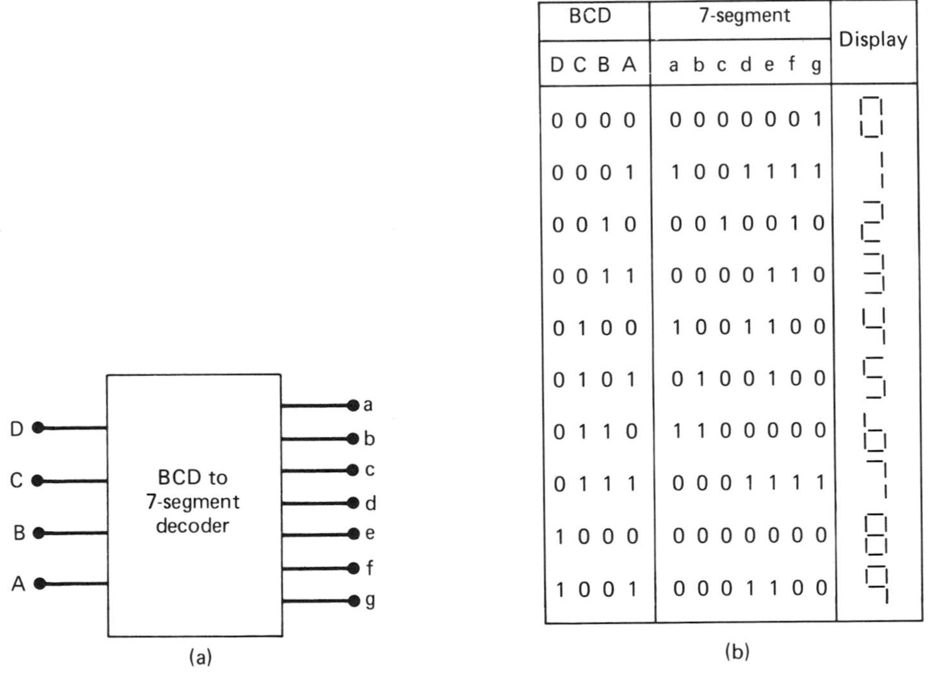

BCD	7-segment	Display
D C B A	a b c d e f g	
0 0 0 0	0 0 0 0 0 0 1	0
0 0 0 1	1 0 0 1 1 1 1	1
0 0 1 0	0 0 1 0 0 1 0	2
0 0 1 1	0 0 0 0 1 1 0	3
0 1 0 0	1 0 0 1 1 0 0	4
0 1 0 1	0 1 0 0 1 0 0	5
0 1 1 0	1 1 0 0 0 0 0	6
0 1 1 1	0 0 0 1 1 1 1	7
1 0 0 0	0 0 0 0 0 0 0	8
1 0 0 1	0 0 0 1 1 0 0	9

Figure 11.8 (a) Decoder block; (b) Seven-segment code.

output codes of the device are given in Fig. 11.8(b). In this table, we see that the outputs corresponding to the segments to be lit are made logic 0. The segment representations of numbers 0 through 9 for a seven-segment code are also indicated in this table.

11.6 THE INTEGRATED SEVEN-SEGMENT DECODER/DRIVER

The block diagram of an integrated seven-segment decoder is shown in Fig. 11.9. In this diagram, we see that the decoder has four inputs labeled *A*, *B*, *C*, and *D*. Of these, *D* is the most significant bit. Moreover, there are seven output leads marked *a* through *g*. It has three control leads. These leads are marked *RBI* for *ripple blanking input*, *RBO* for *ripple blanking output*, and *LT* for *lamp test*.

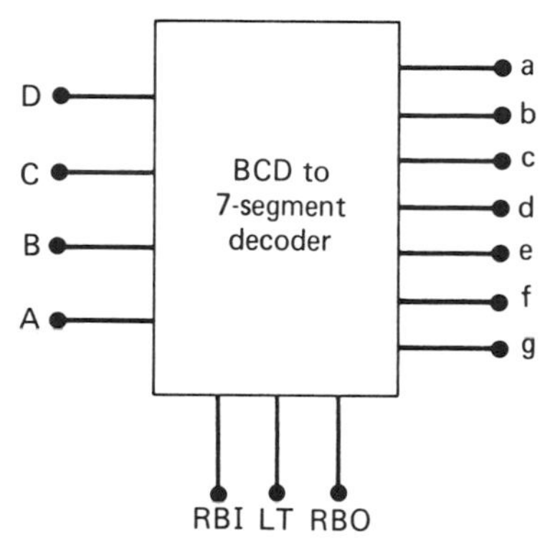

Figure 11.9 IC seven-segment decoder block.

Driving a Seven-Segment Display

The seven-segment decoder is connected to drive the segments of an LED readout as shown in Fig. 11.10. Here the ripple blanking input and lamp test leads are wired to a logic 1 equal to $+V_{cc}$. When connected in this way, the readout will display numbers 0 through 9 as the BCD input changes.

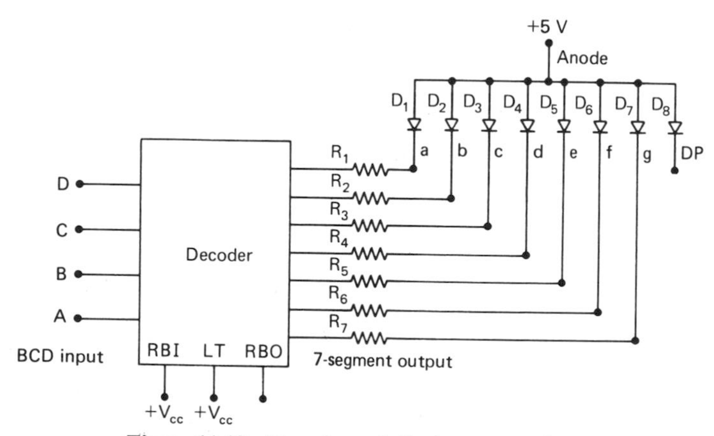

Figure 11.10 Decoder and display connection.

EXAMPLE 11.5

The input to the decoder in Fig. 11.10 is 1001. What is the output of the decoder, and what number is indicated on the display?

SOLUTION:

The BCD input 1001 is the code for number 9. When the input is 9, the table of Fig. 11.8(b) gives the decoder output as

$$abcdefg = 0001100$$

The LED display reads number 9.

If the ripple blanking input is connected to ground or logic 0, the decoder blanks the display of number 0. That is, it will now indicate numbers 1 through 9. For BCD input 0000, the readout stays blank.

The ripple blanking output is used in multidigit applications. In this case, several seven-segment decoders are interconnected with separate seven-segment readouts. If *RBI* is 0 and the *DCBA* input equals 0000, output *RBO* switches to logic 0; however, when this condition does not exist the *RBO* logic level is 1.

On the other hand, the lamp test lead can be used to test the segments of a display. By making it logic 0, all outputs of the decoder switch to logic 0, and each segment on the display lights.

7447 Seven-Segment Decoder/Driver

The 7447 integrated circuit is a standard seven-segment decoder/driver. This device is called a decoder/driver because its outputs are buffered to a voltage rating of 15 V and current of 20 mA. This enables it to directly drive an LED display. Its pin layout is given in Fig. 11.11.

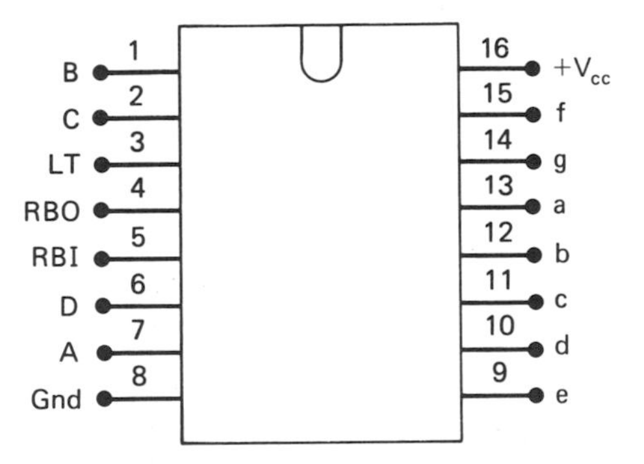

Figure 11.11 7447 pin layout.

Seven-segment decoders are available with different output voltage and current ratings. Moreover, devices with complemented outputs are available to drive common cathode LED readouts.

ASSIGNMENTS

Section 11.2

1. What is another name for a 10-digit numeric indicator tube?
2. About how much voltage is needed to operate a numeric indicator tube?
3. To turn on a number in the numeric indicator tube, is the corresponding cathode supplied with a positive voltage or ground?

Section 11.3

4. What type of decoder circuit is used to interface a BCD calculating circuit to a numeric indicator tube?
5. Does a decoder circuit made to drive a numeric indicator display source or sink the cathodes of the tube?
6. What kind of electronic device is used in the output buffer circuitry of a 74141 device to hold voltage to a safe level?
7. The input of a 74141 decoder is 1001. Find the logic levels of its outputs. If the decoder is wired to a numeric indicator tube, what number lights on the display?

Section 11.4

8. Give three types of seven-segment display devices.
9. Are the segments of a seven-segment display lettered in a clockwise or counterclockwise direction?
10. Which segments must light to make the number 5?
11. How much current is drawn when the number 8 is on an LED display?
12. Can an LED display operate off a 5-V dc power supply?
13. Make a drawing of the light-emitting diodes in a common cathode display.
14. Which diodes in the array of Fig. 11.5(a) are ON if the display reads 2?
15. How much current is supplied to a four-digit liquid crystal display that reads the number 8888?

Section 11.5

16. What type of decoder is used to interface an LED readout to calculating circuitry?
17. Does a decoder source or sink the segments of a common anode LED display?
18. Which number is displayed when the output of the decoder driving a common cathode seven-segment display is $abcdefg = 1100000$?
19. When the input to a seven-segment decoder is 0100, what is the code supplied to the cathodes on the display, and what number is on the display?

Section 11.6

20. What do the notations *RBI*, *RBO*, and *LT* stand for?

21. To display numbers 0 through 9, what logic levels must be set at the *RBI* and *LT* control inputs?

22. The input to the decoder in Fig. 11.10 is 0110. What is the decoder output, and what number lights on the display?

23. Is the ripple blanking circuitry in a 7447 device enabled by making the *RBI* lead logic 0 or 1?

24. In Fig. 11.2 a single fixed resistor is used to limit current through a nixie tube; however, in Fig. 11.10 separate resistors are used for each cathode of an LED display. Why does the LED display need a resistor in each cathode instead of one resistor at the anode?

Digital Multiplexing 12

12.1 INTRODUCTION

Another important use of combinational logic networks is to make *digital multiplexing circuitry*. A multiplexing circuit is an array of logic gates that is a lot like a decoder; however, it is used to switch or select digital data.

In this chapter, we shall study integrated multiplexing devices and circuitry. The topics covered are the following:

1. The digital multiplexer and integrated multiplexer devices
2. The digital demultiplexer and integrated demultiplexer devices
3. The data selector

12.2 THE DIGITAL MULTIPLEXER

The block diagram of a *digital multiplexer* is shown in Fig. 12.1(a). In this diagram, we see that the device has four inputs marked I_1, I_2, I_3, and I_4. On the other hand, it has a single output labeled $Ø_1$.

The input/output operation of the multiplexer is controlled by the logic levels at leads A and B. A truth table of circuit operation is shown in Fig. 12.1(b). Here, we see that output $Ø_1$ is equal to the data from input I_1 when both A and B are logic 0.

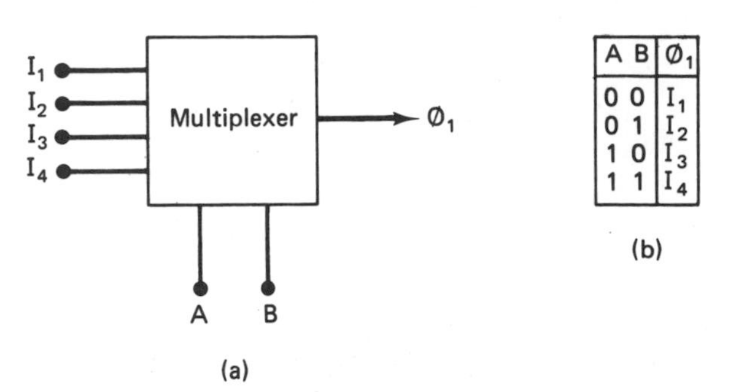

Figure 12.1 (a) Multiplexer block; (b) Input/output operation.

However, by changing the control inputs to $AB = 01$ the data at the I_2 input, instead of at I_1, are switched to output $Ø_1$. In the same way, by making the control signals AB equal to 10 the output becomes input I_3, and when both control signals are logic 1 I_4 is passed through to the output of the circuit.

In this way, we see that a digital multiplexer can be used to select one of several input signals and send it to circuitry connected at its output. The selection of the signal to be outputted is based on the binary combination at a set of control leads.

12.3 INTEGRATED MULTIPLEXER DEVICES

A variety of MSI multiplexer integrated circuits is available. Multiplexer ICs are named in terms of the number of input and output lines they have. Some of the standard device configurations are the 4-line to 1-line multiplexer, 8-line to 1-line multiplexer, and 16-line to 1-line multiplexer.

74153 4-Line to 1-Line Multiplexer

The 74153 integrated circuit is a dual 4-line to 1-line multiplexer device. A logic diagram of one of the two circuits within the 74153 is shown in Fig. 12.2(a). Its pin layout is given in Fig. 12.2(b).

Looking at the logic circuit, we see that it has four data inputs labeled D_0, D_1, D_2, and D_3. Moreover, the output lead is marked F. The second circuit inside the 74153 is identical, and both are controlled by the same set of control leads A and B. In the logic diagram, it is seen that the control inputs are buffered by gate 1 through gate 4 to limit loading on circuits supplying these leads.

Each circuit has one extra control lead marked ST for strobe. The strobe signal is used to enable or disable the operation of the multiplexer circuit. When ST is logic 0, the output of gate 5 is at the 1 level and AND gates 6 through 9 are enabled for operation.

On the other hand, setting ST to logic 1 makes one input on each AND gate logic 0, and there outputs are all held at logic 0. For this reason, all inputs on gate

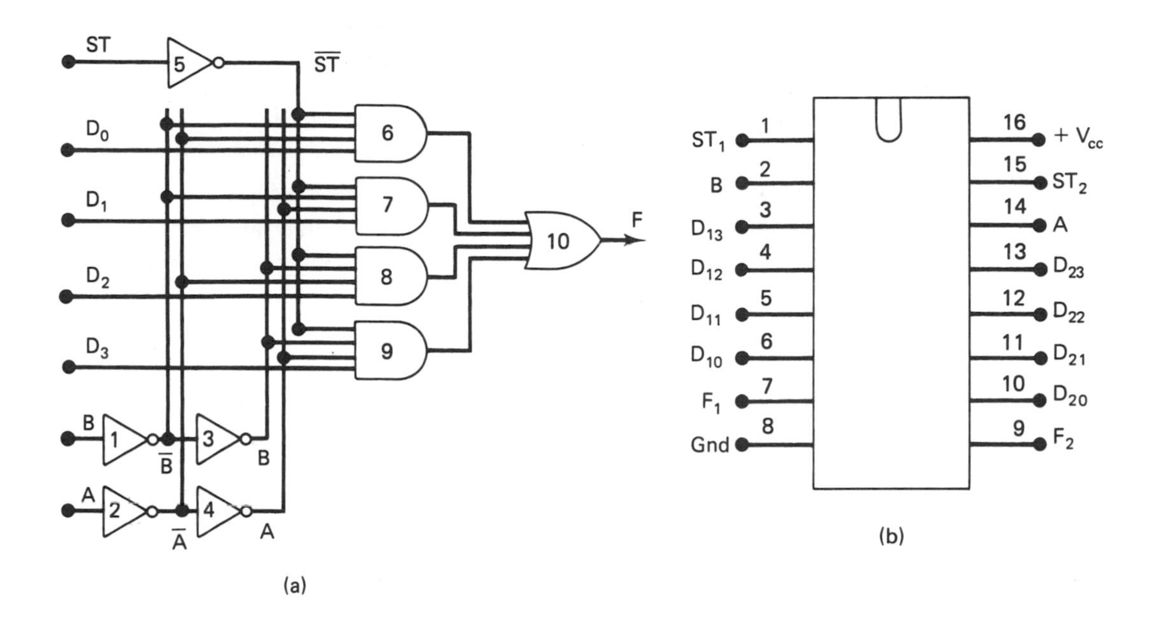

B	A	F
0	0	D_0
0	1	D_1
1	0	D_2
1	1	D_3

(c)

Figure 12.2 (a) Multiplexer logic diagram; (b) 74153 pin layout; (c) Truth table.

10 are 0, and output F of this OR gate is also at the 0 logic level. In this state, the multiplexer is disabled.

Let us illustrate the operation of the multiplexer for control input BA equal to 00 and the strobe input set at 0 to enable the multiplexer for operation. This control input makes the $\bar{B}$ output of gate 1 and $\bar{A}$ output from gate 2 equal to 1. Therefore, the $\bar{A}$, $\bar{B}$, and $\overline{ST}$ inputs on gate 6 are all at the 1 logic level.

As the D_0 input switches between 0 and 1, the output of AND gate 6 takes on the same logic level. When D_0 is logic 1, all inputs on gate 6 are logic 1, and its output goes to the 1 level. This makes one input on the OR gate a logic 1, and the F output of the multiplexer also switches to the 1 level. However, for a logic 0 at D_0 output F stays at logic 0. In this way, we see that the F output follows the logic level at the D_0 input when the control signals are set to $BA = 00$.

At the same time, at least one control input on each of the other three AND gates is at the 0 logic level, and their outputs are all held at logic 0.

If the control inputs are changed from $BA = 00$ to 01, the operation is similar. However, gate 7 is now enabled because $\overline{ST}$, A, and $\bar{B}$ are logic 1. For this case, the signal at input D_1 is passed through to the OR gate and F output. A general description of 74153 operation is given in the truth table of Fig. 12.2(c).

The 74151 Multiplexer

Another standard multiplexer device is the 74151 integrated circuit. The 74151 is an 8-line to 1-line digital multiplexer. Figure 12.3(a) shows its pin layout and Fig. 12.3(b) gives a truth table of its operation.

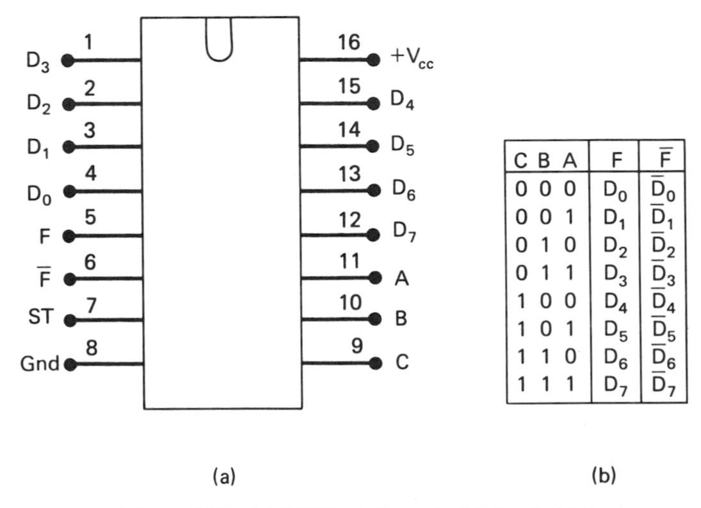

C B A	F	$\bar{F}$
0 0 0	D_0	$\bar{D}_0$
0 0 1	D_1	$\bar{D}_1$
0 1 0	D_2	$\bar{D}_2$
0 1 1	D_3	$\bar{D}_3$
1 0 0	D_4	$\bar{D}_4$
1 0 1	D_5	$\bar{D}_5$
1 1 0	D_6	$\bar{D}_6$
1 1 1	D_7	$\bar{D}_7$

Figure 12.3 (a) 74151 pin layout; (b) Truth table.

From the pin layout, we see that the eight data inputs are called D_0 through D_7, and the circuit provides both output F and its complement $\bar{F}$. As the CBA input is changed, different inputs are switched to the F output, and the complement of this input is available at the $\bar{F}$ output. Again, a 0 logic level is needed at the strobe input to enable the multiplexer for operation.

12.4 MULTIPLEXER APPLICATIONS

The digital multiplexer device can be used in many applications. For instance, it can be made to implement a logic function expressed as a truth table. A second common use is as a frequency selector. Let us illustrate these applications with examples.

EXAMPLE 12.1

Show how to implement the truth table function in Fig. 12.4(a) with a 74151 8-line to 1-line multiplexer device.

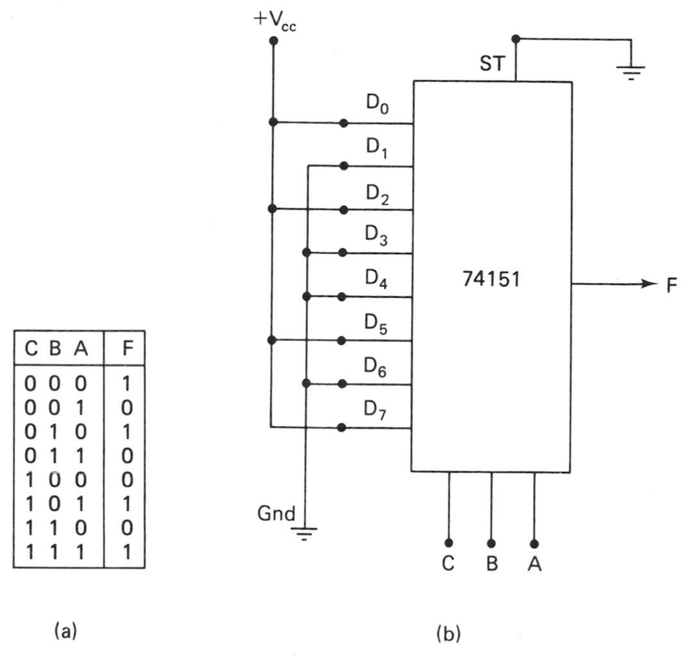

C B A	F
0 0 0	1
0 0 1	0
0 1 0	1
0 1 1	0
1 0 0	0
1 0 1	1
1 1 0	0
1 1 1	1

Figure 12.4 (a) Truth table function; (b) Multiplexer implementation.

SOLUTION:

To implement this truth table with a multiplexer, we must set the data input corresponding to each *CBA* binary combination to the logic level indicated in the *F* column of the truth table.

For example, when the control signals are $CBA = 000$ the D_0 input is sent to the *F* output. Therefore, the D_0 input is wired to a logic 1, and the *F* output of the multiplexer is consistent with the logic level under *F* in the table.

Similarly, we must set the remaining data inputs as follows:

$$D_1 = D_3 = D_4 = D_6 = 0$$
$$D_2 = D_5 = D_7 = 1$$

Besides this, the strobe input must be wired to ground or logic 0. This permanently enables the device.

A block diagram of this hookup with a 74151 multiplexer is shown in Fig. 12.4(b).

EXAMPLE 12.2

Figure 12.5(a) shows the connection of a 74153 4-line to 1-line multiplexer as a frequency selector. The four inputs D_0, D_1, D_2, and D_3 are supplied with different frequency signals. These waveforms are given in Fig. 12.5(b). What signal is passed to the output when the control inputs are set to $BA = 10$?

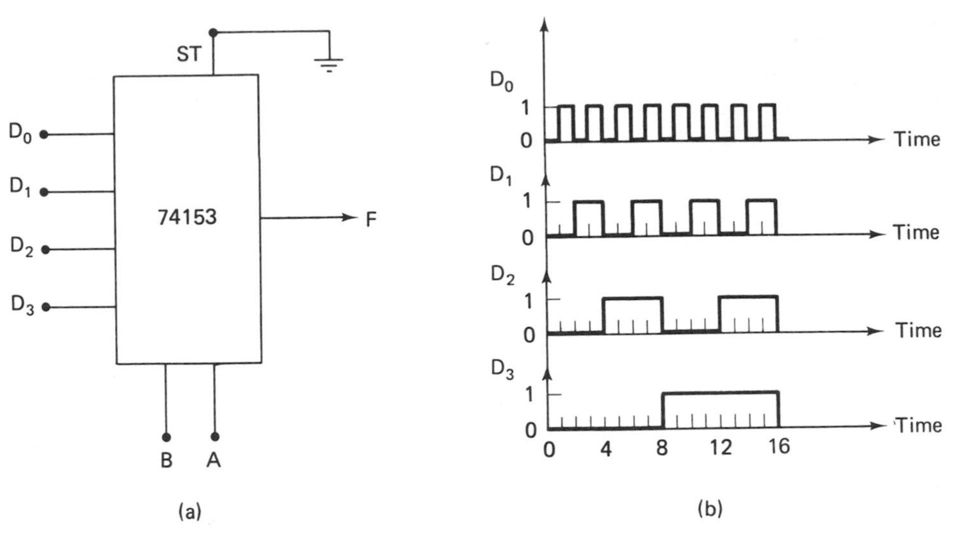

Figure 12.5 (a) Frequency selector; (b) Input signals.

SOLUTION:

The input signal sent to the output is selected by control inputs B and A. When BA equals 10, the waveform at input D_2 appears at the F output.

12.5 THE DIGITAL DEMULTIPLEXER

The *digital demultiplexer* works the opposite way of a multiplexer. That is, it switches data from one input line to one of several different output lines. For instance, Fig. 12.6(a) shows a block diagram of a 1-line to 4-line demultiplexer.

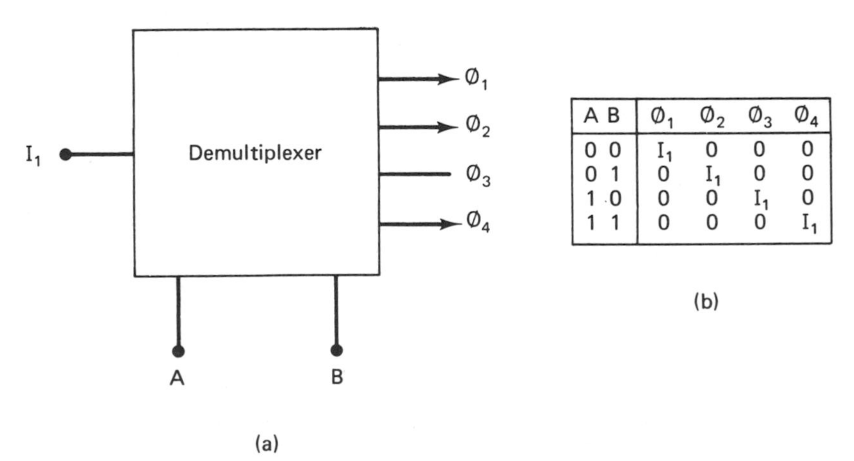

A B	$Ø_1$	$Ø_2$	$Ø_3$	$Ø_4$
0 0	I_1	0	0	0
0 1	0	I_1	0	0
1 0	0	0	I_1	0
1 1	0	0	0	I_1

Figure 12.6 (a) Demultiplexer block diagram; (b) Input/output operation.

Here, we notice that the circuit has a single data input marked I_1 and four data outputs $Ø_1$, $Ø_2$, $Ø_3$, and $Ø_4$. Furthermore, there are two control inputs A and B.

A truth table of demultiplexer operation is shown in Fig. 12.6(b). In this table, we see that the data at input I_1 are sent to a different output line for each binary combination at the AB control inputs.

For example, when the control input at AB is 11 the data at input I_1 are passed through the demultiplexer circuit to output $Ø_4$. If the control input is changed to $AB = 01$, input data are switched to the $Ø_2$ output.

12.6 INTEGRATED DEMULTIPLEXER DEVICES

As with multiplexers, digital demultiplexers are available as standard MSI circuits. Moreover, a demultiplexer circuit is also described in terms of its input and output lines. For instance, the 1-line to 4-line configuration is a standard demultiplexer device.

74155 1-Line to 4-Line Demultiplexer

A 74155 integrated circuit contains two 1-line to 4-line demultiplexer circuits. Each circuit is equivalent to the logic diagram shown in Fig. 12.7(a). A pin layout of the 74155 is given in Fig. 12.7(b).

From the network in Fig. 12.7(a), we find that the logic circuit has a single data input D_1 and four outputs $\bar{F}_1$, $\bar{F}_2$, $\bar{F}_3$, and $\bar{F}_4$. Here, we notice that complemented outputs are produced. So data are inverted between input and output of the demultiplexer. Besides this, there is a strobe lead that is used to enable the input gate. For the input to be enabled for operation, the ST lead must be made logic 0. If ST is at the 1 logic level, all outputs are held at the 1 logic level.

Both circuits within the 74155 device are identical, and their operation is controlled by a single set of control leads marked B and A. By changing the binary combination BA, the complement of input data D_1 is switched from one output to the next. A truth table of circuit operation is shown in Fig. 12.7(c).

Using the logic circuit in Fig. 12.7(a), we can describe the operation of the demultiplexer circuit. Let us assume that the control inputs A and B are both logic 0. This makes the $\bar{A}$ and $\bar{B}$ inverted inputs both logic 1. Looking at the output NAND gates, we see that two inputs on gate 7 are at the 1 level. At the same time, all other NAND gates have at least one input that is logic 0. Therefore, NAND gate 7 is the only gate whose output can change logic levels. The outputs of NAND gates 8 through 10 are held at the 1 level.

The remaining input to gate 7 depends on the logic levels of the strobe signal and data input. When the demultiplexer is enabled, the ST input is held at logic 0. This makes one input on gate 6 logic 0, and the other is the data input D_1 inverted. If D_1 is logic 1, both inputs to the NOR gate are 0, and its output switches to the 1 level. Now, all inputs on gate 7 are 1, and the $\bar{F}_1$ output switches to logic 0. In this

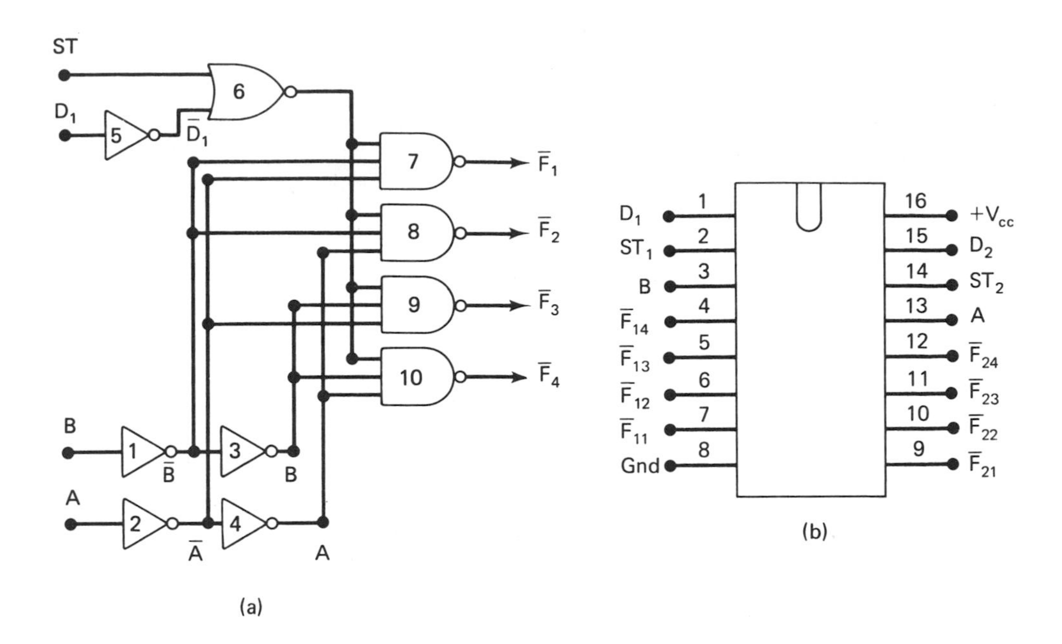

B A	$\bar{F}_1$	$\bar{F}_2$	$\bar{F}_3$	$\bar{F}_4$
0 0	$\bar{D}_1$	1	1	1
0 1	1	$\bar{D}_1$	1	1
1 0	1	1	$\bar{D}_1$	1
1 1	1	1	1	$\bar{D}_1$

(c)

Figure 12.7 (a) Demultiplexer logic diagram; (b) 74155 pin layout; (c) Truth table.

way, we find the $\bar{F}_1$ output corresponding to $BA = 00$, and it equals the complement of the data input $\bar{D}_1$. All other outputs are held at the 1 logic level.

However, if the D_1 input is 0, one input on the NOR gate is logic 1. This makes the output of gate 6 switch to the 0 logic level. Therefore, two inputs of NAND gate 7 are 1, and the other is 0. For this reason, its output is logic 1. This level is again the complement of the D_1 data input.

By changing the logic levels at BA to 01, NAND gate 8 has two inputs set to 1, and data D_1 can be sent through to output $\bar{F}_2$. In the same way, data can be passed to the $\bar{F}_3$ and $\bar{F}_4$ outputs.

EXAMPLE 12.3

Figure 12.8 shows the use of a 1-line to 4-line demultiplexer to control the switching of the output signal of circuit 1 to one of four circuits at the demul-

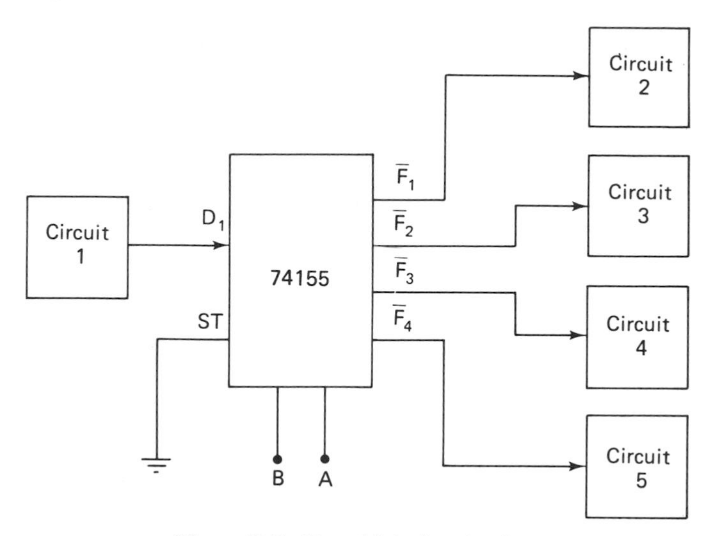

Figure 12.8 Demultiplexing circuitry.

tiplexer output. When the control signals BA are set to 11, which output receives the signal from circuit 1?

SOLUTION:

From the block diagram, we see that the strobe input of the demultiplexer is wired to ground. This logic 0 permanently enables the 74155 demultiplexer device.

For both control inputs equal to 1, the signal from circuit 1 is passed from the D_1 input of the demultiplexer to the $\bar{F}_4$ output. This output goes to the input on circuit 5.

Here, we also notice that the signal is inverted as it is sent from circuit 1 to circuit 5.

EXAMPLE 12.4

If the waveform supplied to the D_1 input of the demultiplexer in Fig. 12.8 is that shown in Fig. 12.9, draw output waveforms $\bar{F}_1$, $\bar{F}_2$, $\bar{F}_3$, and $\bar{F}_4$. The control input is again $BA = 11$.

SOLUTION:

In this demultiplexer circuit, the $\bar{F}_1$, $\bar{F}_2$, and $\bar{F}_3$ outputs are held at logic 1 as shown in the waveforms of Fig. 12.9. However, the signal at output $\bar{F}_4$ is the D_1 data input inverted.

12.7 DATA SELECTOR

A data selector is another type of digital multiplexing circuit. The data selector differs from the earlier devices in that it is used to select a complete data word to be passed to the output.

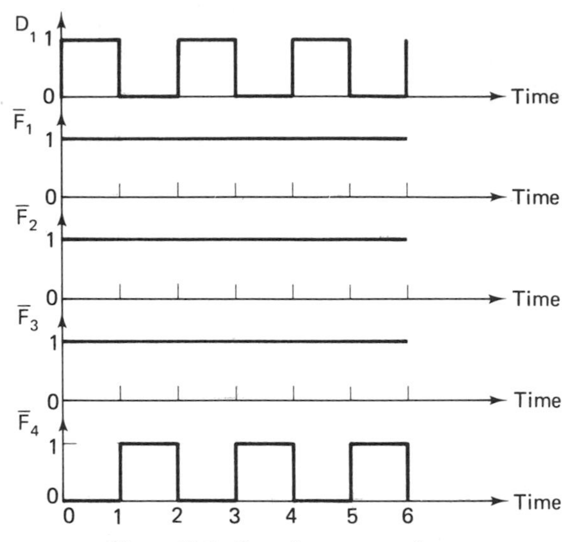

Figure 12.9 Input/output waveforms.

For example, the data selector block of Fig. 12.10 has two 4-bit input words. These words are

$$A = A_3A_2A_1A_0$$
$$B = B_3B_2B_1B_0$$

By changing the logic level at the select control lead S, the 4-bit A or B word is selected and sent to the 4-bit output F:

$$F = F_3F_2F_1F_0$$

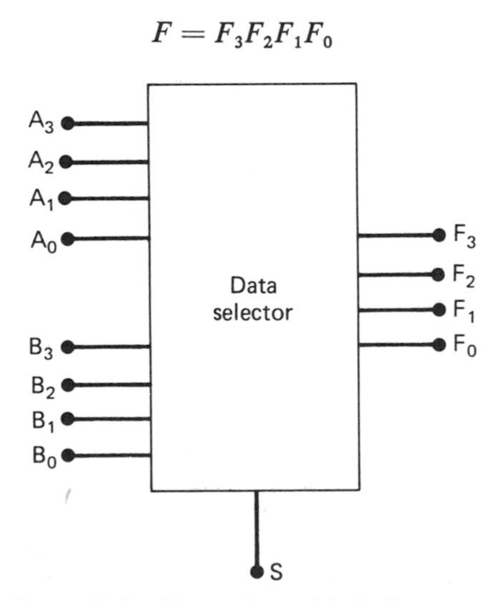

Figure 12.10 Data selector block diagram.

The 74157 Data Selector

A standard MSI data selector integrated circuit is the 74157 device. Figure 12.11(a) shows a pin layout of this 4-bit data selector, and its operation is given by the table in Fig. 12.11(b).

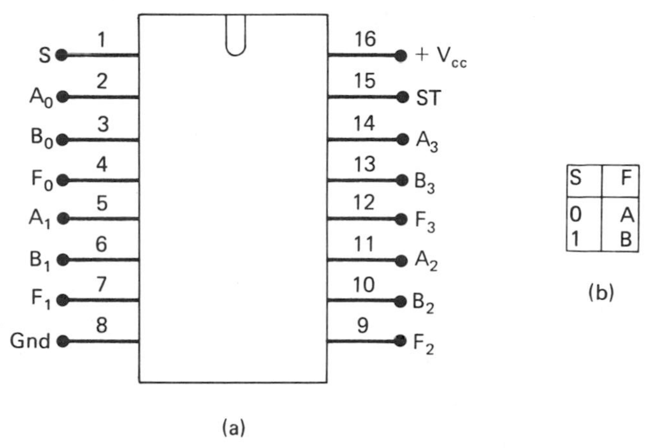

Figure 12.11 (a) 74157 pin layout; (b) Truth table.

In the pin layout, we see that the device has both a strobe (*ST*) and select (*S*) input. The strobe input is used to enable the circuitry for operation. When *ST* is logic 0, the device will work as a 4-bit data selector.

From the truth table in Fig. 12.11(b), we find how the select lead controls the transfer of data words between input and output. A 0 at the *S* lead selects the 4-bit *A* word to be switched to the *F* output lines and a 1 at *S* selects the *B* word.

EXAMPLE 12.5

Make a block diagram of how two 74157 data selectors can be interconnected to make an 8-bit data selector.

SOLUTION:

To make a single 8-bit data selector from two 74157 4-bit data selector devices, the strobe leads are connected together and wired to ground. This enables both devices. In the same way, the select leads are wired together to make a single select control *S*.

Now the *A* inputs of the two devices work together as an 8-bit input, and the same is true of the *B* inputs. Moreover, the two sets of *F* outputs are combined to make an 8-bit output.

$$A = A_7A_6A_5A_4A_3A_2A_1A_0$$

$$B = B_7B_6B_5B_4B_3B_2B_1B_0$$

$$F = F_7F_6F_5F_4F_3F_2F_1F_0$$

The block diagram of this connection is shown in Fig. 12.12.

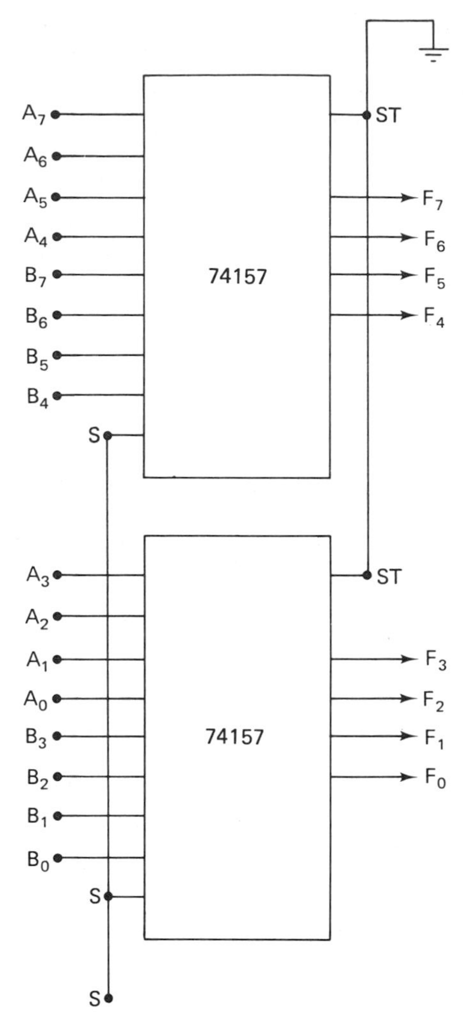

Figure 12.12 Eight-bit data selector.

ASSIGNMENTS

Section 12.2

1. Give a general description of digital multiplexer operation.

Section 12.3

2. What type of multiplexer is the 74153 device?
3. Perform a truth table analysis of the 74153 multiplexer circuit in Fig.

12.2(a) when $ST = 0$, data inputs are $D_3D_2D_1D_0 = 1010$, and the *BA* control input changes from 00 to 01, 10, and 11.

4. Repeat Problem 3 with $ST = 1$.

5. Make a block drawing like that in Fig. 12.1(a) for a 74151 device.

Section 12.4

6. Make a block drawing like that in Fig. 12.4(b) to show how a 74153 multiplexer can be used to implement the exclusive-OR logic function.

7. Draw the output waveform of the frequency selector circuit and input/output waveforms in Figs. 12.5(a) and (b). The control signals *BA* are equal to 11.

8. What is the output *F* in the circuit of Fig. 12.4(b) when the input is $CBA = 101$?

Section 12.5

9. Give a general description of demultiplexer operation.

Section 12.6

10. What type of demultiplexer is the 74155 device?

11. Make a truth table analysis of the demultiplexer circuit in Fig. 12.7(a) for $ST = 0$, $D_1 = 1$, and *BA* changes from 00, to 01, 10, and 11.

12. Figure 12.13 shows the control input waveforms of a 74155 demultiplexer. The data input D_1 is wired to logic 1, and the strobe lead (*ST*) is wired to logic 0. Draw the waveforms at outputs $\bar{F}_1$, $\bar{F}_2$, $\bar{F}_3$, and $\bar{F}_4$.

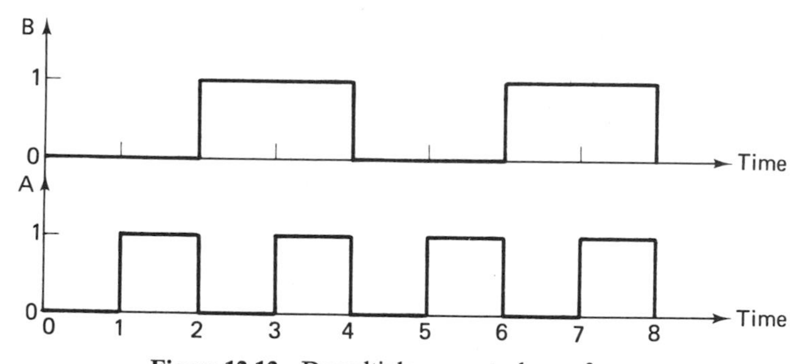

Figure 12.13 Demultiplexer control waveforms.

13. The input code to the demultiplexer in Fig. 12.8 is changed to 01. To which output is the input waveform of Fig. 12.9 sent? Draw the output waveform.

Section 12.7

14. Make a general description of data selector operation.

15. For the 74157 data selector, what is the output word of the device when the select lead S equals 1 and the circuitry is enabled by ST equals 0?

16. What is the output word of the 8-bit data selector in Fig. 12.12 when the S lead is set to logic 0?

Arithmetic Circuitry 13

13.1 INTRODUCTION

Arithmetic operations such as *addition* and *subtraction* are performed in digital equipment using numbers expressed in binary form. In this chapter, addition and subtraction of binary numbers are introduced. Along with these topics, combinational logic networks are developed to implement these mathematical processes with circuitry.

The topics covered in this chapter are as follows:

1. Addition and subtraction of binary numbers
2. Half-adder and full-adder circuits
3. Integrated adder devices
4. Subtracter circuits
5. 1's and 2's complement subtraction circuitry

13.2 ADDITION OF BINARY NUMBERS AND THE HALF-ADDER CIRCUIT

Our study of digital arithmetic circuitry begins with addition of binary numbers. When adding binary numbers, we can add the following combinations of

0s and 1s:

$$\begin{array}{cccc} 0 & 0 & 1 & 1 \\ \underline{+0} & \underline{+1} & \underline{+0} & \underline{+1} \\ 0 & 1 & 1 & 0 \text{ \& carry} \end{array}$$

Looking at these additions, we see three different results. First, the addition $0 + 0$ gives a sum of binary 0. On the other hand, combining binary 0 with a 1 by addition results in 1. The last result is obtained by adding 1 to 1. This should give 2, but in binary form 2 is written as 10.

Another way of describing the answer to $1 + 1$ is to say 0 and a carry of 1 to the next more significant bit. For the first three additions, there is no carry. No carry can be indicated with a binary 0. Therefore, the result of $1 + 0$ can also be written as 01. This is read as 1 and a carry of 0.

Addition Truth Table

In the addition of binary numbers, we must consider two results. One is the sum and the other a carry. In Fig. 13.1, a truth table is used to make a general

A B	S	C
0 0	0	0
0 1	1	0
1 0	1	0
1 1	0	1

Figure 13.1 Single-bit addition truth table.

description of binary addition. However, this table is only correct for binary numbers A and B both with 1 bit. The result of adding number A to B is the *sum* indicated by S and *carry* C. The four logic states indicated in this table are the same as the four basic binary additions just introduced.

Half-Adder Circuit

To implement the truth table function in Fig. 13.1 with a combinational logic network, we use the A and B data as inputs and S and C for two separate outputs. The circuit implementing this table is known as a *half-adder* and is indicated with a block like that shown in Fig. 13.2(a).

From the truth table in Fig. 13.1, we can write two Boolean equations to mathematically describe half-adder operation. The first expression is to describe the S output in terms of inputs A and B. The other expression gives the carry function. With these equations we can construct a logic diagram of a half-adder circuit.

Using the sum-of-products form, the Boolean equation for the sum output S is written as

$$S = \bar{A}B + A\bar{B} = A \oplus B$$

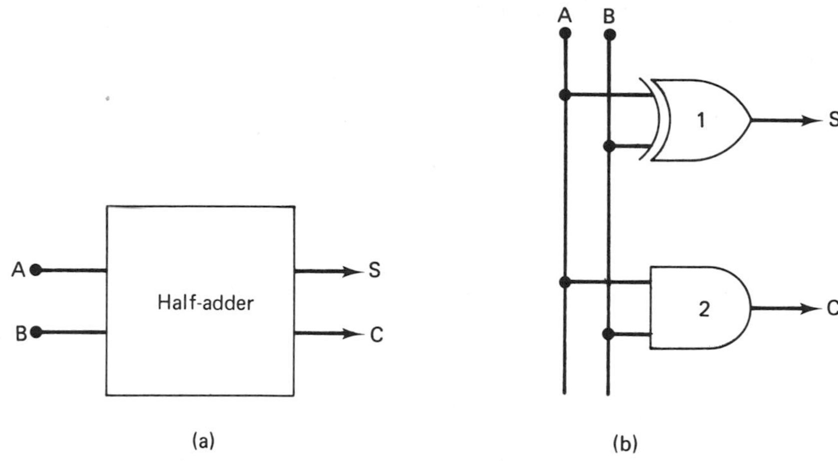

Figure 13.2 (a) Half-adder block; (b) Half-adder circuit.

Furthermore, we find that the equation of the carry output C is the AND function of inputs A and B:

$$C = AB$$

Looking at the sum equation, we notice it is just the exclusive-OR function of binary inputs A and B. For this reason, a single exclusive-OR gate is needed to provide the S output. This gate is marked 1 in the half-adder logic network of Fig. 13.2(b). Moreover, the carry output is provided by the AND gate labeled 2 in the logic diagram.

EXAMPLE 13.1

The input to the half-adder circuit in Fig. 13.2(b) is $AB = 01$. Find the logic level of the S and C outputs.

SOLUTION:

When $A = 0$ and $B = 1$, the S output of the exclusive-OR gate is logic 1:

$$S = 1$$

On the other hand, the C output on the AND gate is at binary 0:

$$C = 0$$

This result says that the sum of binary 0 and 1 is 1 with no carry and is consistent with the table of Fig. 13.1.

13.3 THE FULL-ADDER CIRCUIT

In real addition problems, multibit binary numbers must be used. For instance, the two numbers A and B could each have 4 bits. In this case, the addition can be expressed in general as

$$\begin{array}{r} A_3A_2A_1A_0 \\ +\ B_3B_2B_1B_0 \\ \hline S_4S_3S_2S_1S_0 \end{array}$$

Here we see that the sum of two 4-bit binary numbers can have 5 bits. The fifth bit is marked S_4 and is actually the carry from the fourth bit sum.

Let us give an example of this type of addition:

$$\begin{array}{rl} 1\,1\,1 & \text{carry} \\ 1010 & A \\ +\ \ 1110 & B \\ \hline 11000 & \text{sum} \end{array}$$

This problem is the binary addition of the decimal number A equals 10 to B equals 14. The sum we obtained is the binary equivalent of decimal number 24.

General Addition

Looking at the third bit in the addition example, we see a more general type of addition. In this bit, three binary numbers are added. The carry from the addition in the second bit is added to the sum of the A_2 and B_2 bit values. The result is the sum and a carry to the next bit.

This more general addition includes both a *carry in* C_i and *carry out* C_o. The sum is given by the equation

$$C_i + A + B = S \;\&\; C_o$$

An addition like this occurs in all bits other than the least significant bit of a multi-bit problem.

Full-Adder Circuit

The circuit used to implement this general addition is known as a *full-adder*. A block diagram of a full-adder is shown in Fig. 13.3(a). From this block, we see that the circuit has as its inputs C_i, A, and B. On the other hand, S and C_o are outputs.

A truth table of full-adder operation is shown in Fig. 13.3(b). In this table, we find that the same binary combinations of inputs A and B used for the half-adder circuit are listed for the carry input equal to 0 and repeated for C_1 equal to 1.

This table includes all possible binary combinations of inputs C_i, A, and B. For each input, the logic level of the sum output S and carry output C_o is indicated.

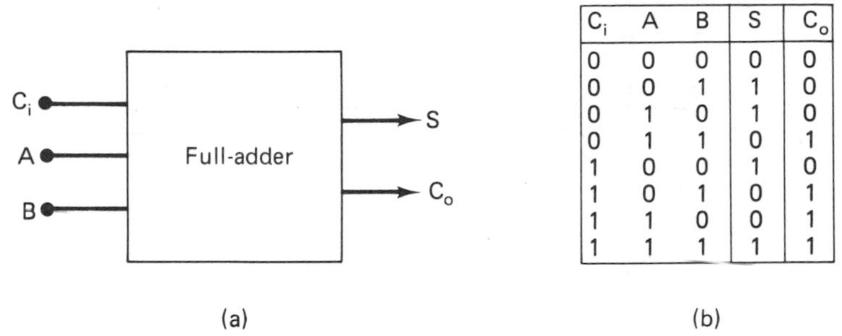

C_i	A	B	S	C_o
0	0	0	0	0
0	0	1	1	0
0	1	0	1	0
0	1	1	0	1
1	0	0	1	0
1	0	1	0	1
1	1	0	0	1
1	1	1	1	1

(a) (b)

Figure 13.3 (a) Full-adder block; (b) Truth table.

Boolean Equation of Full-Adder Operation

Once again, we must write a separate Boolean equation to describe the sum output and carry output of the full-adder. From the S and C_o columns of the truth table, we get the sum-of-products equations

$$S = \bar{C}_i\bar{A}B + \bar{C}_iA\bar{B} + C_i\bar{A}\bar{B} + C_iAB$$
$$C_o = \bar{C}_iAB + C_i\bar{A}B + C_iA\bar{B} + C_iAB$$

Before a logic network is constructed for the full-adder circuit, both equations must be simplified using the theorems introduced in Chapter 7.

The sum equation is simplified using theorem 8a. This gives the expression

$$S = \bar{C}_i(\bar{A}B + A\bar{B}) + C_i(\bar{A}\bar{B} + AB)$$

Looking at the AB terms within the parentheses, we find that the first can be reduced using the exclusive-OR of A and B. On the other hand, the second term contains the complement of the exclusive-OR function. Rewriting the expression with exclusive-OR notation results in

$$S = \bar{C}_i(A \oplus B) + C_i(\overline{A \oplus B})$$

Now we find that the equation represents the exclusive-OR of C_i with $A \oplus B$, and we get the final simplified form

$$S = C_i \oplus (A \oplus B)$$

In a similar way, the carry equation can be simplified to give

$$C_o = AB + C_i(A \oplus B)$$

Implementing the Full-Adder Equations

With the simplified Boolean equations for S and C_0, we can implement logic circuitry for a full-adder:

$$S = C_i \oplus (A \oplus B)$$
$$C_o = AB + C_i(A \oplus B)$$

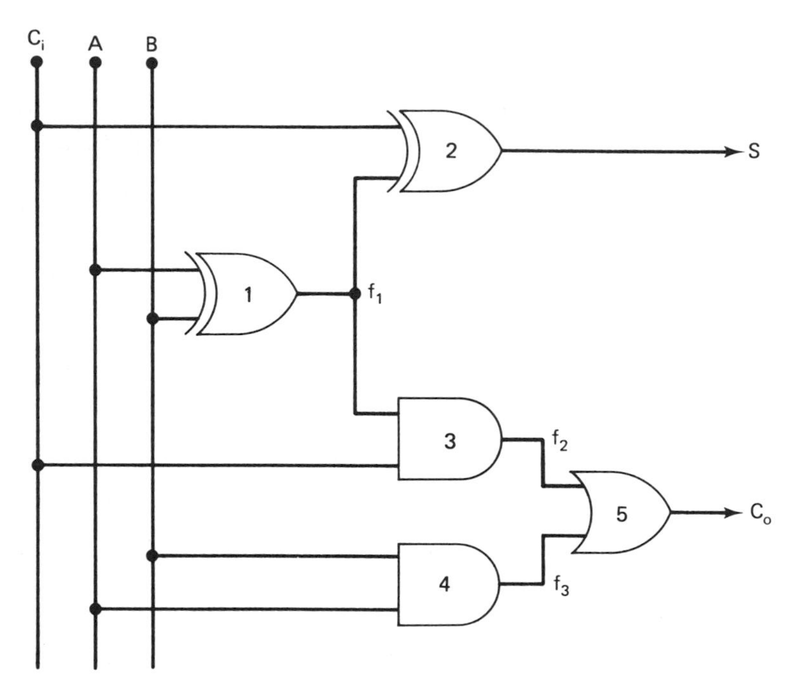

Figure 13.4 Full-adder logic network.

A full-adder circuit is constructed from exclusive-OR gates, AND gates, and an OR gate, as shown in Fig. 13.4.

In this full-adder network, two exclusive-OR gates are used to implement the sum expression. Moreover, the carry out function is formed with two AND gates and one OR gate. Looking at the carry out circuitry, we see that a further simplification is obtained by using $A \oplus B$ from an exclusive-OR gate in the sum circuitry.

EXAMPLE 13.2

What is the sum of the binary numbers $A = 01100101$ and $B = 10010111$? Find the decimal value of the sum.

SOLUTION:

$$\begin{array}{rl} 111 & C \\ 01100101 & A \\ +10010111 & B \\ \hline 11111100 & S \end{array}$$

The decimal value of $S = 11111100$ is found as follows:

$$S = 128(1) + 64(1) + 32(1) + 16(1) + 8(1) + 4(1)$$

$$S = 252$$

EXAMPLE 13.3

Find the logic level at the output of each gate in the full-adder circuit of Fig. 13.4 when the inputs are $C_i = 1$, $A = 0$, and $B = 0$.

SOLUTION:

The inputs to the exclusive-OR gate 1 are $AB = 00$. Therefore, output f_1 is logic 0:

$$f_1 = 0$$

Now the inputs on gate 2 are $C_i f_1 = 10$, and the sum output switches to the 1 level:

$$S = 1$$

For the carry circuitry, the inputs of AND gate 3 are $C_i f_1 = 10$, and its output stays at logic 0:

$$f_2 = 0$$

Moreover, both inputs on gate 4 are logic 0, and f_3 is also held at logic 0:

$$f_3 = 0$$

The carry output is produced by OR gate 5, and its inputs are $f_2 f_3 = 00$. For this reason, the carry output is binary 0:

$$C_o = 0$$

The outputs $S = 1$ and $C_o = 0$ are the same as given in the table of Fig. 13.3(b) for this input condition.

13.4 INTEGRATED ADDER DEVICES

Adder circuitry can be obtained in standard integrated circuits. Both full-adder circuits and *multibit adders* are available. We shall now turn our interest to these devices.

7480 Gated Full-Adder

A block diagram of the circuit within a 7480 integrated circuit is shown in Fig. 13.5(a). Looking at this diagram, we notice the circuit has three outputs. These outputs are the sum $\sum$, its complement $\overline{\sum}$, and the complement of the carry output $\bar{C}_o$.

On the input side, the full-adder has a single carry input marked C_i; four A inputs marked A_c, A^*, A_1, and A_2; and four B inputs B_c, B^*, B_1, and B_2. These two groups of inputs are gated to make A and B for the input columns of the truth table in Fig. 13.5(b).

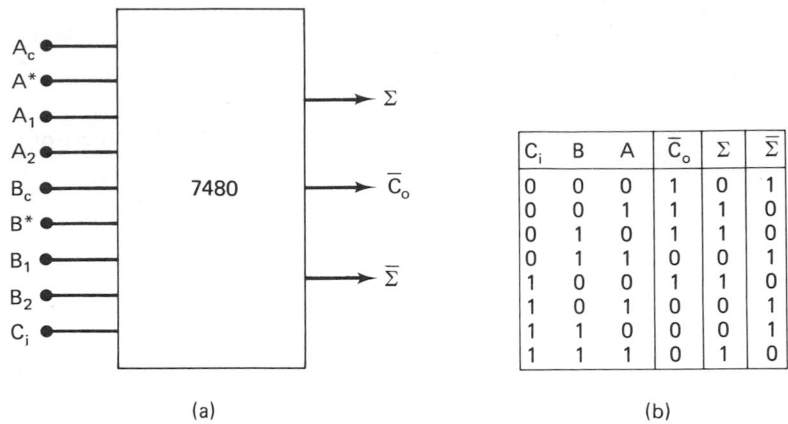

C_i	B	A	$\bar{C}_o$	Σ	$\bar{\Sigma}$
0	0	0	1	0	1
0	0	1	1	1	0
0	1	0	1	1	0
0	1	1	0	0	1
1	0	0	1	1	0
1	0	1	0	0	1
1	1	0	0	0	1
1	1	1	0	1	0

Figure 13.5 (a) 7480 block diagram; (b) Truth table.

The logic level of the *A* input in the table of Fig. 13.5(b) is related to the four *A* data inputs by the equation

$$A = \bar{A}_c + \bar{A}^* + A_1 \cdot A_2$$

In the same way, the *B* input is formed by the *B* data inputs and the equation

$$B = \bar{B}_c + \bar{B}^* + B_1 \cdot B_2$$

The 7480 can have some of its *A* and *B* data inputs wired to fixed logic levels to give simple full-adder operation. For instance, we can use the A_1 and B_1 leads as data inputs. Then the remaining leads must be set to the logic levels that follow:

$$A_c = B_c = 1$$
$$A^* = B^* = 1$$
$$A_2 = B_2 = 1$$

Now the *A* and *B* input equations simplify to give

$$A = 0 + 0 + A_1 \cdot 1 = A_1$$
$$B = 0 + 0 + B_1 \cdot 1 = B_1$$

or

$$A = A_1$$
$$B = B_1$$

This connection of the 7480 is shown in Fig. 13.6(a) and its pin layout in Fig. 13.6(b).

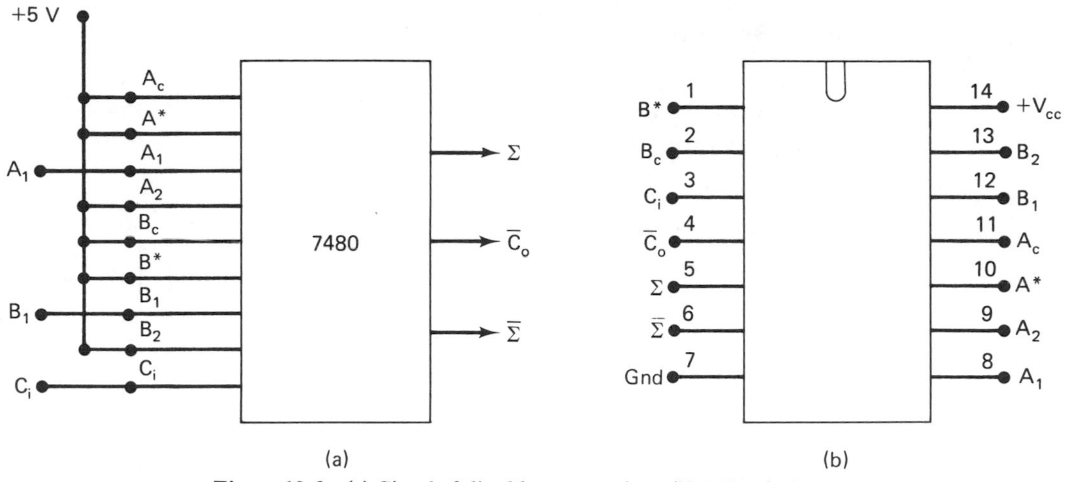

Figure 13.6 (a) Simple full-adder connection; (b) 7480 pin layout.

EXAMPLE 13.4

The data inputs on a 7480 full-adder circuit are as follows:

$$A_c = B_c = 1$$
$$A^* = B^* = 1$$
$$A_1 = 1, \qquad A_2 = 1$$
$$B_1 = 0, \qquad B_2 = 1$$
$$C_i = 1$$

What are the logic levels of the Σ, $\bar{\Sigma}$, and $\bar{C}_o$ outputs?

SOLUTION:

The A and B inputs can be found by using the input equations

$$A = \bar{A}_c + \bar{A}^* + A_1 \cdot A_2 = 0 + 0 + 1 \cdot 1 = 1$$
$$B = \bar{B}_c + \bar{B}^* + B_1 \cdot B_2 = 0 + 0 + 0 \cdot 1 = 0$$

In this way, we find that the inputs to the adder circuit are

$$A = 1$$
$$B = 0$$
$$C_i = 1$$

For this input, the truth table of Fig. 13.5(b) tells us that the full-adder outputs are

$$\bar{C}_o = 0$$
$$\Sigma = 0$$
$$\bar{\Sigma} = 1$$

7483 4-Bit Adder

A 4-bit binary adder circuit is available in the 7483 integrated circuit. Figure 13.7(a) shows the connection of the full-adders within the 7483 device, and Fig. 13.7(b) gives its pin layout.

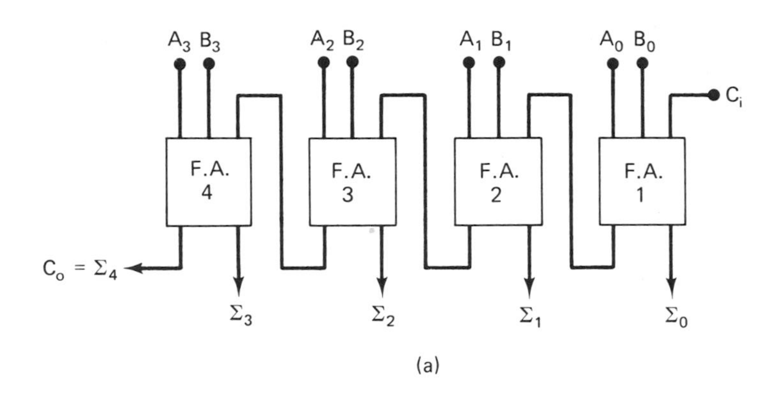

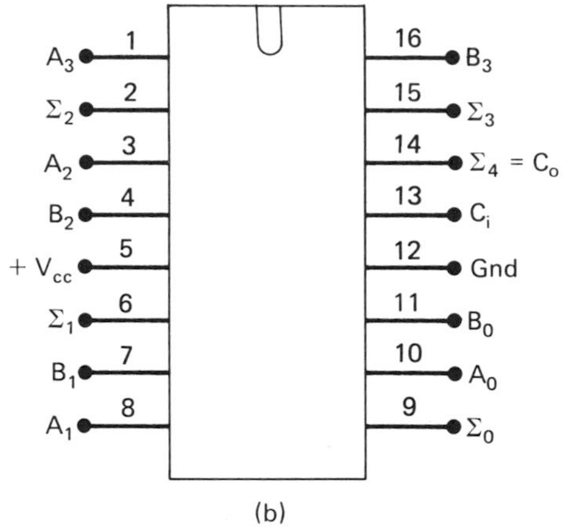

Figure 13.7 (a) 7483 block diagram; (b) 7483 pin layout.

Here we find that the 7483 has two 4-bit data inputs A and B with a 5-bit sum output $\sum$:

$$A = A_3A_2A_1A_0$$
$$B = B_3B_2B_1B_0$$
$$\sum = \sum_4\sum_3\sum_2\sum_1\sum_0$$

The most significant bit of the sum output is actually the carry bit on the fourth full-adder.

EXAMPLE 13.5

The inputs to the 4-bit adder circuit in Fig. 13.7(a) are $A = 0101$ and $B = 1011$. What is the sum output if the carry input is 1?

SOLUTION:

When set in this way, the 4-bit numbers A and B are added together. However, the logic 1 at the carry input C_i is added to the LSB:

$$\begin{array}{rl} 1 & C_i \\ 0101 & A \\ +\ 1011 & B \\ \hline 10001 & \Sigma \end{array}$$

13.5 SUBTRACTION OF BINARY NUMBERS AND THE HALF-SUBTRACTER CIRCUIT

The subtraction arithmetic process can also be directly implemented with combinational logic circuitry. As with addition, subtraction of numbers in digital equipment is performed with numbers expressed in binary form. The basic subtractions are as follows:

$$\begin{array}{r} 0 \\ -\ 0 \\ \hline 0 \end{array} \qquad \begin{array}{r} 0 \\ -\ 1 \\ \hline 1 \end{array} \text{ \& borrow} \qquad \begin{array}{r} 1 \\ -\ 0 \\ \hline 1 \end{array} \qquad \begin{array}{r} 1 \\ -\ 1 \\ \hline 0 \end{array}$$

From these subtractions, we find that subtracting binary 1 from 0 requires a borrow of 1 from the next more significant bit. When a 1 is borrowed, it is brought back as 1 + 1, and subtracting we get 1. This result is expressed as 1 and a borrow of 1.

Looking at the other three subtractions, we find that each can be performed without a borrow. This condition can be indicated as a borrow of 0.

The circuit needed to implement a 1-bit binary subtraction is known as a *half-subtracter circuit*. Figure 13.8(a) gives a block diagram of a half-subtracter and Fig. 13.8(b) a truth table of binary subtraction $A - B$.

Here the data inputs of the circuit are labeled A and B. Furthermore, there are two outputs. These outputs are marked D and Br for *difference* and *borrow*, respectively.

Using the truth table in Fig. 13.8(b), Boolean equations can be written for the D and Br outputs. Doing this, we get

$$D = \bar{A}B + A\bar{B} = A \oplus B$$

$$Br = \bar{A}B$$

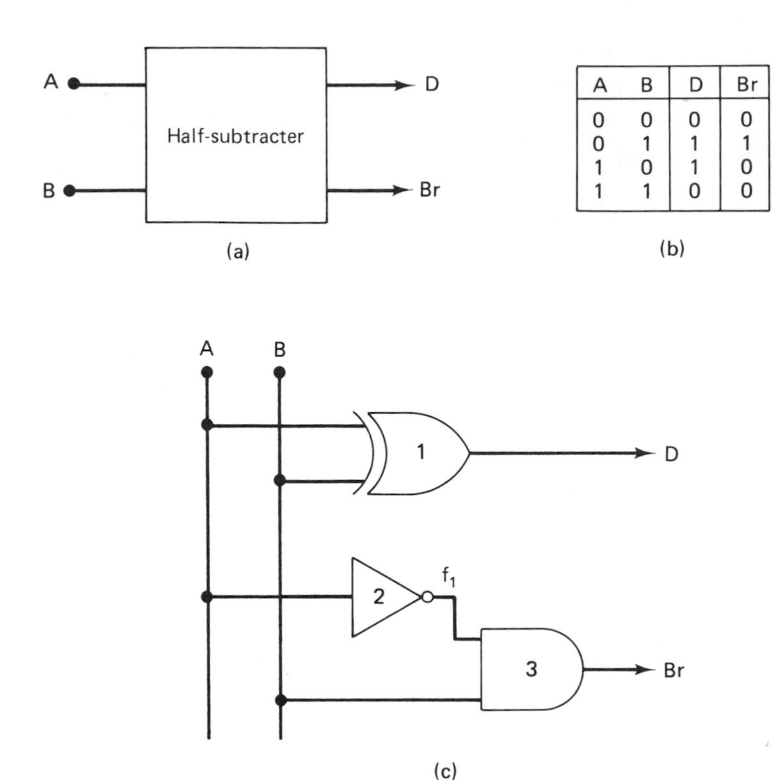

A	B	D	Br
0	0	0	0
0	1	1	1
1	0	1	0
1	1	0	0

Figure 13.8 (a) Half-subtracter block diagram; (b) Truth table; (c) Logic diagram.

The difference expression is the exclusive-OR function of inputs A and B. On the other hand, borrow is the AND function of the complement of input A with input B. The logic diagram of a half-subtracter is shown in Fig. 13.8(c).

EXAMPLE 13.6

The inputs to the logic network in Fig. 13.8(c) are $A = 0$ and $B = 1$. Find the logic level at the output of each gate.

SOLUTION:

The inputs on the exclusive-OR gate are $AB = 01$, and its output is logic 1:

$$D = A \oplus B = 1$$

At the inverter, input A is 0. This makes the f_1 output logic 1:

$$f_1 = \bar{A} = 1$$

Finally, the inputs on the AND gate are both 1, and the Br output is also logic 1:

$$Br = \bar{A}B = 1$$

13.6 THE FULL-SUBTRACTER

As with digital adder circuits, subtracter circuits normally involve more than 1 bit. For this type of operation, a full-subtracter is needed in addition to a half-subtracter circuit.

For example, we could subtract a 4-bit binary number B from another 4-bit number A. This results in a 4-bit *difference* D:

$$\begin{array}{r} A_3A_2A_1A_0 \\ -\quad B_3B_2B_1B_0 \\ \hline D_3D_2D_1D_0 \end{array}$$

To illustrate the subtraction process, let us do an example problem:

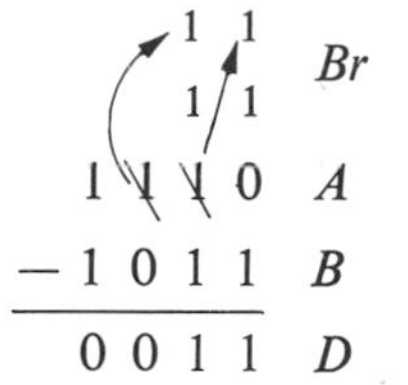

$$\begin{array}{rcccccl} & & 1 & 1 & & & Br \\ & & 1 & 1 & & & \\ & 1 & 1 & 1 & 0 & & A \\ - & 1 & 0 & 1 & 1 & & B \\ \hline & 0 & 0 & 1 & 1 & & D \end{array}$$

The subtraction in the A_0 bit cannot be performed directly. Instead, 1 must be borrowed from the A_1 bit and brought back as two 1s in the A_0 bit. Now subtraction can be done, and the result is 1.

Moving to the A_1 bit, we find that 1 has already been borrowed to leave a 0. For this reason, the subtraction cannot be done without borrowing. Bringing 1 back from the A_2 bit and subtracting, we obtain a difference of 1. The next 2 bits can be subtracted without borrowing, and both give binary 0.

Looking at the subtraction of B_1 from A_1, we notice that two borrows have been performed. First, a 1 is borrowed from the A_1 bit and returned to the A_0 bit. This is called a *borrow in* for the A_1 bit. Moreover, a 1 was borrowed from the A_2 bit and returned to the A_1 bit so the subtraction could be performed. This type of borrow is known as a *borrow out* for the A_1 bit. The subtraction in the A_1 bit can be expressed in general as

$$A - B - Br_i = D \,\&\, Br_o$$

This is known as a *full subtraction*, and it is performed by a *full-subtracter*

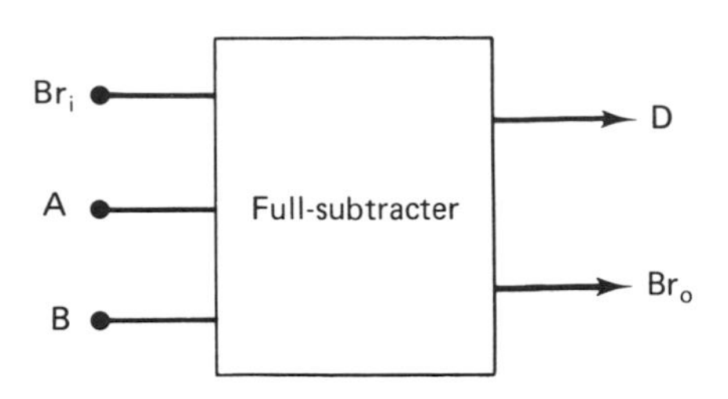

Br_i	A	B	D	Br_o
0	0	0	0	0
0	0	1	1	1
0	1	0	1	0
0	1	1	0	0
1	0	0	1	1
1	0	1	0	1
1	1	0	0	0
1	1	1	1	1

(a) (b)

Figure 13.9 (a) Full-subtracter block diagram; (b) Truth table.

circuit. Figure 13.9(a) shows a block diagram of a full-subtracter circuit, and its truth table is given in Fig. 13.9(b).

From this table, we can write a sum-of-products Boolean equation for the difference and borrow outputs. This results in the equations that follow:

$$
\begin{aligned}
D &= \overline{Br_i}\bar{A}B + \overline{Br_i}A\bar{B} + Br_i\bar{A}\bar{B} + Br_iAB \\
&= \overline{Br_i}(\bar{A}B + A\bar{B}) + Br_i(\bar{A}\bar{B} + AB) \\
&= \overline{Br_i}(A \oplus B) + Br_i(\overline{A \oplus B}) \\
D &= Br_i \oplus (A \oplus B) \\
Br_o &= \overline{Br_i}\bar{A}B + Br_i\bar{A}\bar{B} + Br_i\bar{A}B + Br_iAB \\
&= \bar{A}B(\overline{Br_i} + Br_i) + Br_i(\bar{A}\bar{B} + AB) \\
Br_o &= \bar{A}B + Br_i(\overline{A \oplus B})
\end{aligned}
$$

Implementing the simplified difference and borrow equations, we get the full-subtracter circuit in Fig. 13.10.

EXAMPLE 13.7

Find the difference $A - B$ for the numbers $A = 10010111$ and $B = 01100101$.

SOLUTION:

```
 1 1
 ⁄1 1        Br
⁄10010111    A
−01100101    B
---------
 00110010    D
```

13.7 SUBTRACTION BY COMPLEMENTS

An alternative approach used in binary subtraction is to use *complement methods.* With complements, we can find the difference of two binary numbers by an addition process instead of directly through subtraction. The most widely used complements in digital equipment are the *1's complement* and *2's complement.*

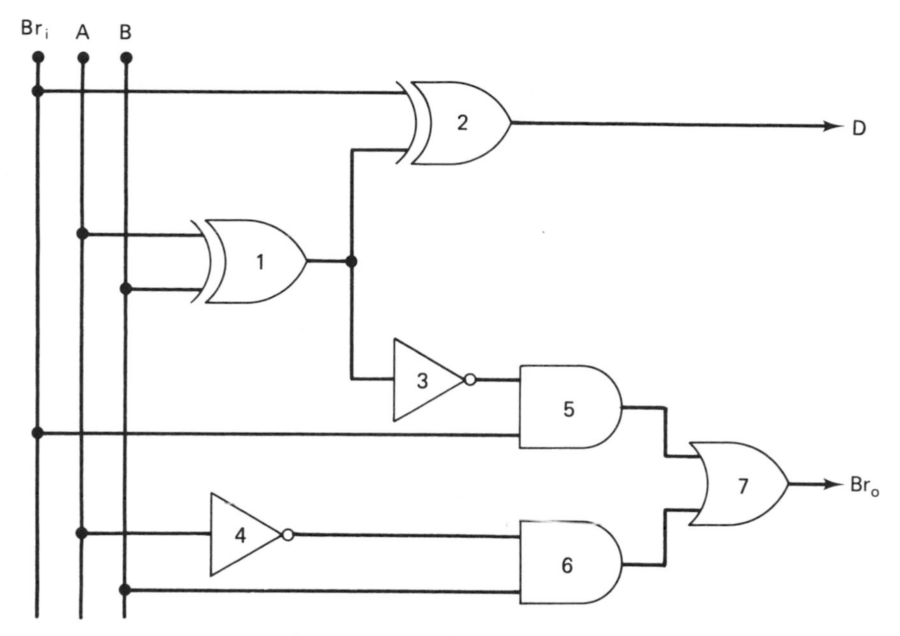

Figure 13.10 Full-subtracter circuit.

In this section, we shall show how to form the 1's and 2's complements of a binary number and how they are used in the subtraction of binary numbers.

1's and 2's Complements of a Binary Number

Before we can begin to subtract binary numbers with the complement methods, we must learn how to form the 1's complement and 2's complement of a binary number.

To form the 1's complement of a binary number, we just change all 1s in the number to 0s and all 0s to 1s. For instance, the 1's complement of the binary number 1010 is 0101.

On the other hand, the 2's complement of a binary number can be formed by first finding its 1's complement and then adding binary 1 to the LSB. As an example, let us continue our example introduced for the 1's complement by forming its 2's complement. We already found the 1's complement to be 0101, and adding 1 results in the 2's complement as 0110.

EXAMPLE 13.8

What are the 1's complement and 2's complement of the binary number 111000?

SOLUTION:

The 1's complement is obtained by changing all 1s to 0s and 0s to 1s:

111000

000111 1's complement

Now to get the 2's complement 1 is added to the LSB of the 1's complement:

```
000111
    +1
------
001000    2's complement
```

Subtraction Using the 1's Complement

First we shall show how the 1's complement is used to perform binary subtraction. To do this, the minuend of the subtraction problem is written in its normal binary form. On the other hand, the subtrahend is replaced with its 1's complement. After this, we add the two binary numbers. One more step is needed to complete the 1's complement subtraction. It is called an *end around carry* and is performed by taking the carry from the MSB of the sum and adding it to the LSB of the sum. The number that results after this is the difference between the two binary numbers.

To better understand this procedure, let us take an example problem. As an example, let us subtract the number 0111 from 1010:

```
minuend       1010
subtrahend  - 0111
            ------
```

We start the subtraction by replacing the subtrahend with its 1's complement and adding. This gives

```
minuend       1010
subtrahend  + 1000
            ------
             10010
```

To get the difference, we must still perform the end around carry. The carry from the MSB is 1, and it is added to the LSB of the sum to give the difference:

```
             10010
                +1
             -----
difference    0011
```

We can check our subtraction by converting the minuend, original subtrahend, and difference to decimal form:

```
minuend        1010 =  10
subtrahend   - 0111 = -7
             ------   ---
difference     0011 =   3
```

This gives 10 minus 7 equal to 3, and the subtraction is correct.

EXAMPLE 13.9

Find the difference 111100 − 010101 using the 1's complement subtraction method.

SOLUTION:

$$\begin{array}{r} 111100 \\ -010101 \\ \hline \end{array}$$

Changing the subtrahend to 1's complement form and adding results in

$$\begin{array}{r} 111100 \\ +101010 \\ \hline 1100110 \end{array}$$

Performing the end around carry, we get

$$\begin{array}{r} 1100110 \\ +1 \\ \hline 100111 \end{array}$$

The difference is equal to 100111.

Subtraction Using the 2's Complement

The 2's complement of the subtrahend can be used instead of the 1's complement in the subtraction of binary numbers. In this case, we just replace the subtrahend with its 2's complement, add the two numbers, and then discard the carry from the MSB.

To illustrate the 2's complement procedure, let us repeat the subtraction 1010 − 0111 we did with the 1's complement method:

$$\begin{array}{lr} \text{minuend} & 1010 \\ \text{subtrahend} & -\ 0111 \\ \hline \end{array}$$

The 2's complement of the subtrahend is found by changing all 1s to 0s and 0s to 1s to give the 1's complement and then adding 1. For the subtrahend 0111, we get

$$1000 + 1 = 1001$$

By substituting the 2's complement into the subtrahend and adding, the result is

$$\begin{array}{lr} \text{minuend} & 1010 \\ \text{subtrahend} & +\ 1001 \\ \hline \text{difference} & \not{1}0011 \end{array}$$

Eliminating the carry from the MSB, we get the difference as 0011. Comparing this answer to that found by the 1's complement method, we see that both results are the same.

EXAMPLE 13.10

Use the 2's complement method of subtraction to find the difference 101010 − 000111.

SOLUTION:

$$\begin{array}{r} 101010 \\ -000111 \\ \hline \end{array}$$

The 2's complement of the subtrahend is found to be

$$111000 + 1 = 111001$$

By replacing the subtrahend with the 2's complement, the difference is obtained by adding:

$$\begin{array}{r} 101010 \\ +111001 \\ \hline \not{1}100011 \end{array}$$

The difference is 100011.

13.8 SUBTRACTION CIRCUITRY USING COMPLEMENT METHODS

Circuitry can be made that performs subtraction by the complement methods instead of with full-subtraction circuits. In the previous section, we found that binary subtraction could be performed by using the 1's complement or 2's complement and an addition procedure. The advantage of the complements method from a circuit point of view is that subtraction can be performed with an integrated adder device.

1's Complement Subtraction Circuit

The procedure we used to subtract binary numbers by the 1's complement method starts with the formation of the 1's complement of the subtrahend binary number. After this, the complement of the subtrahend is added to the minuend instead of subtracting. The 1's complement subtraction is completed by performing and end around carry. This is done by adding the carry from the MSB in the sum to the LSB of the sum to get the difference.

A circuit of a 4-bit *1's complement subtracter* is shown in Fig. 13.11. To form the 1's complement of a number, we change all 0s to 1s and all 1s to 0s. In the subtracter circuit, inverters are used to produce the 1's complements of the subtrahend inputs $B_3B_2B_1B_0$. The complements are labeled $\bar{B}_3\bar{B}_2\bar{B}_1\bar{B}_0$. A 7483 4-bit binary

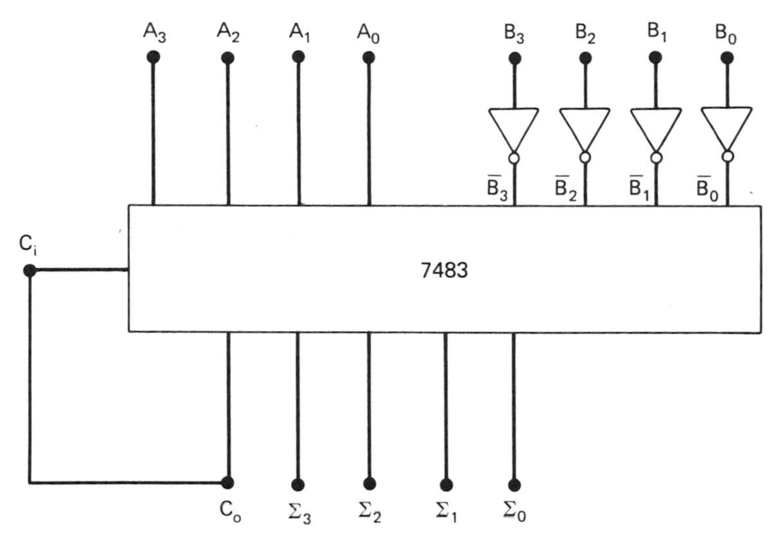

Figure 13.11 1's complement subtraction circuit.

adder is used to produce the sum of the minuend $A_3A_2A_1A_0$ and 1's complement of the subtrahend. The outputs of the adder are $\sum_3\sum_2\sum_1\sum_0$, and the carry from the MSB adder is C_o. This carry output is sent back to the carry input C_i of the LSB in the adder. This return connection of C_o to C_i represents the end around carry operation. After the C_o bit is input to C_i it is added once more to A_0 and $\bar{B}_0$ to give the LSB of the difference. The remaining bits $A_3A_2A_1$ and $\bar{B}_3\bar{B}_2\bar{B}_1$ are again added to give the final difference.

EXAMPLE 13.11

The inputs to the subtracter circuit in Fig. 13.11 are $A_3A_2A_1A_0 = 1100$ and $B_3B_2B_1B_0 = 0011$. Find the difference output of the circuit.

SOLUTION:

The outputs of the inverters are

$$\bar{B}_3\bar{B}_2\bar{B}_1\bar{B}_0 = 1100$$

The 4-bit adder combines its $\bar{B}$ inputs to the A inputs to give

$$\begin{array}{rl} 1100 & A \\ +\ 1100 & B \\ \hline \not{1}1000 & \Sigma \end{array}$$

The carry output C_0 is the most significant bit of the sum output:

$$C_o = 1$$

This logic 1 is returned to the C_i input of the adder device, and the addition is repeated. The result is

$$\begin{array}{rl} 1 & C_i \\ 1100 & A \\ +\ 1100 & B \\ \hline \not{1}1001 & \Sigma \end{array}$$

The C_o bit of the sum is eliminated to give the difference by the 1's complement method:

$$\text{difference} = 1001$$

2's Complement Subtraction Circuit

Figure 13.12 shows a circuit similar to the 1's complement subtracter in Fig. 13.11; however, this 7483 connection performs subtraction by the 2's complement method.

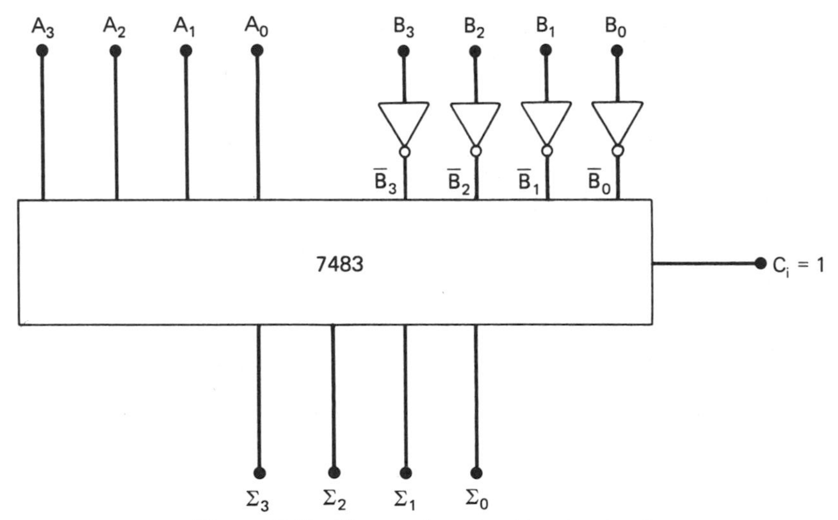

Figure 13.12 2's complement subtracter circuit.

In the *2's complement subtracter circuit,* inverters change the subtrahend inputs $B_3B_2B_1B_0$ to 1's complement form at outputs $\bar{B}_3\bar{B}_2\bar{B}_1\bar{B}_0$. We also notice that the carry input C_i to the LSB of the adder is held at the 1 logic level. This adds 1 to the LSB input and produces the 2's complement of the subtrahend.

After this, the 7483 adds the minuend $A_3A_2A_1A_0$ to the 2's complement of subtrahend $B_3B_2B_1B_0$. The output produced at lines $\Sigma_3\Sigma_2\Sigma_1\Sigma_0$ is the difference. The carry output bit C_o of the adder is not used.

EXAMPLE 13.12

The 2's complement subtracter circuit in Fig. 13.12 is used to subtract the number $B_3B_2B_1B_0 = 1000$ from $A_3A_2A_1A_0 = 1111$. What is the difference output of the circuit?

SOLUTION:

The outputs of the inverters are

$$\bar{B}_3\bar{B}_2\bar{B}_1\bar{B}_0 = 0111$$

With the 7483 adder device, the minuend and inverted subtrahend are added along with $C_i = 1$:

$$\begin{array}{rl} 1 & C_i \\ 1111 & A \\ +\ 0111 & \bar{B} \\ \hline \not{1}0111 & \end{array}$$

Eliminating the C_o bit, we get the difference:

$$\text{difference} = 0111$$

ASSIGNMENTS

Section 13.2

1. The half-adder circuit in Fig. 13.2(b) has input $AB = 11$. What are the logic levels of the S and C outputs?

2. Show how to implement carry out in the circuit of Fig. 13.2(b) with NAND gates.

3. Make a complete truth table analysis of the half-adder circuit in Fig. 13.13(a).

4. Perform a truth table analysis of the half-adder circuit in Fig. 13.13(b).

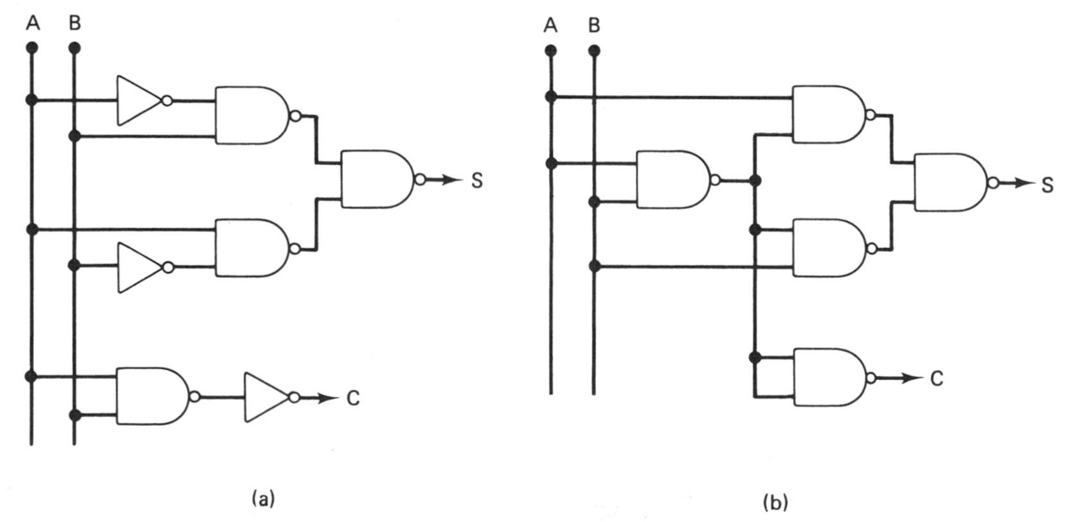

Figure 13.13 Half-adder circuits.

5. Write a product-of-sums Boolean equation for the S output of a half-adder. Implement the sum and carry outputs with NOR gate circuitry.

Section 13.3

6. Find the binary sum in each of the following problems.
(a) $10101 + 11000 =$ (b) $01001111 + 11011100 =$

7. Add the binary number 11000 to 11011. Convert the binary answer to decimal form and express with 8 bits of BCD code.

8. Show how the carry equation $C_o = \bar{C}_iAB + C_i\bar{A}B + C_iA\bar{B} + C_iAB$ is simplified to the form $C_o = AB + C_i(A \oplus B)$.

9. What is the logic level at the output of each gate in the circuit of Fig. 13.4 if the inputs are $C_i = 1$, $A = 1$, and $B = 1$.

10. Perform a complete truth table analysis of the full-adder circuit in Fig. 13.14.

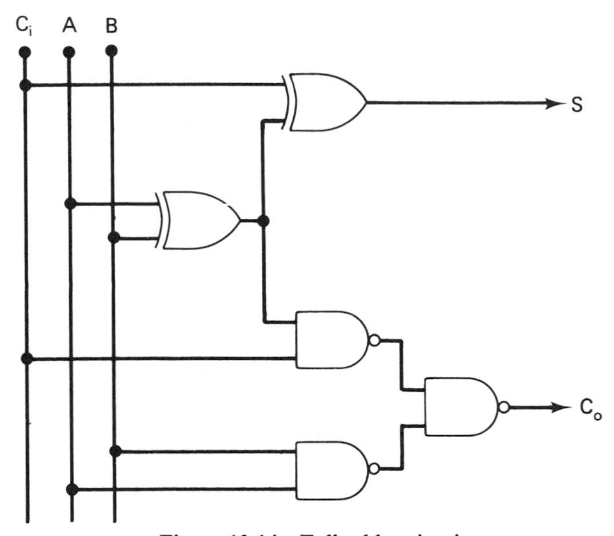

Figure 13.14 Full-adder circuit.

11. Draw a full-adder circuit using just NAND gates.

Section 13.4

12. The data inputs to the 7480 full-adder circuit in Fig. 13.6(a) are $A_1 = 1$, $B_1 = 0$, and $C_i = 0$. What are the logic levels of outputs $\sum$, $\overline{\sum}$, and $\bar{C}_o$?

13. The following signals are the inputs of a 7480 full-adder circuit. Find the $\sum$, $\overline{\sum}$, and $\bar{C}_o$ outputs.

$$A_c = B_c = 1$$
$$A^* = B^* = 1$$
$$A_1 = 1, \quad A_2 = 1$$
$$B_1 = 1, \quad B_2 = 1$$
$$C_i = 0$$

14. A 7483 integrated circuit is set up to work as a 4-bit adder with no carry input. The inputs are $A = 1010$ and $B = 1010$. What is the $\sum$ output in binary form?

Section 13.5

15. The inputs of the half-subtracter block in Fig. 13.8(a) are $AB = 00$. What are the logic levels at the D and Br outputs?

16. Redraw the half-subtracter circuit of Fig. 13.8(c) using NAND gates for the borrow circuitry.

17. Find the logic level at the output of each gate in the circuit of Fig. 13.8(c) for input $AB = 11$.

18. Make a complete truth table analysis of the half-subtracter circuit in Fig. 13.15.

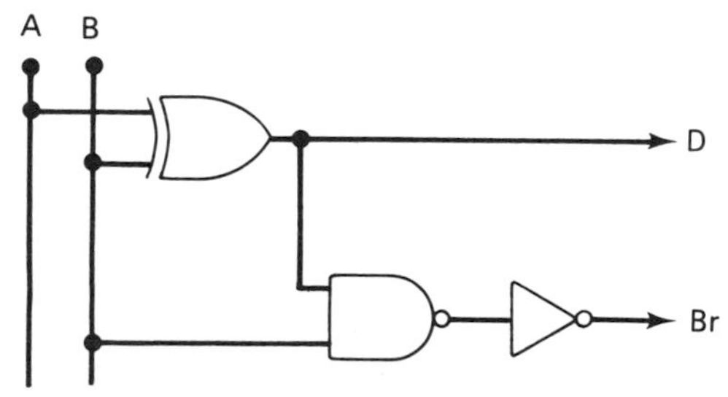

Figure 13.15 Half-subtracter circuit.

Section 13.6

19. Perform the binary subtractions that follow.
(a) 11000 − 10101 = (b) 11011100 − 01001111 =

20. Perform a complete truth table analysis of the full-subtracter circuit in Fig. 13.10.

21. Redraw the subtraction circuit in Fig. 13.10 using NAND gates for the borrow circuitry.

22. Draw a block diagram showing how four full subtracter circuits can be connected to make a 4-bit subtracter.

Section 13.7

23. Find the 1's complement of each number that follows.
(a) 0010111 (b) 101000000

24. What is the 2's complement of each number that follows.
(a) 1110111 (b) 011101101

25. Do the subtractions that follow by the 1's complement method.
(a) 11100 – 01010 = (b) 101011100 – 011111111 =

26. Use the 2's complement method to subtract the numbers that follow.
(a) 10000 – 00111 = (b) 1111000 – 0001111 =

27. Subtract $B = 1001$ from $A = 1010$ with the 1's complement method.

28. Using the 2's complement method, subtract $B = 010001$ from $A = 100000$.

Section 13.8

29. Repeat Example 13.11 for inputs $A = 1110$ and $B = 0100$ to the 1's complement subtracter circuit in Fig. 13.11.

30. The inputs to the subtracter circuit of Fig. 13.12 are $A = 0111$ and $B = 0101$. What are the outputs of the inverters and the difference output of the 7483 device?

Pulse Circuits 14

14.1 INTRODUCTION

In digital equipment, *pulse circuits* are used to generate pulse signals. Two of the most common signals are known as the *pulse* and *square wave*.

In this chapter, we shall look into the characteristics of these waveforms and the circuits used to make each waveshape. The topics covered are as follows:

1. The clock
2. Astable multivibrator
3. Monostable multivibrator
4. Integrated monostable devices
5. The 555 universal timer

14.2 THE CLOCK

The operation of the circuits in many digital systems is controlled or synchronized with a timing signal. This signal is known as the *clock* and is generated by a *clock circuit*. The output produced by a clock circuit is normally a square wave.

A clock circuit can be represented with a block like that shown in Fig. 14.1.

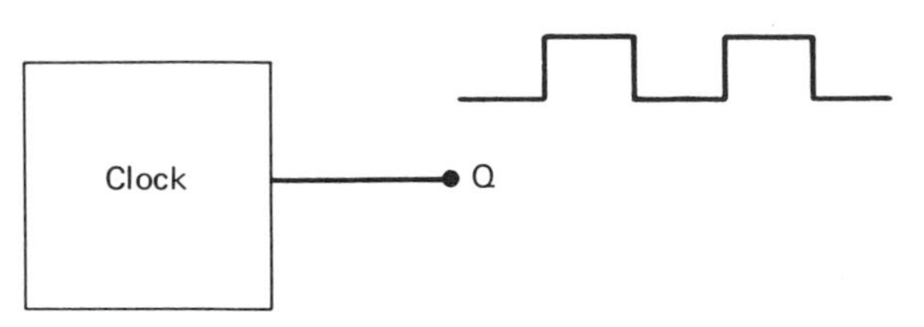

Figure 14.1 The clock circuit.

Here, we see that the output terminal of the circuit is marked Q, and a square wave signal is drawn at this lead.

The Square Wave

A square wave signal switches back and forth between two dc voltage levels. Figure 14.2 shows the waveform of a typical square wave. In this diagram, the voltage levels are marked binary 1 and 0. The high voltage level represents a logic 1 and the low level logic 0.

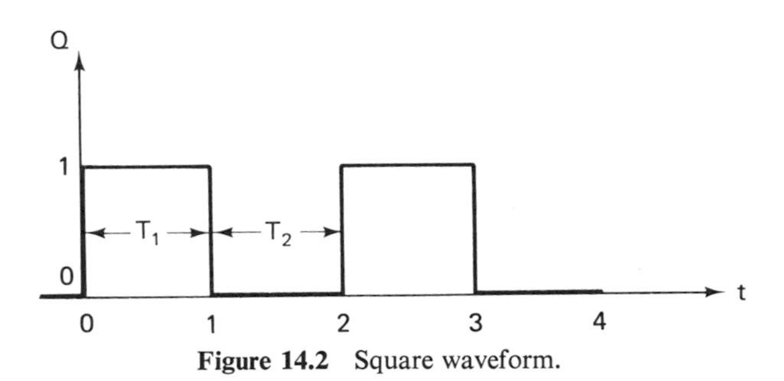

Figure 14.2 Square waveform.

Looking at this square wave, we see that Q stays at the 1 logic level during the time interval from 0 to 1. The amount of time during which the waveform is at the 1 logic level is marked T_1 and called the *pulse width* of the square wave.

On the other hand, Q switches to logic 0 for the time interval from 1 to 2. The amount of time a square wave stays at logic 0 is known as the *pulse interval*, which is labeled T_2.

The values of pulse width T_1 and pulse interval T_2 are measured in the unit seconds (s). However, the short duration of time common to switching waveforms requires the use of the millisecond, microsecond, and nanosecond units. These units are abbreviated ms, μs, and ns, respectively.

The points at which a square wave switches between logic levels are the edges of the waveforms. The change from logic 0 to 1 is known as the *leading edge* of the waveform, and the return from logic 1 to 0 is called the *trailing edge*.

In the waveform of Fig. 14.2, the changes between 0 and 1 are shown to occur instantly. However, in actual circuitry a short delay occurs. The time needed to switch from logic 0 to 1 is the *rise time* of the signal, and the time to switch from 1 to 0 is its *fall time*. Rise time and fall time actually represent the amount of time it takes to go between the 10% and 90% levels of the signals.

Symmetrical and Asymmetrical Square Waves

The output of a clock circuit can be a *symmetrical* or *asymmetrical* square wave. The waveform of Fig. 14.3(a) is symmetrical. From this wave, we see that

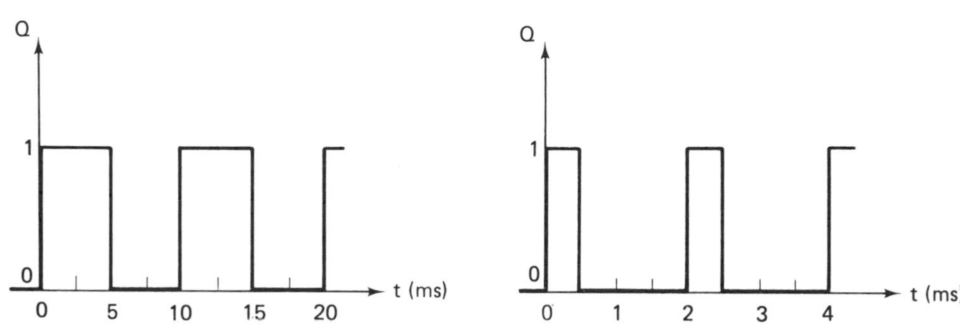

Figure 14.3 (a) Symmetrical square wave; (b) Asymmetrical square wave.

the pulse width and pulse interval are the same length of time. This condition for a symmetrical square wave is expressed as

$$T_1 = T_2 \qquad \text{symmetrical}$$

On the other hand, an asymmetrical square wave has a different value of pulse width and pulse interval. This case is stated mathematically as

$$T_1 \neq T_2 \qquad \text{asymmetrical}$$

An asymmetrical waveform is shown in Fig. 14.3(b).

EXAMPLE 14.1

How long are the pulse width and pulse interval in the waveform of Fig. 14.3(a)?

SOLUTION:

In Fig. 14.3(a), we find the pulse width and pulse interval both equal to 5 ms:

$$T_1 = T_2 = 5 \text{ ms}$$

Properties of a Square Wave

Two important properties of a square wave are its period and frequency. From Fig. 14.4 we see that the square waveshape repeats every .2 s. This interval

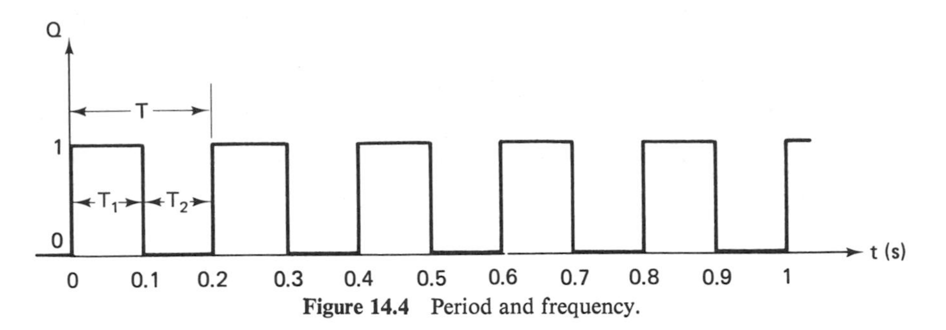

Figure 14.4 Period and frequency.

of time is called the *period* of the square wave and is indicated by the symbol T:

$$T = .2 \text{ s}$$

The period is also measured in the basic unit of seconds. This property of a square wave is sometimes called the *pulse repetition time* and labeled *PRT*.

The period of a square wave can be related to its pulse width and pulse interval. Looking at Fig. 14.4, we notice that T is equal to the sum of T_1 and T_2. This relationship is given by the equation

$$T = T_1 + T_2$$

From the symmetrical square wave of Fig. 14.3(a), we find that the pulse width and pulse interval are equal. For this reason, they can be found by dividing the period of the waveform by 2:

$$T_1 = T_2 = \frac{T}{2} \qquad \text{symmetrical}$$

The *frequency* of a square wave tells the number of times the basic waveform repeats in a 1-s time interval. Frequency is labeled with the symbol f and measured in pulses per second or hertz. When the hertz unit is in use, it is abbreviated Hz. The frequency property is sometimes called the *pulse repetition rate* or *PRR*.

For instance, the square wave in Fig. 14.4 repeats every .2 s or five times in the 1-s time interval. This gives a frequency of 5 pulses/s or 5 Hz:

$$f = 5 \text{ Hz}$$

The frequency of a square wave can be calculated as the reciprocal of its period. This relation is given mathematically by the equation

$$f = \frac{1}{T}$$

Let us check the use of this expression by calculating the frequency of the square wave in Fig. 14.4. Earlier, we said the period T is .2 s. Using the frequency equation, we get

$$f = \frac{1}{T} = \frac{1}{.2\text{s}} = 5 \text{ Hz}$$

$$f = 5 \text{ Hz}$$

EXAMPLE 14.2

Find the period and frequency of the asymmetrical square wave in Fig. 14.3(b).

SOLUTION:

From the waveform, we see that the pulse width is .5 ms and the pulse interval 1.5 ms:

$$T_1 = .5 \text{ ms}$$

$$T_2 = 1.5 \text{ ms}$$

Now the period and frequency are found to be

$$T = T_1 + T_2 = .5 \text{ ms} + 1.5 \text{ ms} = 2 \text{ ms}$$

$$T = 2 \text{ ms}$$

$$f = \frac{1}{T} = \frac{1}{2 \text{ ms}} = 500 \text{ Hz}$$

$$f = 500 \text{ Hz}$$

14.3 THE ASTABLE MULTIVIBRATOR CIRCUIT

The basic circuit used to generate a clock signal is the *astable multivibrator*. An astable circuit can be made with logic gates, as shown in Fig. 14.5(a). Typical waveforms for this circuit are drawn in Fig. 14.5(b). Here, we find the voltage V_c

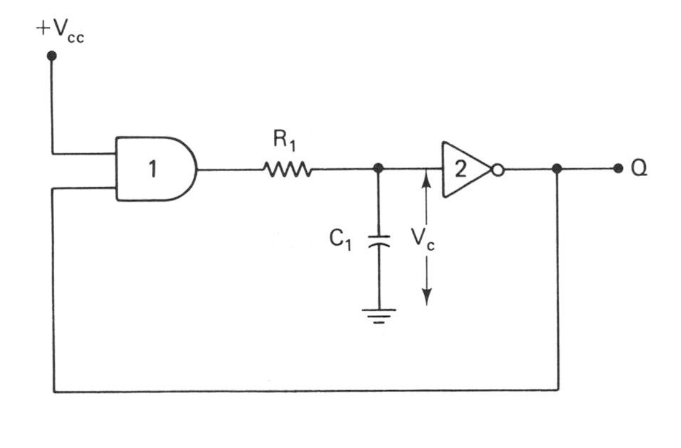

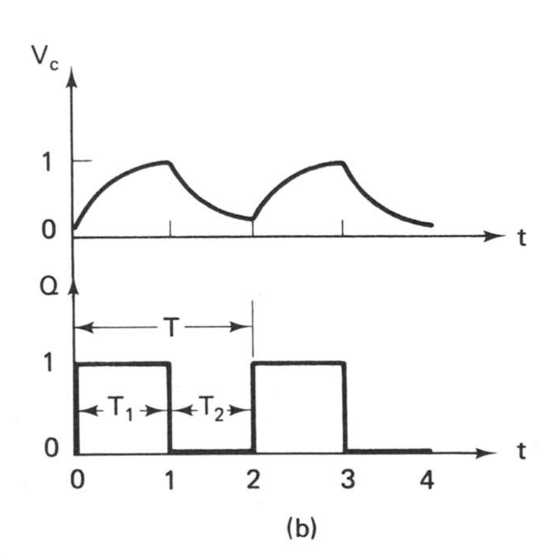

Figure 14.5 (a) Logic gate astable multivibrator; (b) Waveforms.

across capacitor C_1 and the square wave at output Q. The name astable multivibrator means the output of the circuit has no stable state. Looking at the Q waveform, we see that it switches back and forth between the 1 and 0 logic levels at regular time intervals.

In the logic diagram of Fig. 14.5(a), we see that the circuit is made with one AND gate and an inverter. The output of the AND gate is connected to the input of the inverter through an RC circuit. These devices, R_1 and C_1, are used to set the timing and frequency of the square wave at the Q output.

The output of the inverter gate is returned to one input on the AND gate, and the other input of gate 1 is wired to the $+V_{cc}$ supply.

Looking at the waveforms in Fig. 14.5(b), we find that the Q output is logic 1 when the capacitor voltage V_c is increasing toward the 1 logic level of the inverter input. This corresponds to the pulse width T_1 of the output square wave.

On the other hand, as V_c decreases from the 1 level back toward the 0 logic level Q stays in the 0 logic state. This interval of circuit operation corresponds to the output pulse interval T_2. When V_c reaches the 0 level of the inverter input, the output switches back to logic 1. Afterwards, circuit operation repeats with period T.

Circuit Operation

When the power supply voltage is applied to the astable circuit in Fig. 14.6(a), capacitor C_1 is initially discharged. Therefore, C_1 acts like a short circuit and makes the input of the inverter gate logic 0 or ground. For this reason, the Q output starts at the 1 logic level.

This Q output is returned to one input on the AND gate and makes this input logic 1. The other input on gate 1 is already at $+V_{cc}$ or logic 1. Since both inputs are 1, the AND gate output is also logic 1. These logic levels are marked in the circuit diagram of Fig. 14.6(a).

As time continues, the circuit stays in this state with C_1 charging through the output circuitry of gate 1 and resistor R_1. The increasing capacitor voltage waveform is shown between time interval 0 to 1 in Fig. 14.6(c). During the charging time interval, the Q output remains at the 1 logic level.

When V_c reaches the 1 state voltage value of the input on gate 2, its input logic level becomes logic 1. This input causes the Q output to switch to the 0 logic level. At the same time, one input on the AND gate goes to logic 0. Therefore, the output of gate 1 changes to logic 0. Now the circuit has settled into a new state with logic levels as indicated in Fig. 14.6(b).

In this circuit, the output of gate 1 is logic 0. This connects one end of resistor R_1 to ground, and C_1 begins to discharge. As shown in Fig. 14.6(c), V_c decreases back toward the 0 state level of the inverter input during the interval of time from 1 to 2.

This completes the first cycle of circuit operation. With time, the same operating cycle repeats, and the output of the circuit settles to a continuous square wave.

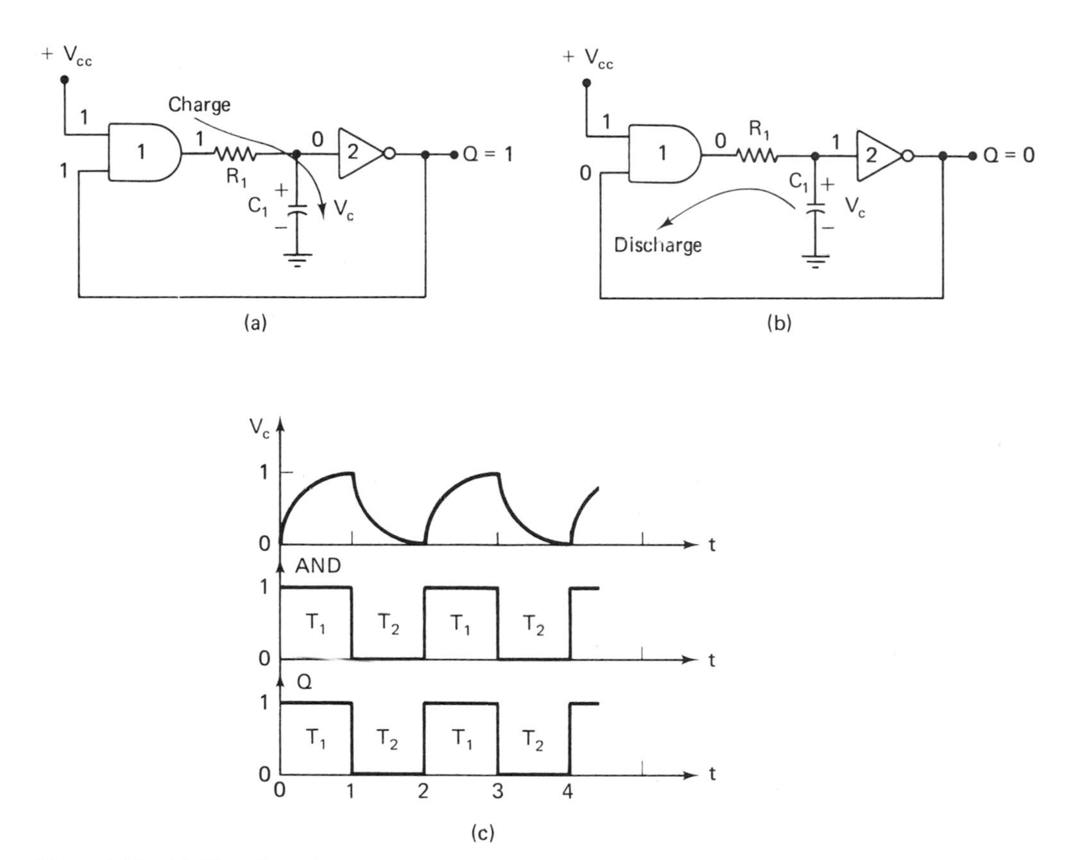

Figure 14.6 (a) Charging of $C1$; (b) Discharging of $C1$; (c) Waveforms.

Square Wave Timing

The duration of the pulse width and pulse interval in the square wave at the Q output of an astable circuit depends on the *RC time constant*. This time constant is labeled with the Greek symbol tau (τ) and is measured in the unit of seconds. The time constant of the multivibrator circuit in Fig. 14.6(a) is found as the product of the values of timing resistor R_1 and capacitor C_1:

$$\tau = R_1 C_1$$

Ideal astable waveforms are shown in Fig. 14.7. Here, the pulse width and pulse interval are the same length. Moreover, the logic levels are assumed to be 1 equals +5 V and 0 equals 0 V. In this case, the output is symmetrical, and the values of T_1 and T_2 can be calculated with the equations

$$T_1 = .69 R_1 C_1$$

$$T_2 = T_1$$

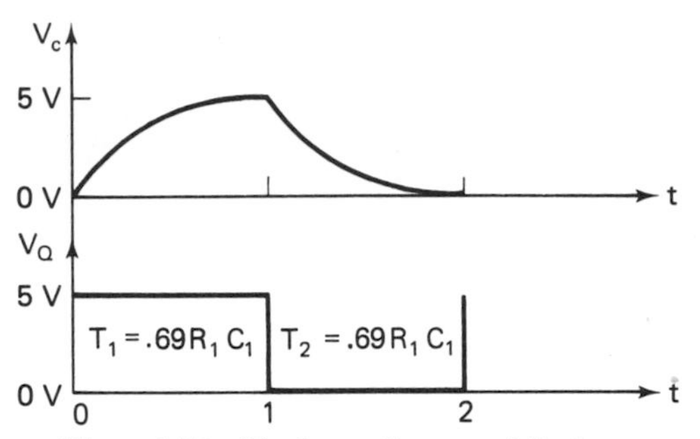

Figure 14.7 Ideal waveforms and timing.

However, the output circuitry of the AND gate and practical logic levels both cause the timing of the output square wave to change.

Knowing the pulse width and interval, we can find the period and frequency of the clock signal. These properties can be found with the equations that follow:

$$T = T_1 + T_2$$

$$f = \frac{1}{T}$$

EXAMPLE 14.3

What are the pulse width, pulse interval, period, and frequency of the square wave at the output of the astable multivibrator in Fig. 14.5(a)? The device values are $R_1 = 300\ \Omega$ and $C_1 = 5000$ pF.

SOLUTION:

Using the timing equations, the properties of the output square wave are found to be

$$T_1 = .69R_1C_1 = (.69)(300\ \Omega)(5000\ \text{pF}) = 1\ \mu\text{s}$$

$$T_1 = 1\ \mu\text{s}$$

$$T_2 = T_1 = 1\ \mu\text{s}$$

$$T = T_1 + T_2 = 1\ \mu\text{s} + 1\ \mu\text{s}$$

$$T = 2\ \mu\text{s}$$

$$f = \frac{1}{T} = \frac{1}{2\ \mu\text{s}} = 500\ \text{kHz}$$

$$f = 500\ \text{kHz}$$

14.4 THE MONOSTABLE MULTIVIBRATOR

Another widely used multivibrator circuit is the *monostable multivibrator*. This circuit is used to generate a single pulse instead of a square wave. For this reason, the monostable multivibrator is also known as a *one-shot circuit*.

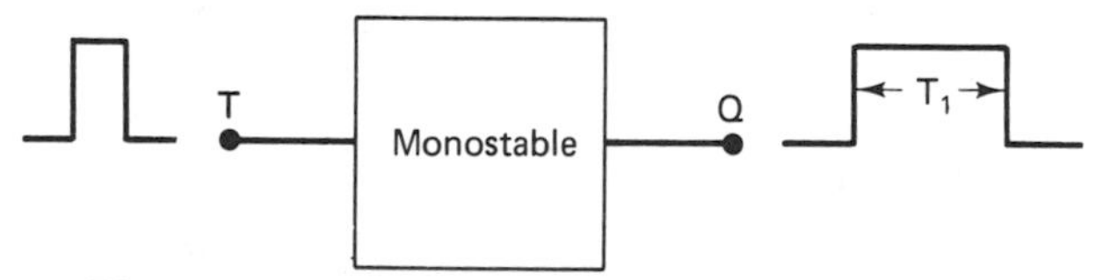

Figure 14.8 Monostable multivibrator block diagram.

A block diagram of a monostable circuit is shown in Fig. 14.8. This circuit has an input lead marked T for trigger and a single output terminal Q. The input/output operation of the monostable is given by the signal waveforms drawn in Fig. 14.8. Here, we see that a short trigger pulse must be applied at the T input to start circuit operation.

When triggered, the Q output switches from the 0 logic level to the 1 state. Q stays at the 1 level for a fixed interval of time after the trigger has ended. As interval T_1 elapses, the output returns to the 0 level and waits for another trigger pulse. In this way, a single pulse with fixed voltage levels and time duration is produced. The monostable multivibrator has a single stable output state, and the circuit waits in this state to be triggered into operation.

Figure 14.8 shows both the input triggered off a pulse to logic 1 and an output that switches to logic 1. However, circuits can also be made to trigger off the 0 logic level or produce an output pulse to logic 0 instead of 1.

Logic Gate Monostable Multivibrator

The monostable multivibrator circuit can be made using logic gate circuitry. Figure 14.9(a) shows a monostable circuit constructed from a NAND gate and inverter. As in the astable, the output of gate 1 is coupled to the input on gate 2 through an RC circuit. However, here the connection of resistor R_1 and capacitor C_1 is reversed.

Input and output waveforms for the monostable are given in Fig. 14.9(b). From these waveforms, we find that the circuit triggers off a pulse to logic 0 at the

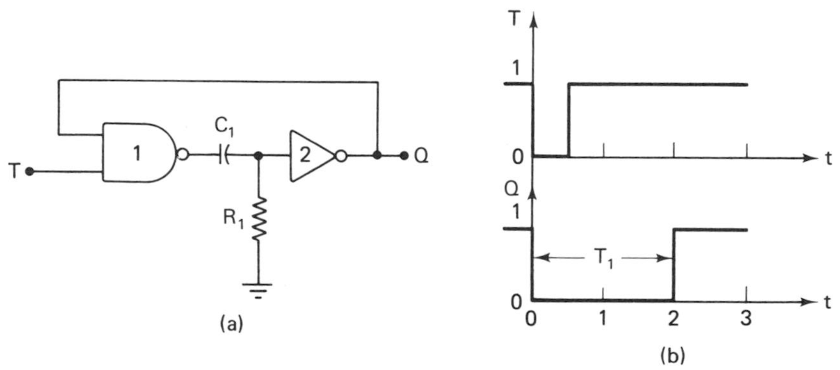

Figure 14.9 (a) Logic gate monostable; (b) Input/output waveforms.

T input and that it produces a pulse to logic 0 at output Q. The width of the output pulse is labeled T_1.

Resting State Operation

The monostable circuit waits to be triggered in the *resting state* of Fig. 14.10(a). In this state, the trigger input T on the NAND gate is held at the 1 level. Resistor R_1 is wired from the input of gate 2 to ground. This makes the input of the inverter

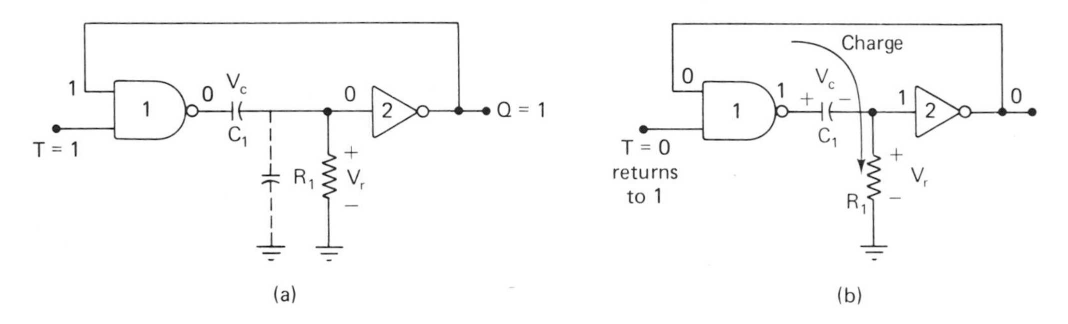

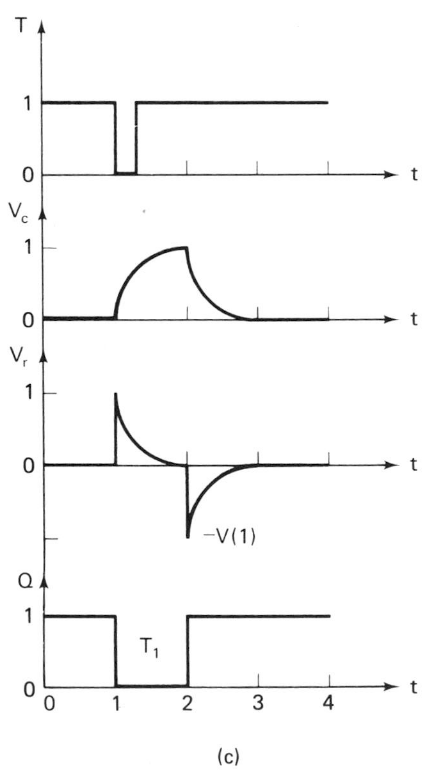

Figure 14.10 (a) Resting state; (b) Triggered state; (c) Waveforms.

logic 0 and the Q output logic 1. This resistor must be a low value, such as several hundred ohms.

The 1 logic level at the Q output makes the second input of the NAND gate logic 1. Now both inputs are logic 1, and its output is at the 0 level. Therefore, capacitor C_1 is effectively tied to ground through the output of gate 1 and is connected across resistor R_1. In this state C_1 is discharged.

This completes the resting state operation of the monostable. Logic levels at each input and output are marked in Fig. 14.10(a).

Triggered Operation

When a trigger pulse to logic 0 occurs at input T on gate 1, one input of the NAND gate is logic 0 and the other 1. These inputs make the output of gate 1 switch to the 1 level. This change in operation is shown to occur at the instant of time marked 1 in the timing diagram of Fig. 14.10(c).

At this moment, capacitor C_1 is discharged and acts like a short. Voltage V_r across resistor R_1 becomes instantly equal to the 1 state voltage level of the NAND gate output, and capacitor C_1 begins to charge with the polarity shown in Fig. 14.10(b). The logic 1 at the inverter input causes output Q to switch to the 0 level.

The 0 at the Q output is returned to one input on gate 1 and holds its output at logic 1 after the trigger pulse is complete. The logic levels for this state of circuit operation are given in Fig. 14.10(b). The circuit stays in this state while the capacitor charges. As voltage V_c across C_1 increases, V_r decreases back toward the 0 logic level of the inverter input. These voltage waveforms are shown in Fig. 14.10(c).

When the resistor voltage drops to the 0 logic level, the output of the inverter returns to the 1 logic level. This occurs at instant 2 in the timing diagram of Fig. 14.10(c). Once both inputs on the NAND gate are logic 1, its output switches back to logic 0.

However, the capacitor is now charged and makes the input voltage of the inverter negative, as shown in Fig. 14.10(c). This voltage decays back toward 0 V with time, and the circuit returns to the resting state shown in Fig. 14.10(a). The monostable should not be retriggered during this recovery interval.

Pulse Timing

The width of the output pulse in the monostable circuit of Fig. 14.9(a) depends on the charging of capacitor C_1 through resistor R_1 and the output of the NAND gate. The rate at which C_1 charges is given by the time constant of the RC circuit. This time constant is given by the expression

$$\tau = R_1 C_1$$

Figure 14.11 shows ideal waveforms for the monostable circuit. As with the astable, the logic levels are assumed to be 0 = 0 V and 1 = +5 V for ideal circuit operation. For this condition, the duration of the output pulse is given by the

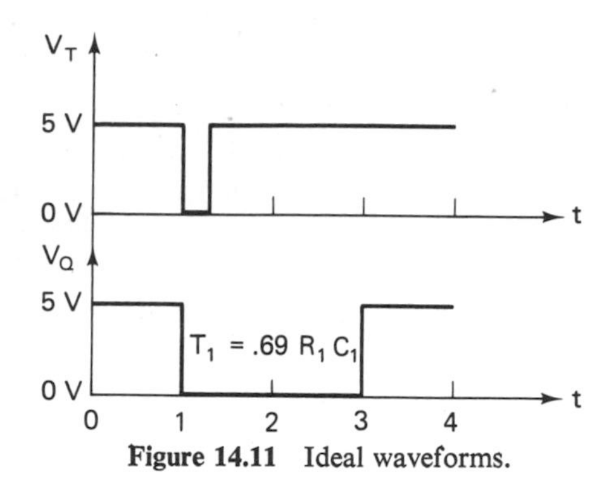

Figure 14.11 Ideal waveforms.

equation

$$T_1 = .69R_1C_1$$

However, the output circuitry of the NAND gate and logic levels of both gates lead to changes in the timing of the output pulse.

EXAMPLE 14.4

Find the ideal output pulse width for a monostable circuit made with a 100-Ω resistor and 10-μF capacitor.

SOLUTION:

The pulse width of the monostable output is found as follows:

$$T_1 = .69R_1C_1 = (.69)(100\ \Omega)(10\ \mu\text{F}) = .69\ \text{ms}$$
$$T_1 = .69\ \text{ms}$$

14.5 INTEGRATED MONOSTABLE MULTIVIBRATOR CIRCUIT

The monostable multivibrator circuit is available as a standard small-scale integrated circuit. In this section, we shall study the operation of the 74123 monostable device.

Integrated Monostable

A block diagram of an integrated monostable structure is shown in Fig. 14.12(a). Here, we see that the circuit has a lead for output Q and also its complement $\bar{Q}$. Moreover, there are two trigger inputs labeled A and B. Triggering of the circuit depends on the logic levels at the BA control inputs.

The leads C_x and R_x/C_x are provided for connection of external timing circuitry. An RC network can be wired to these terminals to set the pulse width of the output.

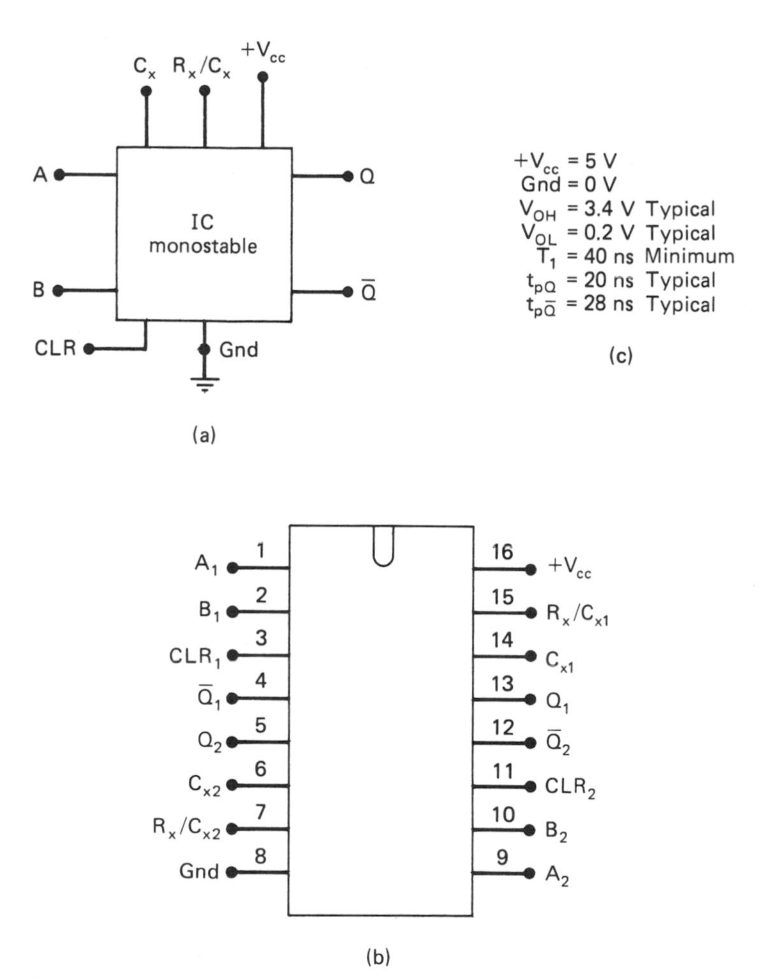

Figure 14.12 (a) Block diagram; (b) 74123 pin layout; (c) Electrical characteristics.

Last, a clear lead marked *CLR* is provided. The *CLR* signal can be used to reset the monostable circuit after the output pulse has started.

The 74123 integrated circuit is a dual retriggerable monostable device. A pin layout of the 74123 is shown in Fig. 14.12(b) and some electrical characteristics in Fig. 14.12(c). Looking at the pin layout, we notice both circuits are the same as the block diagram of Fig. 14.12(a).

Connection for Monostable Operation

To set up a 74123 monostable for operation, the power supply V_{cc} must be connected and an external *RC* circuit must be wired to the timing leads. This connection is shown in Fig. 14.13. From this diagram, we find that a fixed resistor

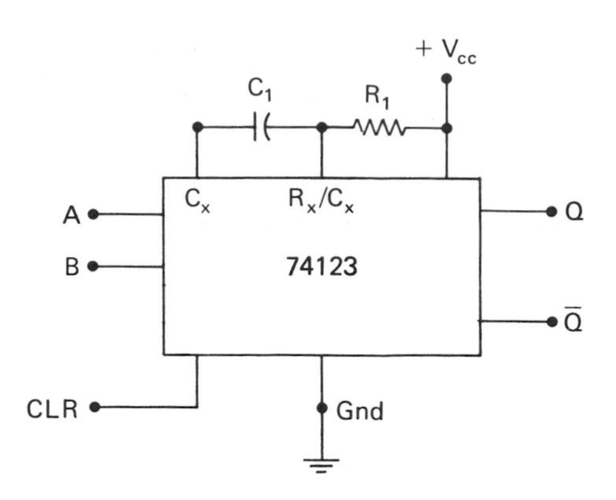

Figure 14.13 74123 connection.

R_1 must be tied from the R_x/C_x lead to the $+V_{cc}$ end of the power supply. Besides this, a capacitor C_1 is needed from the C_x lead to R_x/C_x.

The time constant τ of the external timing network is given by the equation

$$\tau = R_1 C_1$$

This time constant is used to set the pulse width of the output. Pulse width T_1 can be found with the expression

$$T_1 = .28 R_1 C_1 \left(1 + \frac{700}{R_1}\right)$$

The shortest pulse width that can be obtained from the 74123 device is 40 ns, and the duration of the output can be set as long as several seconds.

EXAMPLE 14.5

Find the pulse width of a 74123 monostable when the external timing circuit is made with $R_1 = 10\ \text{k}\Omega$ and $C_1 = 1500\ \text{pF}$.

SOLUTION:

Using the pulse width equation, we get

$$T_1 = .28 R_1 C_1 \left(1 + \frac{700}{R_1}\right) = (.28)(10\ \text{k}\Omega)(1500\ \text{pF})\left(1 + \frac{700}{10\ \text{k}\Omega}\right)$$

$$T_1 = 4.5\ \mu\text{s}$$

To operate, the clear lead of the monostable must be held at the 1 logic level. A table of the different modes of operation for the 74123 device is shown in Fig. 14.14(a). From this table, we see that the device can be triggered into operation with the A or B lead. Moreover, its operation can be inhibited with the control inputs.

When the A input is made logic 0, the monostable input triggers off a pulse to logic 1 at the B input. This gives the positive pulse-triggered monostable circuit

Inputs		Outputs	
B	A	Q	$\bar{Q}$
X	1	0	1
0	X	0	1
↑	0	⎍	⎎
1	↓	⎍	⎎

(a)

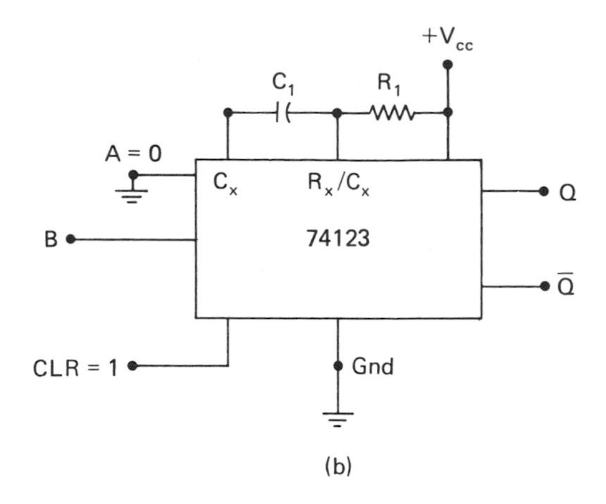

(b)

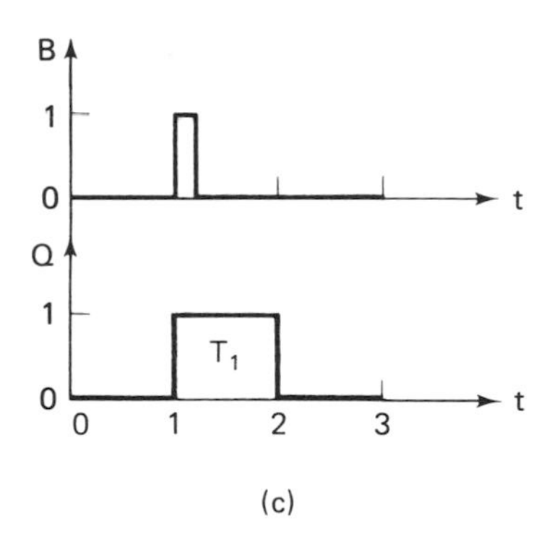

(c)

Figure 14.14 (a) Modes of operation; (b) Positive pulse-triggered monostable; (c) Input/output waveforms.

of Fig. 14.14(b). As the circuit is triggered a pulse to logic 1 or positive pulse occurs at the Q output. These waveforms are shown in Fig. 14.14(c).

At the same time, a pulse to ground appears at the $\bar{Q}$ output. The propagation delay t_p between input and output at Q is 20 ns, and at the $\bar{Q}$ output the delay is 28 ns.

If the A input is used to trigger the monostable, the B input must be held at logic 1. Now a pulse to logic 0 is needed to trigger the circuit into operation. However, the output pulses are the same as when triggered off the B input.

If the B input is 0 or the A input 1, the circuit will not trigger off the other input. For both of these cases, the other trigger lead is indicated as X in the table of Fig. 14.14(a). This means it is a don't care state and can be logic 0 or 1. For each of these conditions, the output stays in the resting state with $Q = 0$ and $\bar{Q} = 1$.

Override Clear and Retriggered Operation

To make the 74123 device more flexible, it is provided with *override clear* and *retriggered features.* With the override clear operation, the length of the normal output pulse can be shortened. On the other hand, retriggered operation is used to lengthen the pulse width.

The waveforms in Fig. 14.15(a) show override clear operation. In this diagram, we see that making the *CLR* lead logic 0 cuts short the pulse at the *Q* output. Retriggered operation is shown in Fig. 14.15(b). Here the input is retriggered before the pulse width is complete. This increases the duration of the output pulse.

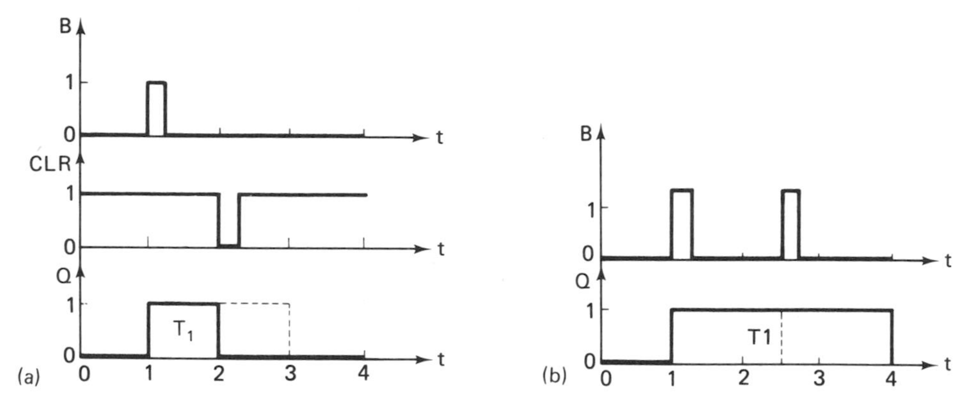

Figure 14.15 (a) Override clear input/output waveforms; (b) Retriggered input/output waveforms.

14.6 THE 555 UNIVERSAL TIMER

The 555 *universal timer* is an integrated circuit made for use as a pulse waveform generator in digital equipment. This device can be made to work as an astable multivibrator to produce a square wave output or a monostable multivibrator to give a one-shot pulse. The logic levels at the output of the 555 are compatible with TTL circuitry. The advantage in the use of the 555 device is that its output frequency or pulse width is very stable.

A block diagram of the 555 timer is shown in Fig. 14.16(a). Looking at this block, we notice the device has six functional leads in addition to the pins for the $+V_{cc}$ and gnd of the power supply. The 555 has a trigger input lead and single output. Moreover, the device has four control leads. These control leads are called *reset*, *threshold*, *discharge*, and *control voltage*.

The 555 will operate off any $+V_{cc}$ supply voltage in the range from +4.5 V dc to a maximum of 16 V dc. However, for TTL compatibility the supply should be made +5 V. The 555 device is packaged in an eight-pin DIP case. Its pin layout is shown in Fig. 14.16(b).

Function of the Leads

When using the 555 as an astable or monostable multivibrator, the most important leads are the trigger input, output, threshold, and discharge leads.

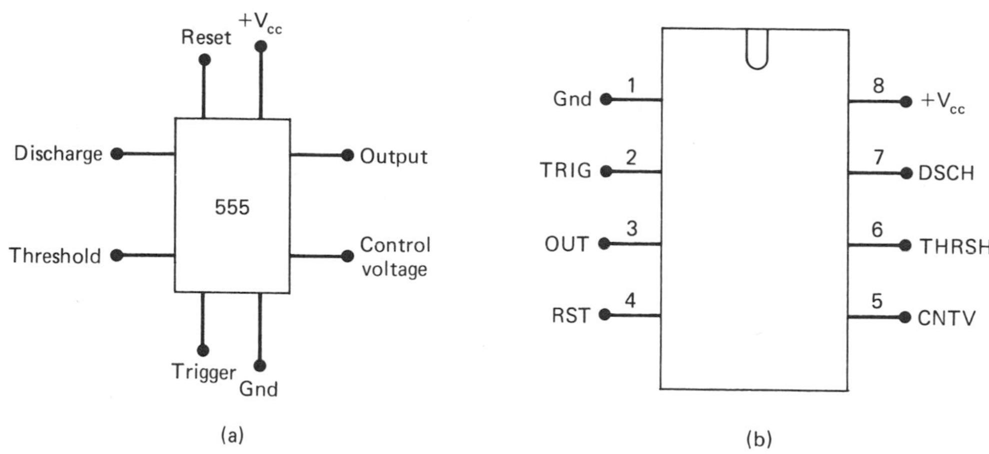

Figure 14.16 (a) Timer block diagram; (b) 555 pin layout.

However, the reset and control voltage terminals can be used to provide extra features for some other applications.

The trigger input is used to start the operation of the timing circuit, and a square wave or pulse is produced at the output. Moreover, the voltage at the threshold lead decides when the output switches between the 0 and 1 logic level. This voltage is developed across a capacitor in an external *RC* circuit. This *RC* combination is used to set the timing of the astable or monostable waveform.

The discharge lead provides a path for the discharge of the capacitor in the *RC* timing circuit. So this lead is involved in the recovery of the circuit for the next output pulse.

The reset lead is not necessarily used for astable or monostable operation. However, reset can be used to terminate a pulse before it naturally times out. This operation is similar to the override clear feature on the 74123 integrated monostable multivibrator. To stop the output pulse, the reset lead must be set to the 0 logic level. If this feature is not in use, the reset pin is wired to the $+V_{cc}$ power supply.

By applying an external dc voltage to the control voltage input, the internal operation can be changed. This type of operation is needed in some applications. When not in use, the control voltage lead should be connected to ground through a .01-μF capacitor. This capacitor is used to stop pickup of stray noise.

Astable Multivibrator Circuit

To make an astable multivibrator with a 555 timer, the circuit connection in Fig. 14.17(a) is used. Its waveforms are shown in Fig. 14.17(b). In this circuit, the timing of the output waveform is set by resistors R_1 and R_2 and capacitor C_1. On the other hand, resistor R_3 is used to improve the shape of the output waveform. The value of this device should be about 1 kΩ. The second capacitor C_2 is added to eliminate noise pickup at the unused control voltage lead.

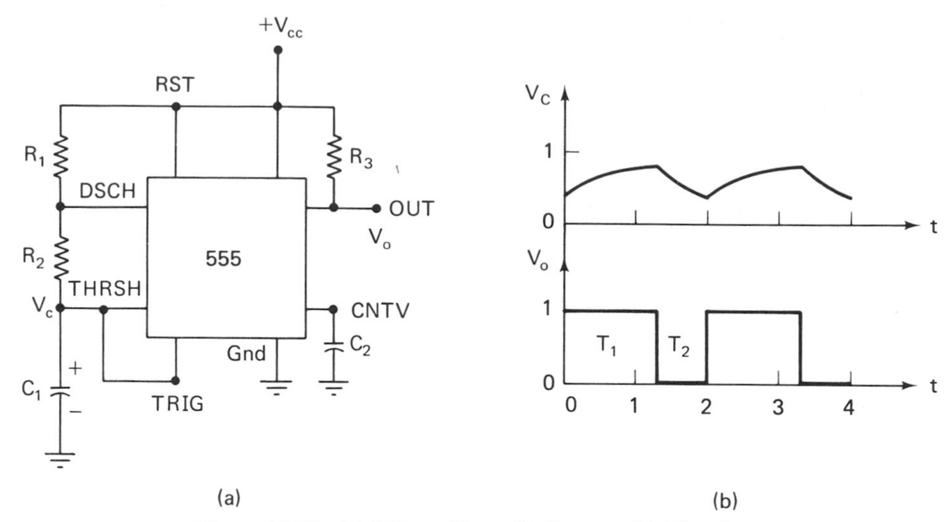

Figure 14.17 (a) 555 astable multivibrator; (b) Waveforms.

Looking at the circuit, we see that capacitor C_1 is connected from the threshold lead to ground. During the first period of time in Fig. 14.17(b), the discharge lead acts like an open circuit, and the capacitor charges through resistors R_1 and R_2 to the trigger voltage level.

The trigger lead is directly wired to the threshold lead. When the capacitor voltage is below the threshold, the output lead is at the 1 logic level. However, as V_c reaches the trigger level the output switches to the 0 state, as shown in Fig. 14.17(b).

When the output goes to logic 0, the internal circuitry of the 555 device switches the discharge lead to logic 0 or ground. Now C_1 begins to discharge through R_2. As V_c reaches a low-level threshold, the output returns to the 1 level.

This switch in the output logic state makes the discharge lead again act like an open circuit. In this way, C_1 begins to recharge, and the timing cycle repeats.

The output of the astable is an asymmetrical square wave. The pulse width T_1 and pulse interval T_2 are given by the equations that follow:

$$T_1 = .69(R_1 + R_2)C_1$$
$$T_2 = .69R_2C_1$$

Using these values, the period and frequency of the output can be calculated.

EXAMPLE 14.6

The timing devices in the astable circuit of Fig. 14.17(a) are $R_1 = R_2 = 1\ \text{k}\Omega$ and $C_1 = .015\ \mu\text{F}$. What are the pulse width, pulse interval, period, and frequency of the output square wave.

SOLUTION:

$$T_1 = .69(R_1 + R_2)C_1 = (.69)(1\ \text{k}\Omega + 1\ \text{k}\Omega)(.015\ \mu\text{F})$$
$$T_1 = 20.7\ \mu\text{s}$$
$$T_2 = .69R_2C_1 = (.69)(1\ \text{k}\Omega)(.015\ \mu\text{F})$$
$$T_2 = 10.4\ \mu\text{s}$$
$$T = T_1 + T_2 = 20.7\ \mu\text{s} + 10.4\ \mu\text{s}$$
$$T = 31.1\ \mu\text{s}$$
$$f = \frac{1}{T} = \frac{1}{31.1\ \mu\text{s}} = 32.1\ \text{kHz}$$

Monostable Multivibrator Circuit

A connection of the 555 timer as a monostable multivibrator is shown in Fig. 14.18(a). The differences between this circuit and the astable in Fig. 14.18(a) are

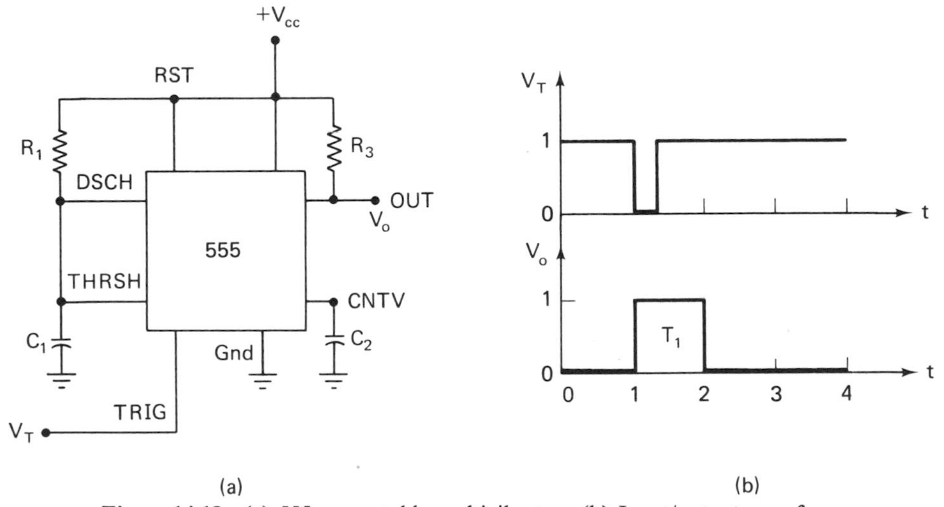

Figure 14.18 (a) 555 monostable multivibrator; (b) Input/output waveform.

the connections of the trigger and threshold leads. Here, we see that the trigger lead is not tied to the threshold lead; instead, it is used as a trigger input. Moreover, resistor R_2 is eliminated, and the threshold and discharge leads are directly tied together.

To start a pulse at the monostable output, the trigger lead must be set to logic 0 for an instant. This makes capacitor C_1 begin to charge through resistor R_1. After the capacitor reaches the threshold voltage of the 1 state, the discharge lead is set to logic 0, and the capacitor immediately discharges. The input and output waveforms are shown in Fig. 14.18(b).

The duration of the output pulse is given by the equation

$$T_1 = 1.1R_1C_1$$

and the recovery time for retriggering is effectively zero.

EXAMPLE 14.7

A 555 monostable circuit is made with $R_1 = 10\ \text{k}\Omega$ and $C_1 = 1\ \mu\text{F}$. Find the pulse width of the output.

SOLUTION:

$$T_1 = 1.1R_1C_1 = (1.1)(10\ \text{k}\Omega)(1\ \mu\text{F}) = 11\ \text{ms}$$

$$T_1 = 11\ \text{ms}$$

ASSIGNMENTS

Section 14.2

1. What type of output signal is produced by a clock circuit?

2. Define the terms *pulse width* and *pulse interval.*

3. How long are the pulse width and pulse interval in the waveform of Fig. 14.19?

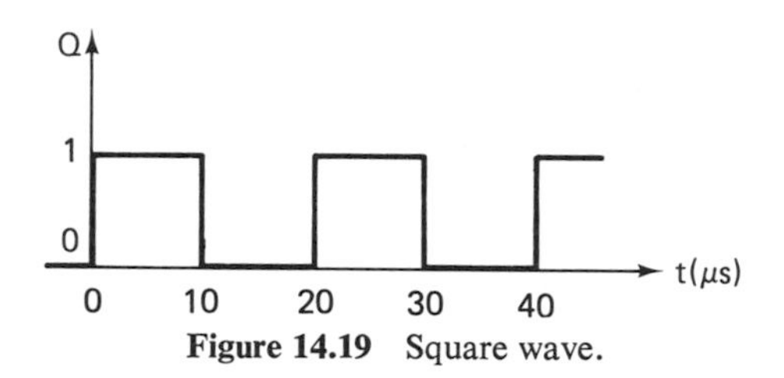

Figure 14.19 Square wave.

4. What is meant by the period and frequency of a square wave?

5. What are the period and frequency of the waveform in Fig. 14.19?

6. Find the period and frequency of the waveform in Fig. 14.3(a).

Section 14.3

7. To change the frequency of the output square wave in the astable circuit of Fig. 14.5(a), the values of which devices can be changed?

8. Make a logic drawing to show how the astable circuit of Fig. 14.5(a) can be made using the NAND gates on a single 7400 integrated circuit.

9. The circuit in Fig. 14.5(a) is made with $R_1 = 500\ \Omega$ and $C_1 = .3\ \mu\text{F}$. Find the ideal pulse width, pulse interval, period, and frequency of the output square wave. Sketch the output waveform.

Section 14.4

10. What kind of output waveform is produced by a monostable multivibrator?

11. Give another name for the monostable circuit.

12. Which devices in the monostable of Fig. 14.9(a) can be used to adjust the duration of the output?

13. Make a logic diagram showing how the gates on a 7400 IC can be used to contruct a monostable like that in Fig. 14.9(a). However, use NAND gates as inverters to make the input trigger off the 1 logic level instead of 0 and the output produce a pulse to logic 1.

14. Calculate the ideal pulse width of the monostable circuit in Fig. 14.9(a). The device values are $R_1 = 150\ \Omega$ and $C_1 = .001\ \mu F$.

Section 14.5

15. What type of integrated circuit is the 74123 device?

16. The circuit in Fig. 14.13 is made with $R_1 = 15\ k\Omega$ and $C_1 = .03\ \mu F$. Find the pulse width T_1.

17. Make a drawing like that in Fig. 14.14(b) of a negative pulse-triggered monostable. Show typical input/output waveforms.

18. What is meant by the override clear operation of the 74123 monostable?

19. Describe the retriggerable feature of the 74123 monostable.

Section 14.6

20. What is the advantage of using a 555 timer device to make an astable or monostable multivibrator?

21. Which devices in the astable circuit of Fig. 14.17(a) can be used to change the value of the pulse width? To adjust the pulse interval?

22. The astable circuit in Fig. 14.17(a) is made with $R_1 = 1\ k\Omega$, $R_2 = 2.2\ k\Omega$, and $C_1 = 1000\ pF$. Find the pulse width, pulse interval, period, and frequency. Sketch the output waveform.

23. In the monostable circuit of Fig. 14.18(a), what devices should be used to change the duration of the output pulse?

24. What is the duration of the output pulse in the circuit of Fig. 14.18(a) for $R_1 = 10\ k\Omega$ and $C_1 = 1000\ pF$?

Flip-flops 15

15.1 INTRODUCTION

In this chapter, we shall begin our study of *sequential digital circuitry*. A sequential circuit is one in which the new output depends on both the present inputs and the previous logic state of the outputs. For this reason, we say a sequential circuit has the ability to store information. That is, it has memory.

The building block of sequential digital circuitry is known as the *flip-flop*. A flip-flop is actually a third type of multivibrator circuit called a *bistable multivibrator*, which means it has two stable states.

Many different flip-flop circuits can be made with gate circuitry and are available as standard ICs. The topics included in this chapter are as follows:

1. Set-reset flip-flop
2. Clocked set-reset flip-flop
3. Master-slave flip-flop
4. *D*-type flip-flop
5. *J-K* flip-flop

15.2 SET-RESET FLIP-FLOP

The simplest flip-flop circuit is known as a *set-reset flip-flop*. A block diagram of this type of circuit is shown in Fig. 15.1(a). Looking at this diagram, we see that

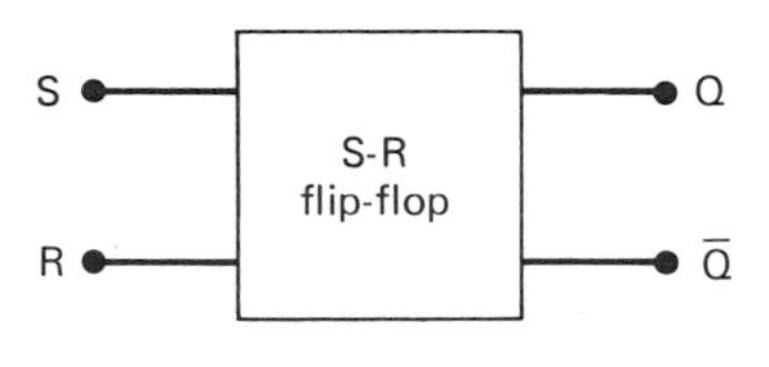

(a)

R	S	Q	$\bar{Q}$	Operation
0	0	Q_0	$\bar{Q}_0$	Unchanged
0	1	1	0	Set
1	0	0	1	Reset
1	1	1	1	Unstable

(b)

Figure 15.1 (a) Set-reset flip-flop block; (b) Truth table.

the block has two inputs marked S and R. The outputs of the flip-flop are Q and its complement $\bar{Q}$. This circuit is also called an *S-R flip-flop*.

A truth table of *S-R* flip-flop operation is given in Fig. 15.1(b). Here, we find that different logic levels at the inputs cause different changes at the outputs. For instance, making both inputs logic 0 will not change the output states. No change in output is denoted as Q equals Q_0 and $\bar{Q}$ equals $\bar{Q}_0$.

However, changing the level at the S input to 1 and leaving R equal to 0 causes the flip-flop to *set*. When in the set state, the flip-flop outputs are always $Q = 1$ and $\bar{Q} = 0$.

If both inputs are returned to the 0 logic level, the flip-flop stays set with outputs $Q = 1$ and $\bar{Q} = 0$. Here, we see that the circuit holds or stores the output state. But setting the inputs to $R = 1$ and $S = 0$ causes the flip-flop to *reset*. As it resets, the outputs become $Q = 0$ and $\bar{Q} = 1$. When both inputs go back to 0, the outputs stay in the reset state.

In this way, we see that the output of the circuit is switched to the set state by making the S input logic 1. It stays in this stable state until the R input is made logic 1 to return the outputs to the reset state. In both cases, the other input must be held at logic 0.

For correct operation, both the S and R inputs of the circuit should not be set to logic 1 together. If they are, the Q and $\bar{Q}$ outputs switch to 1. This is an unstable state, and as the inputs are returned to logic 0 the Q and $\bar{Q}$ outputs can settle into either the set or reset state.

Logic Gate Set-Reset Flip-flop

A circuit that operates as an *S-R* flip-flop can be made using logic gates. Such a circuit is the NAND gate connection in Fig. 15.2(a).

This circuit is made from two two-input NAND gates. The first input on the gate labeled 1 is used as the set input, and its other input is wired to the output of gate 2. On gate 2 one input is marked reset, and the second is tied to the output of gate 1. The output of NAND gate 1 is the Q output of the flip-flop, and $\bar{Q}$ is the output from gate 2.

The operation of the NAND gate flip-flop is shown in the table of Fig. 15.2(b). Here, we notice the circuit works off a logic 0 at the set or reset input.

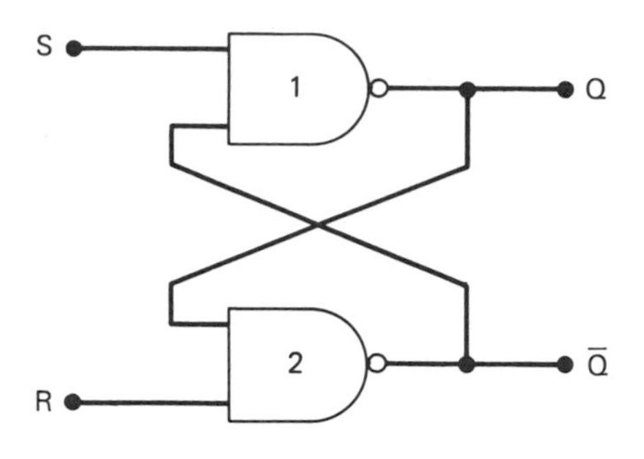

S	R	Q	$\bar{Q}$	Operation
0	0	1	1	Unstable
0	1	1	0	Set
1	0	0	1	Reset
1	1	Q_0	$\bar{Q}_0$	Unchanged

(b)

(a)

Figure 15.2 (a) NAND gate *S-R* flip-flop; (b) Truth table.

Circuit Turn On

When voltage is first applied to the circuit, both the set and reset inputs are made logic 1. After this, one of the logic gates turns on faster than the other, and its output switches to logic 0. This forces the output of the other gate to logic 1.

Let us assume the output of gate 2 is switched to logic 1 first. This gives a logic 1 at the $\bar{Q}$ output. Now both inputs on gate 1 are logic 1, and the Q output is held at logic 0. The flip-flop has settled into the reset state shown in Fig. 15.3(a).

Set Operation

When the S input is made logic 0 and R equals 1, the flip-flop sets as shown in the table of Fig. 15.2(b). Now the Q and $\bar{Q}$ outputs switch logic states to give $Q = 1$ and $\bar{Q} = 0$.

When the S input on gate 1 in Fig. 15.3(a) is switched to 0, the Q output goes to the 1 level. Now both inputs on gate 2 are logic 1, and the $\bar{Q}$ output switches to logic 0. This change in $\bar{Q}$ makes the second input on gate 1 equal to 0. If the set input is returned to logic 1, the 0 level at the $\bar{Q}$ output holds the Q output at logic 1. At the same time, Q equal to 1 makes the $\bar{Q}$ output stay at logic 0. The outputs have settled into the set state of Fig. 15.3(b).

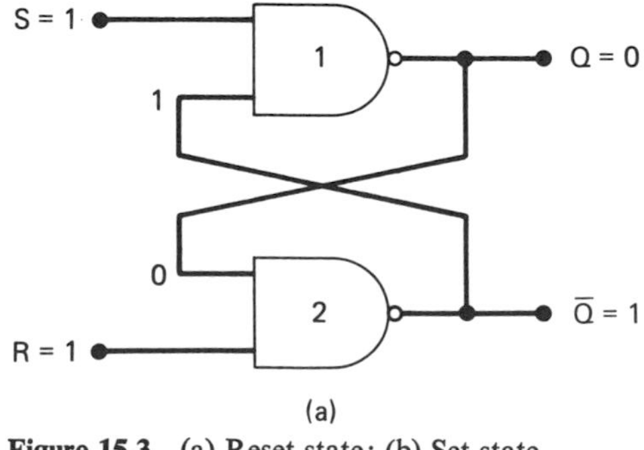

(a)

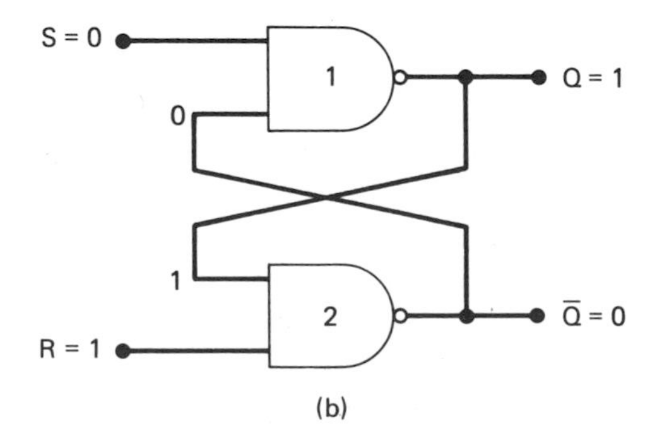

(b)

Figure 15.3 (a) Reset state; (b) Set state.

Reset Operation

The reset operation is started by making the logic level of R go to logic 0. As the level of R switches to logic 0, the $\bar{Q}$ output of gate 2 changes to the 1 level. Now both inputs on gate 1 are logic 1, and output Q switches back to the 0 state.

As the reset input is returned to the 1 level, the circuit is back in the reset state of Fig. 15.3(a).

EXAMPLE 15.1

The inputs on the S-R flip-flop in Fig. 15.2(a) are made $S = 0$ and $R = 1$. What are the states of the Q and $\bar{Q}$ outputs?

SOLUTION:

When the data inputs are $S = 0$ and $R = 1$, the truth table of Fig. 15.2(b) tells us that the flip-flop sets and the outputs are $Q = 1$ and $\bar{Q} = 0$:

$$Q = 1$$
$$\bar{Q} = 0$$

EXAMPLE 15.2

Figure 15.4 shows the S and R data input waveforms for the NAND gate flip-flop in Fig. 15.2(a). Draw the waveform at output Q.

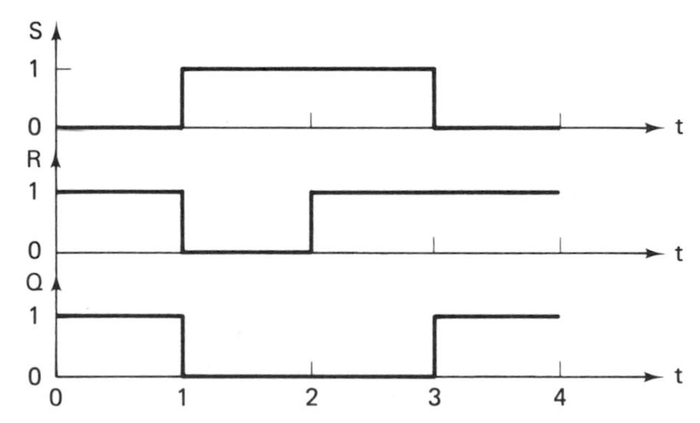

Figure 15.4 Input/output waveforms.

SOLUTION:

During the interval from 0 to 1 on the time axis, the inputs to the circuit are $S = 0$ and $R = 1$. This condition makes the outputs set, and Q switches to logic 1 as shown in Fig. 15.4:

$$S = 0$$
$$R = 1$$
$$Q = 1$$

In the time interval 1 to 2, the inputs change to $S = 1$ and $R = 0$. For these inputs, the flip-flop outputs reset to give $Q = 0$ as shown in the waveform:

$$S = 1$$
$$R = 0$$
$$Q = 0$$

For the next interval, we find that $S = R = 1$, and the output remains reset with $Q = 0$. During the last interval, we again have $S = 0$ and $R = 1$. So the output sets to $Q = 1$.

74279 Integrated *S-R* Flip-flop

The NAND gate *S-R* flip-flop circuit of Fig. 15.2(a) is available as a standard small-scale integrated circuit. A 74279 is a quad set-reset flip-flop device, and its pin layout is given in Fig. 15.5.

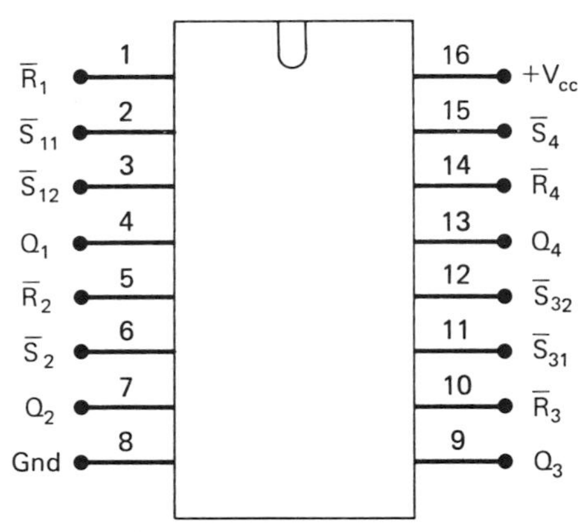

Figure 15.5 74279 pin layout.

Looking at this pin layout diagram, we see that the set and reset inputs are marked $\bar{S}$ and $\bar{R}$, respectively. This notation with the bar over S and R indicates the inputs trigger off the 0 logic level instead of 1.

15.3 THE CLOCKED SET-RESET FLIP-FLOP

A variation on the standard set-reset flip-flop is shown in Fig. 15.6. Here, we see that the circuit has the S and R inputs and Q and $\bar{Q}$ outputs. However, it has one additional input marked *CLK.* This lead is known as the *clock input.* For this reason, the circuit is called a *clocked S-R flip-flop.*

The operation of this circuit is identical to that indicated for the set-reset flip-flop in Fig. 15.2(b). But to operate the circuit must be enabled by setting the clock input to the correct logic level. When this clock input is at the OFF logic

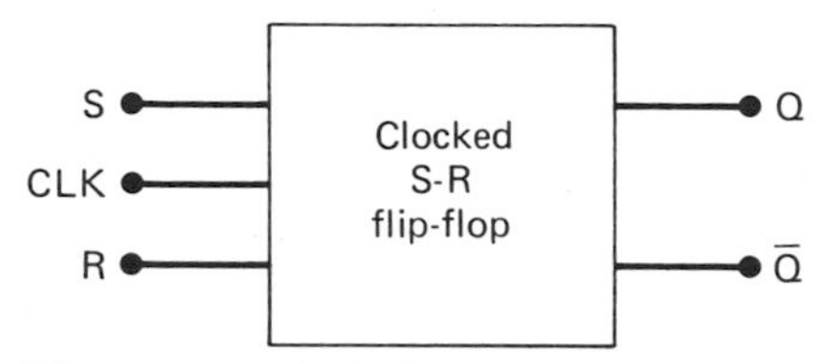

Figure 15.6 Clocked set-reset flip-flop block.

level, data at the S and R inputs cannot enter the circuit, and the outputs remain unchanged.

Edge Triggered

A clocked S-R flip-flop circuit can be made that triggers off a logic 1 or a logic 0 level at the CLK lead. When input data from S and R enter the flip-flop on the change from 0 to 1 at the clock input, the circuit is said to be *positive edge triggered.* On the other hand, a circuit that triggers on the 1 to 0 clock change is called a *negative edge-triggered* flip-flop.

Logic Gate Clocked Set-Reset Flip-flop

The NAND gate connection in Fig. 15.7(a) is a clocked set-reset flip-flop circuit. In this circuit, gate 3 and gate 4 make a standard NAND gate S-R flip-flop like that in Fig. 15.2(a). By adding gates 1 and 2, the entry of data at S and R to the $\bar{S}$ and $\bar{R}$ inputs of the flip-flop is controlled by clock signal CLK.

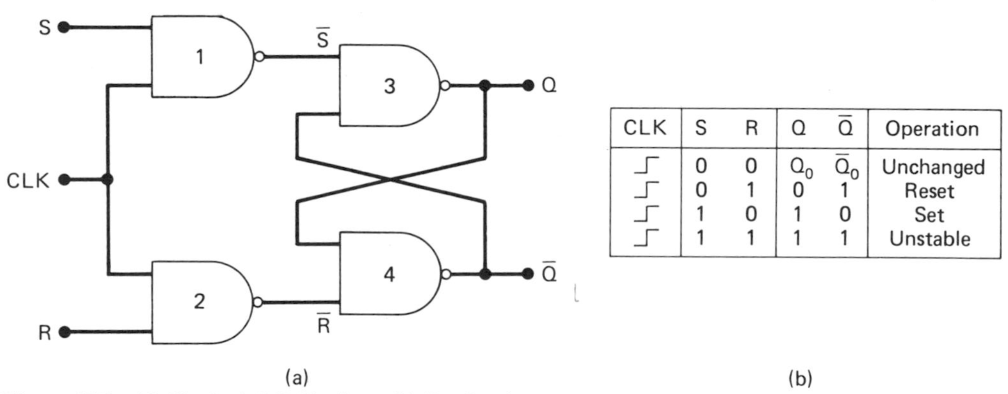

CLK	S	R	Q	$\bar{Q}$	Operation
⎍	0	0	Q_0	$\bar{Q}_0$	Unchanged
⎍	0	1	0	1	Reset
⎍	1	0	1	0	Set
⎍	1	1	1	1	Unstable

Figure 15.7 (a) Clocked S-R flip-flop; (b) Truth table.

If the clock input on the circuit in Fig. 15.7(a) is made logic 0, one input on gate 1 and one on gate 2 are logic 0. This holds the outputs to $\bar{S}$ and $\bar{R}$ both at logic 1, and the Q and $\bar{Q}$ outputs are unchanged.

To enable the input gates, the clock lead must be switched to logic 1 as shown in Fig. 15.7(b). For this reason, the circuit of Fig. 15.7(a) is positive edge triggered. When CLK is at the 1 level, the data at the S and R inputs are inverted and passed to the $\bar{S}$ and $\bar{R}$ leads, respectively. Because of this inverting characteristic of the input circuitry, the logic levels in the truth table of clocked *S-R* flip-flop operation of Fig. 15.7(b) are the opposite of those shown for the standard *S-R* flip-flop in Fig. 15.2(b).

From this table, we see the outputs of the circuit set to $Q = 1$ and $\bar{Q} = 0$ when $S = 1$ and $R = 0$ and the clock is triggered. Moreover, a 1 at *R* and 0 at *S* reset the outputs to $Q = 0$ and $\bar{Q} = 1$ as the clock goes to 1.

Let us look more closely into the operation of the control gates on the clocked *S-R* flip-flop in Fig. 15.7(a). By making the inputs on gate 1 $CLK = 1$ and $S = 1$, the $\bar{S}$ output switches to logic 0. On the other hand, a 0 at *S* with a 1 at *CLK* gives a logic 1 at $\bar{S}$. In this way, we see that the data at the *S* input are inverted as they go through NAND gate 1. Gate 2 at the *R* input operates in the same way.

EXAMPLE 15.3

The data inputs of the clocked set-reset flip-flop in Fig. 15.7(a) are $S = 1$ and $R = 0$. What are the logic levels of the *Q* and $\bar{Q}$ outputs after the *CLK* lead is set to logic 1?

SOLUTION:

The inputs $S = 1$ and $R = 0$ ready the flip-flop to be set on the next clock signal. Therefore, the output sets to $Q = 1$ and $\bar{Q} = 0$ when *CLK* switches to 1.

EXAMPLE 15.4

The *S*, *R*, and *CLK* waveforms of Fig. 15.8 are applied to the clocked set-reset flip-flop in Fig. 15.7(a). Draw the *Q* output waveform.

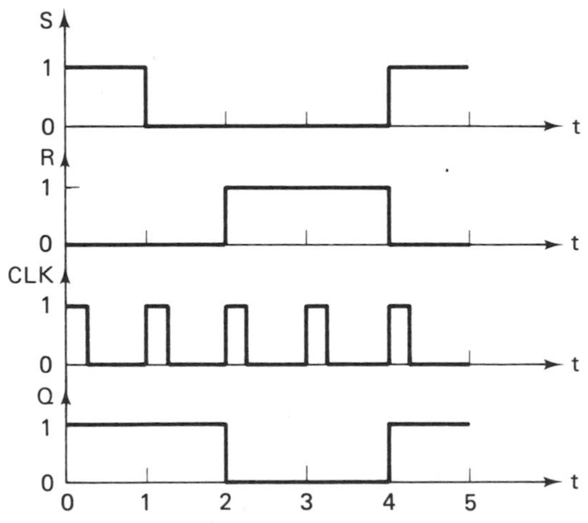

Figure 15.8 Input/output waveforms.

SOLUTION:

Looking at these waveforms, we see that the data at the *S* and *R* inputs are read during each of the five clock pulses. When the first clock pulse occurs, the inputs are $S = 1$ and $R = 0$. This sets the Q output to 1, as shown in Fig. 15.8:

$$S = 1$$
$$R = 0$$
$$CLK = 0 \text{ to } 1 \text{ positive edge triggered}$$
$$Q = 1$$

At the next clock pulse, both $S = 0$ and $R = 0$. So the Q output is unchanged and stays at logic 1:

$$S = 0$$
$$R = 0$$
$$CLK = \text{positive edge triggered}$$
$$Q = 1$$

However, the third clock pulse corresponds to $S = 0$ and $R = 1$. For this input, the Q output resets to logic 0. During the fourth clock, the data inputs are $S = 0$ and $R = 1$. The output remains reset. On the last pulse, the output again returns to the set logic level of $Q = 1$.

15.4 THE DATA LATCH

A circuit that is obtained from the basic clocked set-reset flip-flop is the *data latch* of Fig. 15.9(a). In this diagram, we see that the latch is made by tying an inverter, gate 5, from the set input to the reset input. The single input that results is labeled D for *data input.*

In this way, the set and reset inputs on gates 1 and 2, respectively, are always

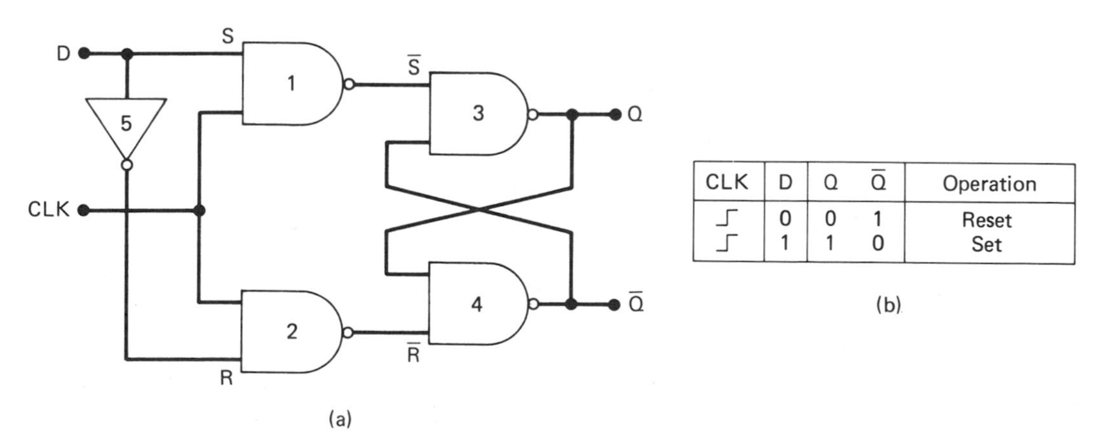

CLK	D	Q	$\overline{Q}$	Operation
↑	0	0	1	Reset
↑	1	1	0	Set

Figure 15.9 (a) Data latch circuit; (b) Truth table.

the opposite logic level. For instance, D equal to 1 gives a 1 logic level at the S input of the flip-flop circuit and a 0 at R. With these inputs, the outputs set to $Q = 1$ and $\bar{Q} = 0$ as the clock input switches to logic 1.

On the other hand, a 0 at the data input makes the set input 0 and reset 1. When clocked, this input condition causes the outputs to reset to $Q = 0$ and $\bar{Q} = 1$.

These two cases are summarized in the table of Fig. 15.9(b). Here, we notice that the Q output of a data latch becomes equal to the logic level at the data input on the positive edge of the clock pulse.

7475 Integrated Data Latch

The pin layout of a 7475 data latch device is given in Fig. 15.10. This IC contains four data latch circuits that work like the circuit shown in Fig. 15.9(a). However, a single clock lead is shared by circuits 1 and 2 and another clock lead by circuits 3 and 4. These inputs are marked CLK_{12} and CLK_{34}, respectively.

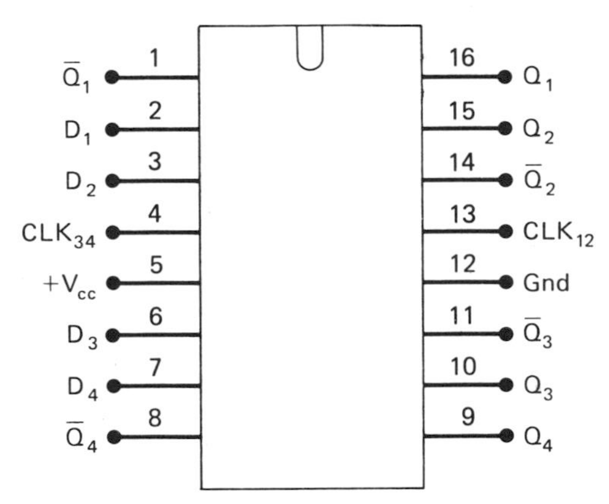

Figure 15.10 7475 pin layout.

15.5 MASTER-SLAVE FLIP-FLOP

Many integrated flip-flop devices use a circuit configuration known as a *master-slave flip-flop* instead of the basic clocked S-R flip-flop in Fig. 15.7(a). The structure of a master-slave circuit is shown in Fig. 15.11. Looking at this circuit, we see that it uses two clocked set-reset flip-flops. The flip-flop on the input side is called the *master* and the other the *slave*. The outputs Q_1 and $\bar{Q}_1$ of the master flip-flop are used as the set and reset data inputs to the slave. The outputs of the circuit are Q_2 and $\bar{Q}_2$ from the slave flip-flop.

Pulse Triggering

Two inverter gates are used to control the triggering of the master and slave flip-flops. The output of gate 1 is used to clock the slave, and gate 2 clocks the master.

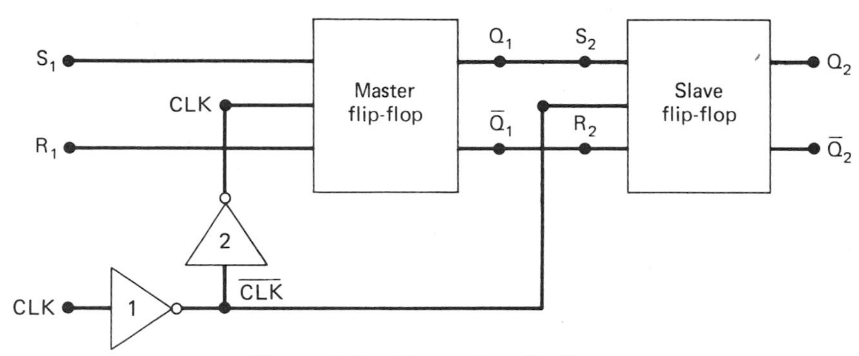

Figure 15.11 Master-slave flip-flop.

Using this circuit connection, we get *pulse-triggered* operation instead of edge triggering. In this case, the outputs of the master flip-flop set or reset on the leading edge or 0 to 1 change of the clock pulse. However, the outputs of the master are not entered into the slave until the clock returns from 1 to 0 at the trailing edge of the pulse.

Operation

At this point, let us look more closely into the operation of the master-slave flip-flop in Fig. 15.11. On the leading edge of the trigger pulse, the *CLK* input switches to logic 1. This makes $\overline{CLK}$ at the output of gate 1 equal to logic 0 and *CLK* from gate 2 equal to 1.

When set in this way, the master flip-flop is enabled to read data inputs S_1 and R_1. In response, outputs Q_1 and $\bar{Q}_1$ switch to the set or reset state. At this moment, the 0 at $\overline{CLK}$ disables the inputs of the slave flip-flop.

As the trailing edge of the clock occurs, $\overline{CLK}$ on gate 1 switches to logic 1 and *CLK* on gate 2 to the 0 level. This disables the inputs of the master flip-flop and enables the slave. In turn, the Q_2 and $\bar{Q}_2$ outputs are set or reset depending on the logic levels held at the master flip-flop outputs.

EXAMPLE 15.5

The inputs to the master-slave flip-flop in Fig. 15.11 are $S_1 = 1$ and $R_1 = 0$. Describe the operation as the *CLK* lead is activated by a positive pulse.

SOLUTION:

On the leading edge of the clock pulse, the master flip-flop is set, and its outputs become $Q_1 = 1$ and $\bar{Q}_1 = 0$:

$$Q_1 = 1$$
$$\bar{Q}_1 = 0$$

This makes the S_2 inputs on the slave flip-flop $S_2 = 1$ and $R_2 = 0$. At the trailing edge of the clock pulse, the slave flip-flop sets, and the outputs become $Q_2 = 1$ and $\bar{Q}_2 = 0$:

$$Q_2 = 1$$
$$\bar{Q}_2 = 0$$

The outputs of the circuit remain in this state until the *S-R* inputs are at the reset logic levels when the clock lead is pulsed.

15.6 *D*-TYPE FLIP-FLOP

The *D*-type or *data flip-flop* is a circuit that operates a lot like the integrated data latch described in this chapter (Section 15.4). However, it uses a master-slave circuit configuration.

A block diagram for a data flip-flop is shown in Fig. 15.12(a). Looking at this block, we see that the circuit has a data input marked *D* and clock input *CLK*. Its outputs, like those on other flip-flops, are marked Q and $\bar{Q}$. Figure 15.12(b) shows a table of *D*-type flip-flop operation. From this table, we find that the Q output becomes equal to the logic level of the *D* input as the clock signal occurs.

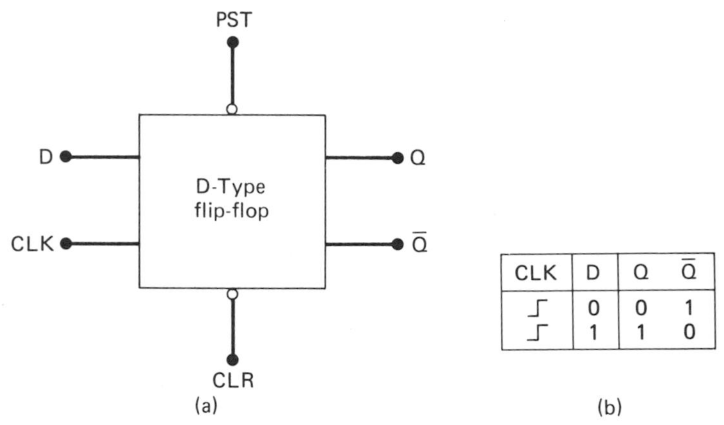

CLK	D	Q	Q̄
⎍	0	0	1
⎍	1	1	0

Figure 15.12 (a) *D*-type flip-flop; (b) Truth table.

From the block, we notice the device has two control leads. One lead is marked *CLR* for clear and is used to reset the output to $Q = 0$ and $\bar{Q} = 1$. The other control is the *preset* lead indicated by *PST*. This lead is used to set $Q = 1$ and $\bar{Q} = 0$.

Each of these leads has a dot where it enters the block. This notation means they are working off the 0 logic level or active low. For normal *D*-type operation, both *CLR* and *PST* should be set to logic 1.

7474 *D*-Type Flip-flop

The 7474 integrated circuit is a dual *D*-type flip-flop device. A pin layout for the 7474 is given in Fig. 15.13(a) and some of its electrical characteristics in Fig. 15.13(b). Here, we see that the electrical properties are essentially the same as for

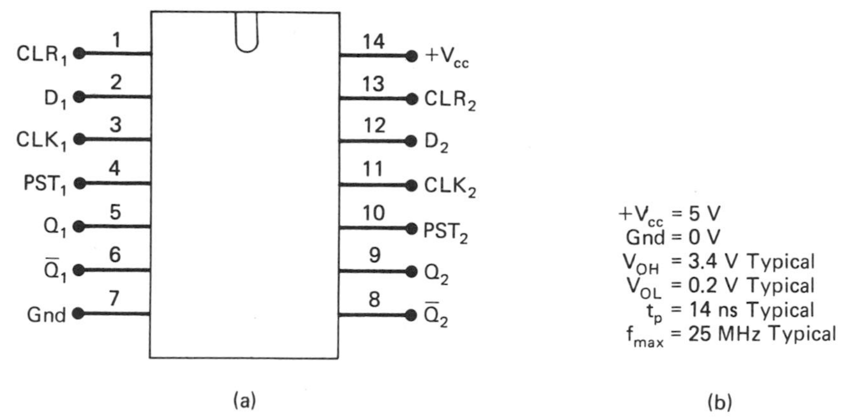

Figure 15.13 (a) 7474 pin layout; (b) Electrical characteristics.

earlier TTL devices. However, for flip-flop devices a new property called the *maximum operating frequency* (f_{max}) is given. The typical value of f_{max} for a 7474 IC is 25 MHz.

EXAMPLE 15.6

The D-type flip-flop in Fig. 15.12(a) has its clear and preset control leads set to logic 1, and the outputs are initially cleared to $Q = 0$ and $\bar{Q} = 1$ by a pulse to logic 0 at CLR. If the D input is at logic 1 when CLK switches from 0 to 1, what are the new output states of Q and $\bar{Q}$?

SOLUTION:

Making CLR and PST equal to 1 sets the flip-flop for normal operation. On the positive edge of the clock pulse, the Q output switches to logic 1 and $\bar{Q}$ to a logic 0:

$$Q = 1$$
$$\bar{Q} = 0$$

EXAMPLE 15.7

A data flip-flop with $CLR = PST = 1$ has the D and CLK waveforms shown in Fig. 15.14. Draw the Q output waveform.

SOLUTION:

In the clock waveform, we see that there are four clock pulses. At the leading edge of the first pulse, $D = 1$, and the Q output switches to logic 1 as shown in Fig. 15.14. For the second pulse, D is at logic 0, and the Q output returns to the 0 level. As the third pulse occurs, D is again 0. Therefore, the Q output stays at 0. However, at the last pulse D is again 1, and Q returns to the 1 level.

Toggle Mode Operation

The D-type flip-flop can be connected to make what is known as a *toggle mode* or *T-type flip-flop*. This connection is shown in Fig. 15.15(a).

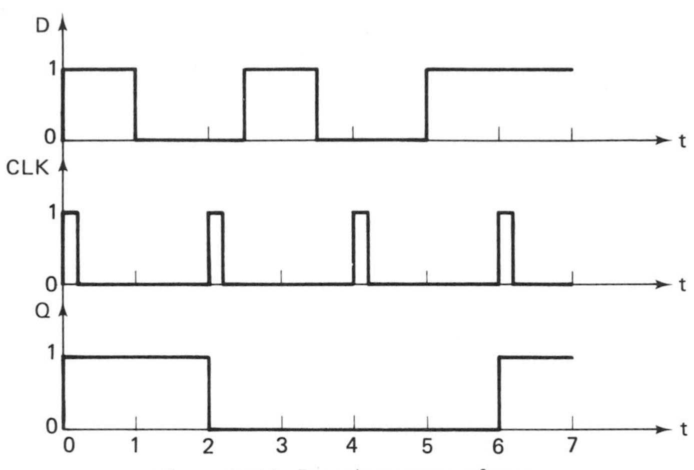

Figure 15.14 Input/output waveforms.

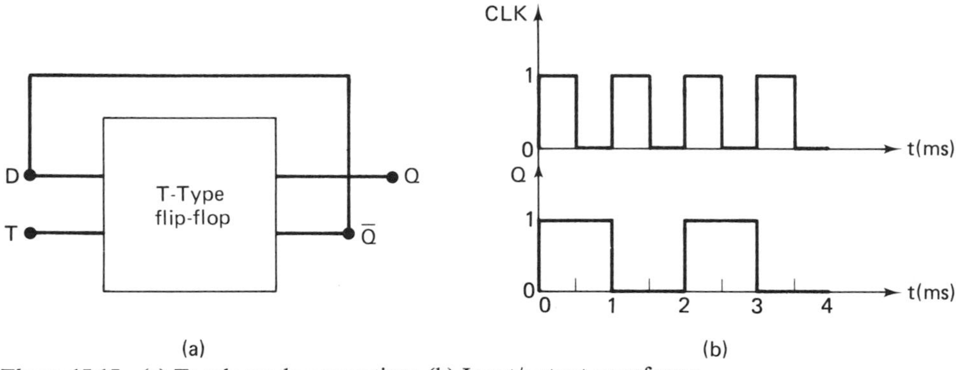

Figure 15.15 (a) Toggle mode connection; (b) Input/output waveforms.

To make a *T*-type circuit, output $\bar{Q}$ is used as the data input. So the $\bar{Q}$ lead is wired to the *D* input. In this way, the data value at *D* is always the opposite logic level of the *Q* output. The output switches logic states on the leading edge of each clock pulse as shown in the input/output waveforms of Fig. 15.15(b).

If a square wave is applied to the clock input of a toggle mode flip-flop, the output is a square wave with half the frequency of the input. For example, in Fig. 15.15(b) the input frequency is 1 kHz, and at the output a 500-Hz square wave is produced.

15.7 *J-K* FLIP-FLOP

Another type of master-slave flip-flop circuit is the *J-K flip-flop* shown in Fig. 15.16(a). From this block diagram, we see that the circuit has two data inputs marked *J* and *K*. Besides this, it has a clock lead marked *CLK*, clear lead *CLR*, and preset lead *PST*. The outputs are once again labeled *Q* and $\bar{Q}$.

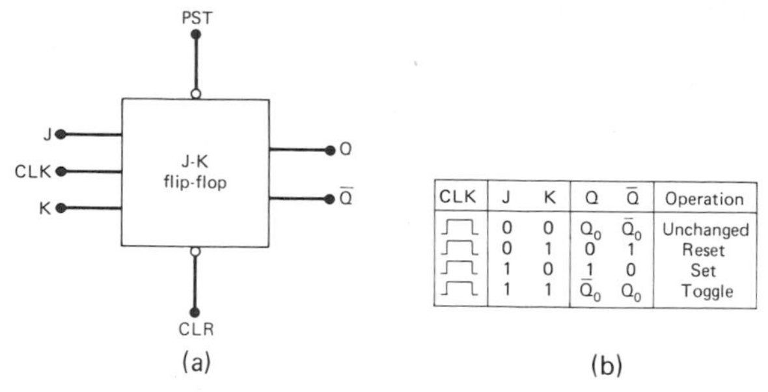

CLK	J	K	Q	$\bar{Q}$	Operation
⎍	0	0	Q_0	$\bar{Q}_0$	Unchanged
⎍	0	1	0	1	Reset
⎍	1	0	1	0	Set
⎍	1	1	$\bar{Q}_0$	Q_0	Toggle

Figure 15.16 (a) *J-K* flip-flop; (b) Truth table.

A truth table of *J-K* flip-flop operation is given in Fig. 15.16(b). Looking at this table, we find that the circuit can work as a set-reset flip-flop or in the toggle mode. Furthermore, it triggers off a positive pulse at the clock input.

The mode of operation is selected by the logic levels at the *J* and *K* data inputs. Let us assume the *J* and *K* data inputs are both set to logic 1. This makes the flip-flop work in the toggle mode, as shown in Fig. 15.16(b). On the other hand, setting $J = K = 0$ stops the outputs from changing logic states.

To operate, both the *CLR* and *PST* leads must be at logic 1. When the clear lead is made logic 0, the outputs of the *J-K* flip-flop reset to $Q = 0$ and $\bar{Q} = 1$. In the same way, the preset input can be used to set the outputs to $Q = 1$ and $\bar{Q} = 0$.

In most applications, the *J-K* flip-flop is used in one of its two modes of operation. For example, both the *J* and *K* inputs can be wired to $+V_{cc}$ to make the device operate as a *T*-type flip-flop.

7473 *J-K* Flip-flop

The pin layout of the 7473 integrated *J-K* flip-flop is shown in Fig. 15.17(a). This device contains two circuits similar to the *J-K* flip-flop of Fig. 15.16(a). These circuits are positive pulse triggered but have no preset lead.

Electrical characteristics of the 7473 are given in Fig. 15.17(b). Looking at

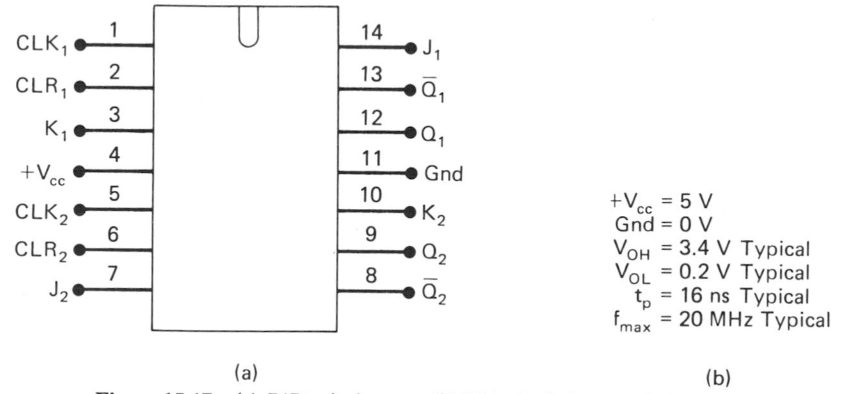

Figure 15.17 (a) 7473 pin layout; (b) Electrical characteristics.

these values, we find that they are close to those of the 7474 D-type flip-flop device. However, the value of the maximum operating frequency is 5 MHz lower than for the 7474.

EXAMPLE 15.8

One circuit on a 7473 IC is in the reset state with outputs $Q = 0$ and $\bar{Q} = 1$. The data inputs are $J = 1$ and $K = 0$ and the CLR lead is logic 1. What will be the logic level of the Q and $\bar{Q}$ outputs after the input is clocked?

SOLUTION:

With inputs $J = 1$, $K = 0$, and $CLR = 1$, the J-K flip-flop is ready to set. As the clock lead is pulsed, the outputs set to $Q = 1$ and $\bar{Q} = 0$:

$$Q = 1$$
$$\bar{Q} = 0$$

EXAMPLE 15.9

A J-K flip-flop has data inputs $J = K = 1$ and the CLR lead is logic 1. If the clock signal is a 10-kHz symmetrical square wave, determine the frequency of the square wave at the Q output and draw the input and output waveforms.

SOLUTION:

When both J and K are set to 1, the flip-flop works in the toggle mode. In this case, the output is a square wave with half the frequency of the input. Therefore, the output is a 5-kHz square wave. To draw the waveforms, we must find the period, pulse width, and pulse interval of both the input and output signals. For the input, we get

$$T = \frac{1}{f} = \frac{1}{10\text{ kHz}} = 100\ \mu\text{s}$$

$$T_1 = T_2 = \frac{T}{2} = \frac{100\ \mu\text{s}}{2} = 50\ \mu\text{s}$$

Now, for the 5-kHz output, we find

$$T = \frac{1}{f} = \frac{1}{5\text{ kHz}} = 200\ \mu\text{s}$$

$$T_1 = T_2 = \frac{T}{2} = \frac{200\ \mu\text{s}}{2} = 100\ \mu\text{s}$$

These waveforms are drawn in Fig. 15.18.

Other *J-K* Flip-flop Devices

Some other J-K flip-flop circuits are available as standard integrated circuits. For instance, the 74103 is a dual J-K flip-flop circuit just like the 7473; however, it is negative edge triggered instead of positive pulse triggered. The 74106 device is similar to the edge-triggered 74103, but it has a preset lead on each flip-flop.

Another variation of the J-K flip-flop is the 7472 (Fig. 15.19). This device has

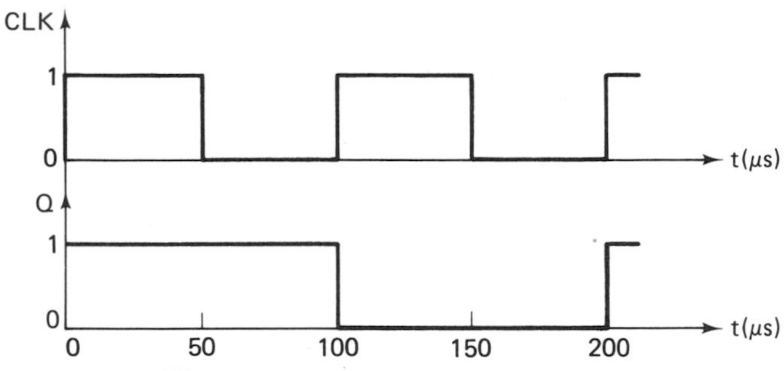

Figure 15.18 Input/output waveforms.

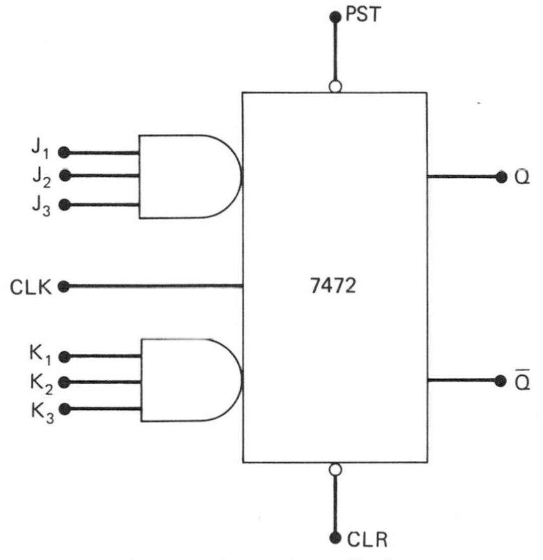

Figure 15.19 7472 *J-K* flip-flop.

gated J and K data inputs. Here, we find that the AND functions $J = J_1 \cdot J_2 \cdot J_3$ and $K = K_1 \cdot K_2 \cdot K_3$ decide if the flip-flop operates in the set, reset, or toggle mode.

ASSIGNMENTS

Section 15.2

1. Make a general description of *S-R* flip-flop operation.

2. Inputs to the NAND gate *S-R* flip-flop circuit in Fig. 15.2(a) are $S = 1$ and $R = 0$. What are the logic states of outputs Q and $\bar{Q}$?

3. Repeat the problem of Example 15.2 for the $\bar{Q}$ output.

4. The input waveforms at pin 5 and 6 of a 74279 IC are shown in Fig. 15.20. Draw the output waveform at pin 7.

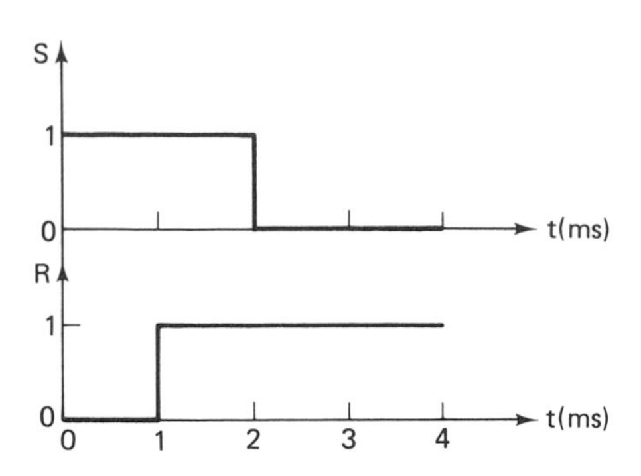

Figure 15.20 Input waveforms.

5. Draw an *S-R* flip-flop circuit using NOR gates.

Section 15.3

6. Describe the difference between the operation of a clocked *S-R* flip-flop and an *S-R* flip-flop.

7. The data inputs to a clocked *S-R* flip-flop are $S = 0$ and $R = 1$. What are the levels of the Q and $\bar{Q}$ outputs after the *CLK* lead is triggered?

8. Draw the $\bar{Q}$ output of the flip-flop in Example 15.4.

Section 15.4

9. Give a general description of data latch operation.

10. The data input at pin 2 of a 7475 data latch is logic 1. What is the logic level of the Q_1 and $\bar{Q}_1$ outputs at pins 16 and 1 after the pin 13 clock lead is triggered?

11. The *D* and *CLK* inputs of a 7475 data latch are shown in Fig. 15.21. Draw the Q output waveform.

Section 15.5

12. Describe the difference between edge triggering and pulse triggering.

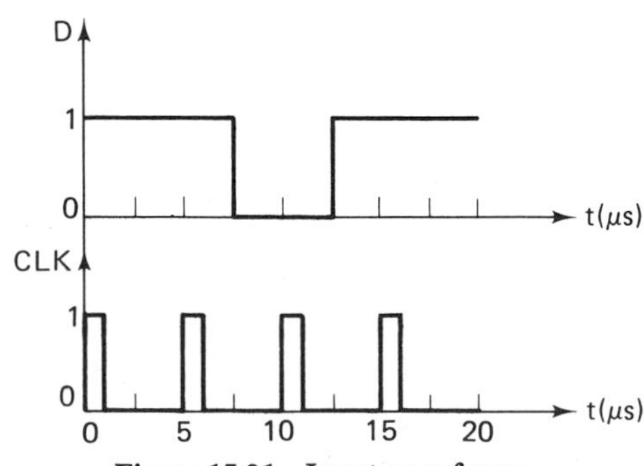

Figure 15.21 Input waveforms.

13. The inputs of the master-slave flip-flop in Fig. 15.11 are $S_1 = 0$ and $R_1 = 1$. Describe the operation of the circuit to a positive trigger pulse at the *CLK* lead.

Section 15.6

14. Give a general description of operation for a positive edge-triggered *D*-type flip-flop.

15. The D_1 input at pin 2 of a 7474 IC is logic 0, and the clear and preset leads are both logic 1. What are the logic states of Q and $\bar{Q}$ after the clock lead is triggered?

16. Draw the $\bar{Q}$ output waveform for the flip-flop in Example 15.7.

17. The D and *CLK* waveforms of a *D*-type flip-flop are shown in Fig. 15.21. Construct the Q and $\bar{Q}$ waveforms. Both *CLR* and *PST* are held at logic 1.

18. A 7474 *D*-type flip-flop is wired for *T*-type operation. The input is the waveform of Fig. 15.22. Draw the Q output.

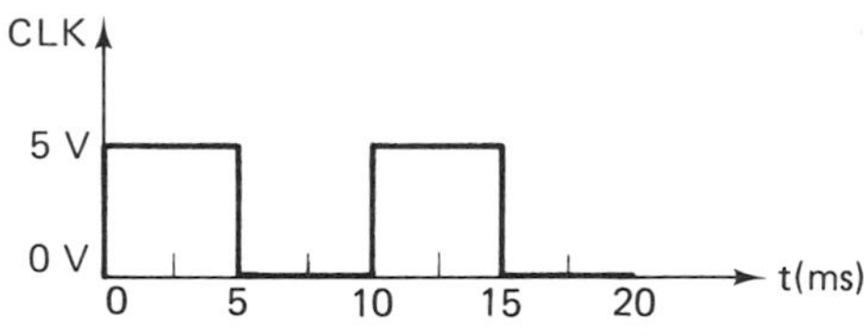

Figure 15.22 Input waveform.

Section 15.7

19. Give a general description of *J-K* flip-flop operation.

20. The inputs on a 7473 *J-K* flip-flop are $J_1 = 1$ and $K_1 = 1$. Besides this, the clear lead is held at logic 1. If the outputs are $Q = 0$ and $\bar{Q} = 1$, what are the new outputs after the circuit is clocked with a positive pulse.

21. The input to a *J-K* flip-flop set for toggle mode operation is a 100-kHz square wave that switches between 0 and +5 V. Evaluate the pulse width, pulse interval, and period of both the input and Q output waveforms. Draw the input and output.

Counters 16

16.1 INTRODUCTION

One of the most important uses of the flip-flop is to construct electronic *counter circuits*. A counter is a digital circuit which can be clocked through a standard sequence of output states.

These outputs usually represent a set of numbers. For instance, a circuit could count from 0 through 15 in binary form. This circuit is known as a *binary counter*. Another example is a counter that advances from binary 0 through 9 instead of 15. Here the output is the BCD code, and the circuit is called a *decade* or *decimal counter*.

In this chapter, we shall introduce the different types of counter circuits and how they are implemented with integrated devices. The topics included for this purpose are as follows:

1. The ripple counter
2. Up- and down-counter circuits
3. Integrated counter devices
4. Synchronous counters

16.2 THE RIPPLE COUNTER

A block diagram of a general counter circuit is shown in Fig. 16.1. On this block, there is a clock input lead marked *CLK* and two control leads. The control inputs are clear (*CLR*) and preset (*PST*). Furthermore, there are four outputs labeled *A*, *B*, *C*, and *D*. Of these, *D* is the most significant bit and *A* the LSB.

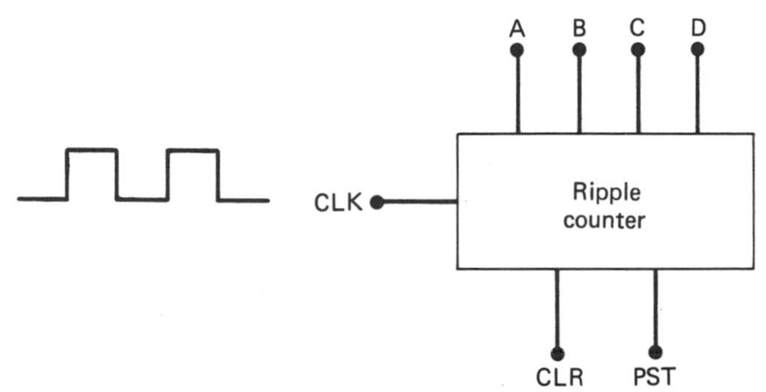

Figure 16.1 Counter block diagram.

As the clock input is pulsed, the output count advances through a standard set of binary combinations. For example, the outputs can be reset to $DCBA = 0000$ by pulsing the *CLR* lead. After this, the first pulse at the clock input advances the output to $DCBA = 0001$ equal to decimal 1. For each additional clock pulse, the output count increments by 1 until it reaches 1111 or 15. On the next clock pulse, the count resets to 0000, and the count sequence repeats.

The *PST* lead can be used to preset all outputs to the 1 logic level. In this way, the count sequence can be started at *DCBA* equal to 1111 instead of 0000.

Ripple Counter Circuit

The simplest type of counter circuit is called a *ripple counter;* it is also one of the most widely used. A ripple counter is formed by interconnecting toggle mode flip-flops as shown in Fig. 16.2.

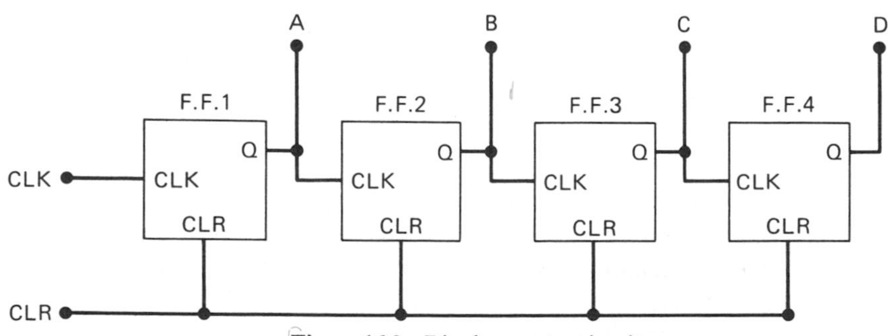

Figure 16.2 Ripple counter circuit.

In this circuit, we find four *T*-type flip-flops connected to make a 4-bit binary counter. The *Q* output of each flip-flop is wired to the *CLK* input of the next flip-flop. The counter output is a 4-bit binary word, with flip-flop 4 producing the MSB output *D* and flip-flop 1 the LSB output *A*.

This circuit configuration is called a ripple counter because data work their way from the *CLK* input through the *A*, *B*, and *C* flip-flops toward the MSB output *D*. When the input is clocked, the output states of the flip-flop change one after the other until a new count is set at the outputs. This results in a propagation delay of up to 100 ns.

16.3 THE UP-COUNTER CIRCUIT

Using the ripple counter configuration, we can construct two basic counter circuits. These circuits are called the *up-counter* and *down-counter*.

Up-counter

A binary up-counter circuit is made to count through the number sequence from 0 through 15. The output is a 4-bit binary number. Figure 16.3(a) gives a truth table of this operation.

A 4-bit binary counter circuit is shown in Fig. 16.3(b). Here, the *A*, *B*, *C*, and *D* outputs are taken from the *Q* outputs of flip-flops 1 through 4, respectively. The clock inputs on flip-flops 2 through 4 are supplied by the *Q* output on the flip-flop of the next least significant bit. For instance, the *Q* output of flip-flop 1 goes to the *CLK* input on flip-flop 2. On the other hand, the clock lead of flip-flop 1 is the *CLK* input of the counter circuit. Clear leads on all flip-flops are wired together to make a single *CLR* input.

When the circuit in Fig. 16.3(b) is turned ON, the outputs can settle into the 0 or 1 logic state. For this reason, the clear input must be pulsed to logic 0 to clear the counter output to $DCBA = 0000$. As shown in the table of Fig. 16.3(a), the first clock pulse sets the *A* output to logic 1. On the other hand, outputs of the other flip-flops stay at the 0 logic level. The result is the second count $DCBA = 0001$ in the table. This change in the *A* output is also shown in the waveforms of Fig. 16.3(c).

On the second clock pulse, flip-flop 1 resets, and the *A* output switches back to logic 0. Now the *A* output has been set from logic 0 to 1 and reset back to 0. This gives a clock pulse at the input of flip-flop 2, and the *B* output sets to the 1 level. The *C* and *D* outputs remain unchanged. The new output state is $DCBA = 0010$, as shown in the table of Fig. 16.3(a) and the waveforms of Fig. 16.3(c).

For the third clock pulse, flip-flop 1 is again set to 1. But the state of the *B* flip-flop is unchanged with output *B* equal to 1. The output count is now $DCBA = 0011$.

As a fourth clock pulse occurs, the outputs of the *A* and *B* flip-flops reset to

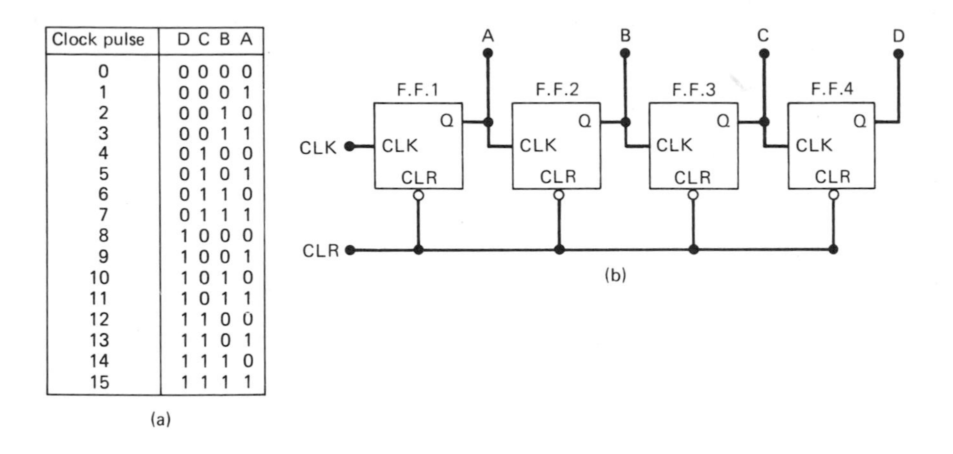

Clock pulse	D C B A
0	0 0 0 0
1	0 0 0 1
2	0 0 1 0
3	0 0 1 1
4	0 1 0 0
5	0 1 0 1
6	0 1 1 0
7	0 1 1 1
8	1 0 0 0
9	1 0 0 1
10	1 0 1 0
11	1 0 1 1
12	1 1 0 0
13	1 1 0 1
14	1 1 1 0
15	1 1 1 1

(a)

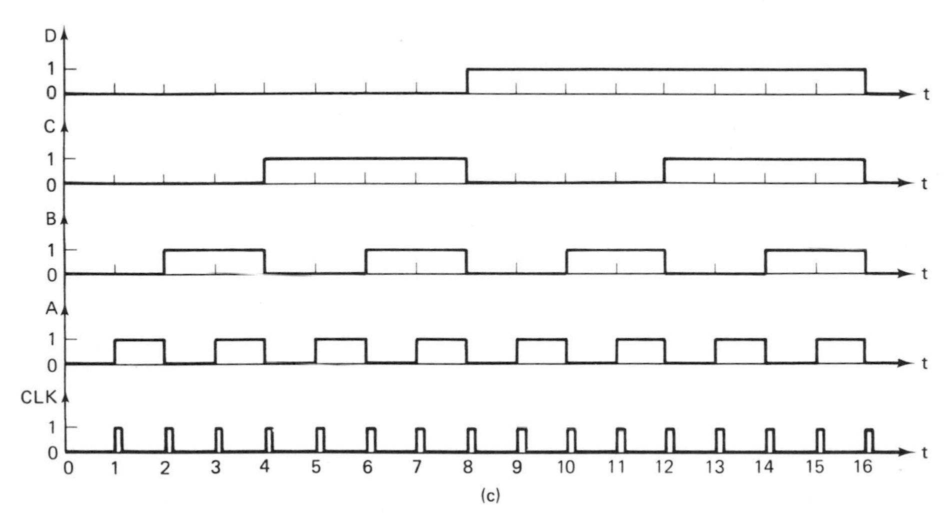

Figure 16.3 (a) Binary up-counter truth table; (b) Up-counter circuit; (c) Input/output waveforms.

logic 0. When the *B* output returns to logic 0, a pulse is completed at the input of flip-flop 3, and output *C* switches to logic 1. This gives the output count *DCBA* = 0100.

For additional clock pulses, the output advances through the binary states from 0101 through 1111. On the sixteenth input pulse, the count automatically resets to *DCBA* = 0000. At any point in the count sequence, the output of the circuit can be reset to 0000 by pulsing the *CLR* lead to logic 0.

EXAMPLE 16.1

What is the count held in the circuit of Fig. 16.3(b) after the sixth clock pulse shown in Fig. 16.3(c)?

SOLUTION:

The waveforms of Fig. 16.3(c) show that the output after the sixth input pulse is the binary combination for number 6:

$$DCBA = 0110$$

EXAMPLE 16.2

Show how two 7473 *J-K* flip-flop devices can be connected to make a 4-bit up-counter. Use a NAND gate to reset the circuit at a count of 10 to give a BCD or decade counter.

SOLUTION:

Each 7473 device contains two *J-K* flip-flop circuits. Their connection into a 4-bit counter is shown in Fig. 16.4. The *J* and *K* inputs are all made logic 1 and can be wired to $+V_{cc}$ through a 1-kΩ resistor to limit noise.

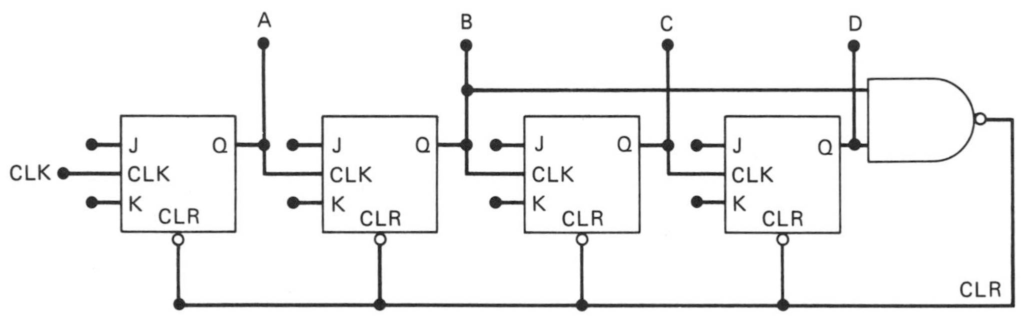

Figure 16.4 *J-K* flip-flop decade counter.

To make a BCD counter, the output of a NAND gate is wired to the clear leads. The inputs of the gate are supplied by the *B* and *D* count outputs. When $DCBA = 1010$, the output of the NAND gate goes to logic 0 and resets the count to $DCBA = 0000$. Now, the NAND gate output is returned to the 1 logic level.

In this way, the count sequence is limited to the BCD range of 0000 to 1001.

Input/Output Frequency

The waveforms of a 4-bit binary up-counter are shown in Fig. 16.5. Looking at these waveforms, we see that the input frequency is divided by 2 by each flip-flop. For this reason, the counter is sometimes called a *frequency divider circuit.*

If the clock input is indicated by *f*, the frequencies of the square waves at the *A*, *B*, *C*, and *D* outputs are given by the equations that follow:

$$f_A = \tfrac{1}{2}f$$
$$f_B = \tfrac{1}{2}f_A = \tfrac{1}{4}f$$
$$f_C = \tfrac{1}{2}f_B = \tfrac{1}{8}f$$
$$f_D = \tfrac{1}{2}f_C = \tfrac{1}{16}f$$

EXAMPLE 16.3

The frequency of the clock input of Fig. 16.5 is 100 kHz. Find the frequency of the *A*, *B*, *C*, and *D* outputs.

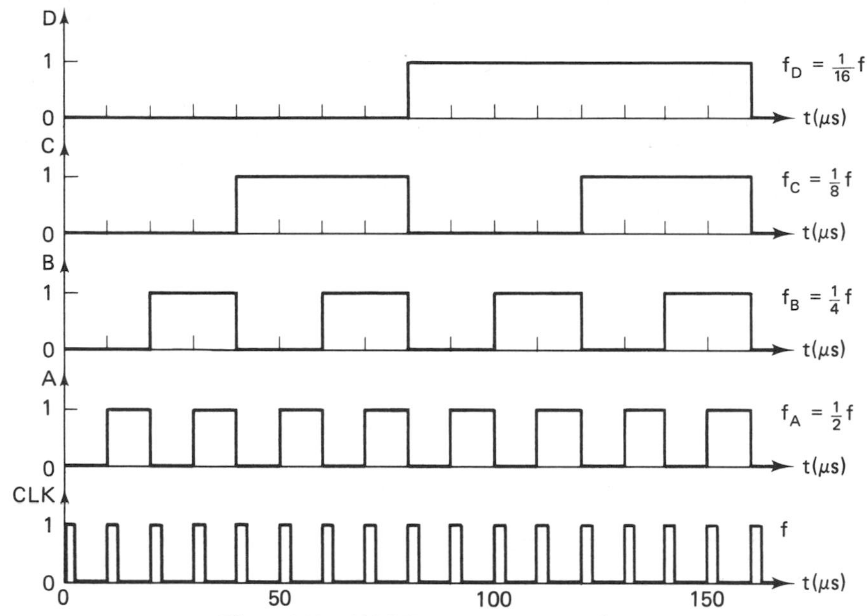

Figure 16.5 100-kHz up-counter waveforms.

SOLUTION:

$$f_A = \tfrac{1}{2}f = \tfrac{1}{2}(100\text{ kHz}) = 50\text{ kHz}$$
$$f_B = \tfrac{1}{4}f = \tfrac{1}{4}(100\text{ kHz}) = 25\text{ kHz}$$
$$f_C = \tfrac{1}{8}f = \tfrac{1}{8}(100\text{ kHz}) = 12.5\text{ kHz}$$
$$f_D = \tfrac{1}{16}f = \tfrac{1}{16}(100\text{ kHz}) = 6.25\text{ kHz}$$

16.4 THE DOWN-COUNTER CIRCUIT

A *down-counter circuit* works the opposite of the up-counter. A truth table of its operation is shown in Fig. 16.6(a). Here, we see that the output starts at a count of $DCBA = 1111$ and is clocked down toward $DCBA = 0000$.

The circuit of a down-counter is given in Fig. 16.6(b). Looking at this circuit, we notice it differs from the up-counter of Fig. 16.3(b) in that the $\bar{Q}$ outputs are used to provide clock signals to the next more significant flip-flop. Moreover, the preset leads, instead of the *CLR* leads, are wired together. In this way, the *PST* input can be pulsed to start the count at $DCBA = 1111$.

Input/output waveforms of down-counter operation are shown in Fig. 16.6(c). Here, we see that the first clock pulse resets the *A* flip-flop and makes the output $DCBA = 1110$. On the next clock pulse, flip-flop *A* sets, and the *B* fl.p-flop is reset. This gives the output state $DCBA = 1101$.

For further clock pulses, the output counts down through the binary com-

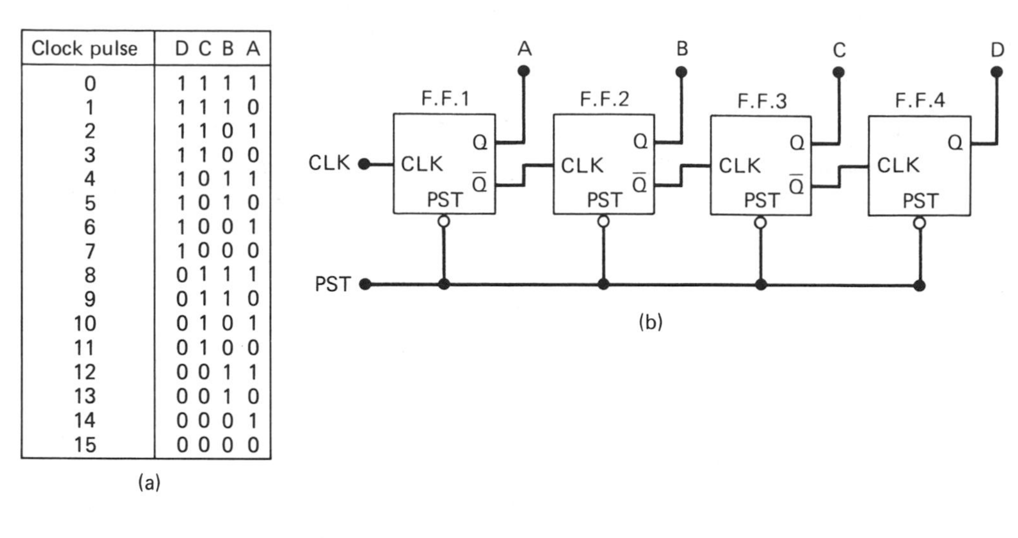

Clock pulse	D C B A
0	1 1 1 1
1	1 1 1 0
2	1 1 0 1
3	1 1 0 0
4	1 0 1 1
5	1 0 1 0
6	1 0 0 1
7	1 0 0 0
8	0 1 1 1
9	0 1 1 0
10	0 1 0 1
11	0 1 0 0
12	0 0 1 1
13	0 0 1 0
14	0 0 0 1
15	0 0 0 0

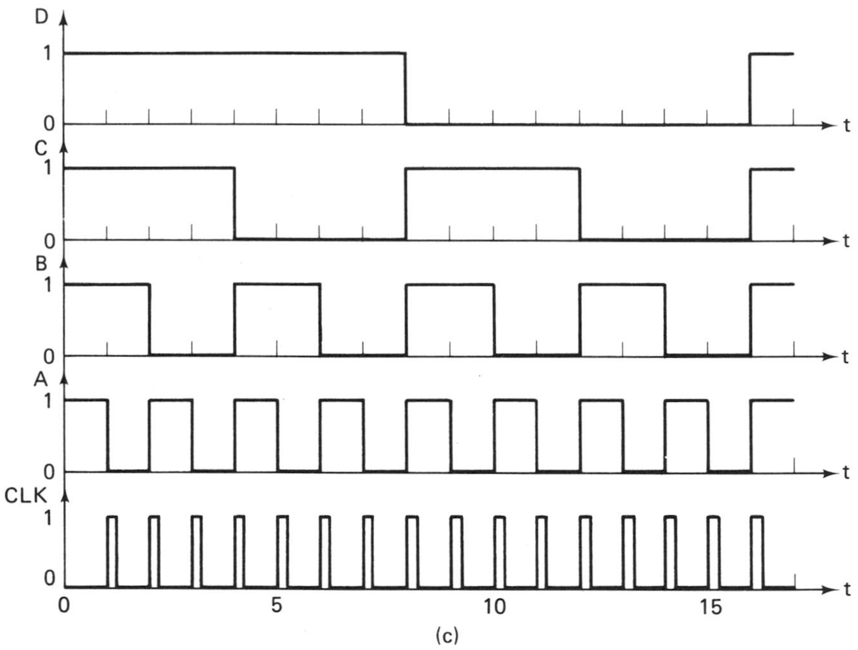

Figure 16.6 (a) Binary down-counter truth table; (b) Down-counter circuit; (c) Input/output waveforms.

binations from 1101 through 0000. As a sixteenth clock pulse is applied to the *CLK* input, the counter output again sets to $DCBA = 1111$.

EXAMPLE 16.4

What count is held in the down-counter circuit of Fig. 16.6(b) after the sixth clock pulse has occurred in Fig. 16.6(c)?

SOLUTION:

From the waveform in Fig. 16.6(c), we find that the count corresponding to the sixth clock pulse is the binary combination for number 9:

$$DCBA = 1001$$

16.5 INTEGRATED RIPPLE COUNTERS

Counter circuits are available in standard medium-scale integrated circuits. Two widely used counter devices are the 7493 binary counter and the 7490 decade counter. In this section, we shall describe both of these devices and their operation.

7493 Binary Counter

The 7493 integrated circuit is made to work as a 4-bit binary counter. A block diagram of this device is shown in Fig. 16.7(a). From this block, we see that the clock input of the circuit is called A_i instead of *CLK*. The outputs on the counter are marked A_o, B_o, C_o, and D_o. Once again, the D_o output is the most significant bit and A_o the LSB. The counter circuit within the 7493 is shown in Fig. 16.7(b). This circuit is negative edge triggered.

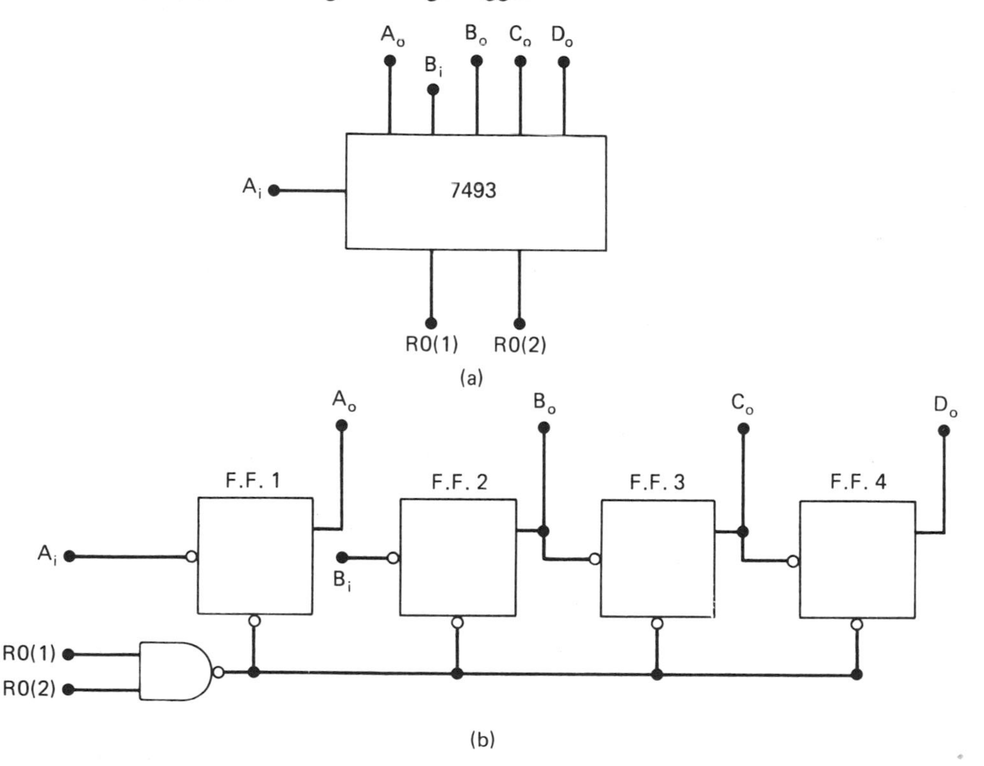

Figure 16.7 (a) 7493 block diagram; (b) Counter circuit.

The 7493 device has an external lead to the input of the B flip-flop. This lead is marked B_i. Besides this, there are two *reset leads* indicated $R0(1)$ and $R0(2)$. The $R0$ leads allow for the reset of the counter at any point in the standard count order.

A pin layout for the 7493 is shown in Fig. 16.8(a) and some of its electrical characteristics in Fig. 16.8(b). From these characteristics, we see that the maximum propagation delay is 70 ns and that the typical value of the maximum operating frequency is 42 MHz.

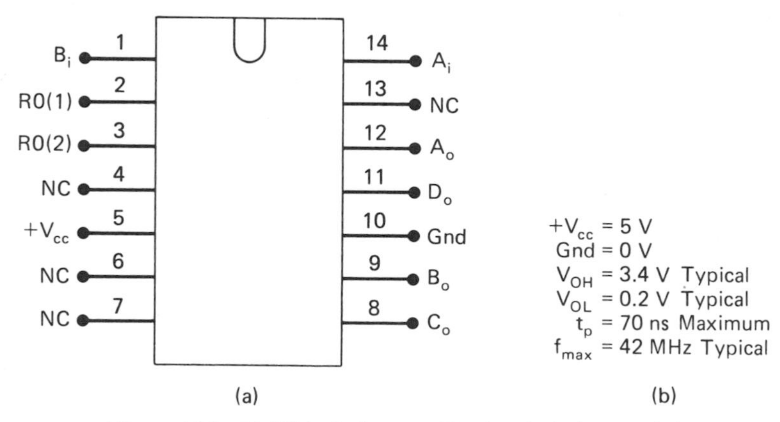

Figure 16.8 (a) 7493 pin layout. (b) Electrical characteristics.

Binary Counter Operation

For operation as a normal binary counter, the 7493 must be connected as shown in Fig. 16.9(a). Looking at this circuit, we see that the A_o output must be tied to the B_i input. This wire completes the connection of the flip-flops into a 4-bit up-counter.

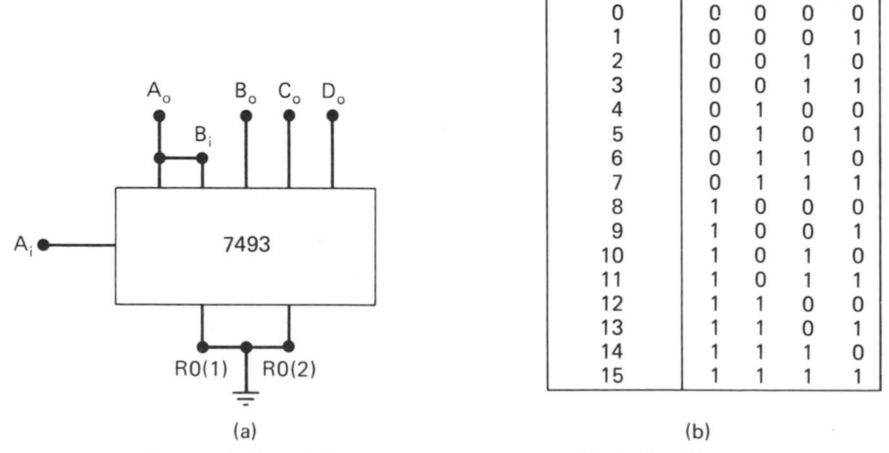

Clock pulse	D_o	C_o	B_o	A_o
0	0	0	0	0
1	0	0	0	1
2	0	0	1	0
3	0	0	1	1
4	0	1	0	0
5	0	1	0	1
6	0	1	1	0
7	0	1	1	1
8	1	0	0	0
9	1	0	0	1
10	1	0	1	0
11	1	0	1	1
12	1	1	0	0
13	1	1	0	1
14	1	1	1	0
15	1	1	1	1

Figure 16.9 (a) Binary counter operation. (b) Truth table.

At least one of the two reset leads must be set to logic 0. For this reason, both $R0(1)$ and $R0(2)$ are shown wired to ground in Fig. 16.9(a).

In this circuit, consecutive clock pulses cause the output $D_oC_oB_oA_o$ to count through the binary sequence from 0000 to 1111. This output is shown in the table of Fig. 16.9(b).

Resetting the 7493 Counter

The $R0(1)$ and $R0(2)$ leads can be connected to the output leads of the counter to reset the count at any point in its normal sequence. When both reset inputs are logic 1, the clear lead on each flip-flop in Fig. 16.7(b) goes to logic 0. This causes the output to reset to $D_oC_oB_oA_o = 0000$.

Figure 16.10 shows the use of the reset leads on a 7493 binary counter to make a BCD or decade counter. In th s circuit, the B_o and D_o outputs are returned to $R0(1)$ and $R0(2)$, respectively. As the output count changes to $D_oC_oB_oA_o = 1010$, $R0(1)$ and $R0(2)$ are logic 1, and the count immediately resets to decimal 0. The new count sequence is from 0000 through 1001.

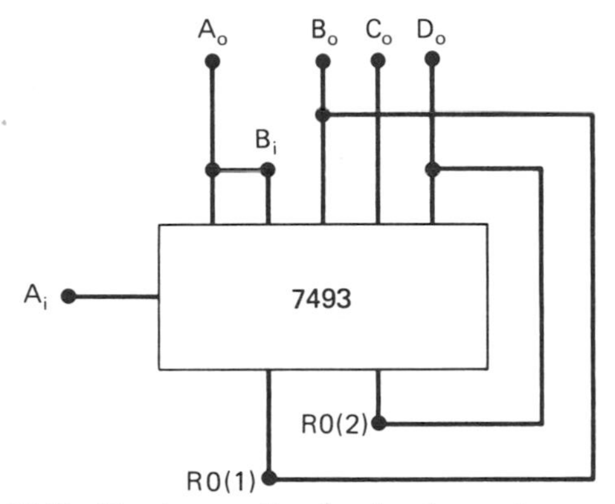

Figure 16.10 Reset connection for decade counter operation.

EXAMPLE 16.5

What is the binary count at the output of the 7493 counter in Fig. 16.9(a) after the eleventh clock pulse?

SOLUTION:

The truth table in Fig. 16.9(b) gives the operation of the binary counter. Looking at this table, we find that the count after the eleventh pulse is the binary number for 11:

$$D_oC_oB_oA_o = 1011$$

EXAMPLE 16.6

What is the count sequence of a 7493 counter if outputs C_o and D_o are returned to reset leads $R0(1)$ and $R0(2)$, respectively?

SOLUTION:

In this circuit, the output resets to $D_oC_oB_oA_o = 0000$ when C_o and D_o go to logic 1 for the first time. This occurs at the binary count $D_oC_oB_oA_o = 1100$ or decimal 12. For this reason, the circuit counts from decimal 0 through 11 in binary form.

7490 Decade Counter

The 7490 integrated counter is a lot like the 7493 device; however, the circuit within the 7490 is constructed to count through a BCD count sequence. For this reason, it is called a decimal or decade counter.

The block diagram of a 7490 counter is shown in Fig. 16.11(a) and its pin layout in Fig. 16.11(b). Looking at the block diagram, we notice that the clock input on the device is again called A_i and its outputs A_o, B_o, C_o, and D_o. But the external input to the B flip-flop is indicated as the BD_i input. This 7490 circuit is also negative edge triggered.

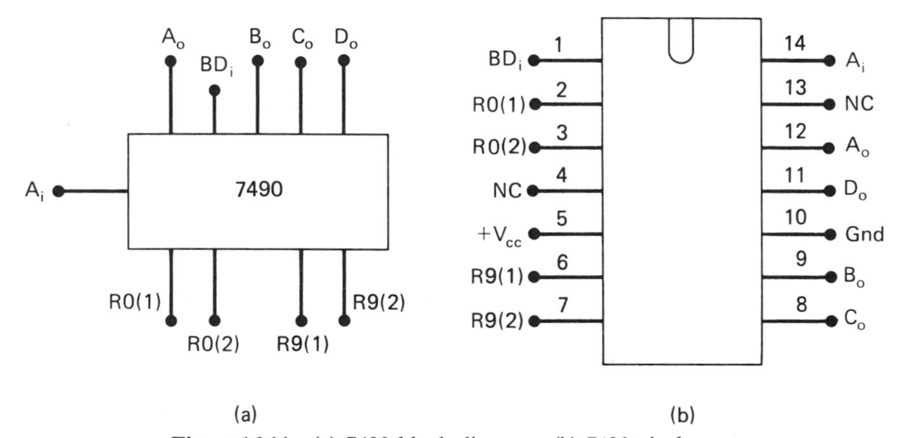

Figure 16.11 (a) 7490 block diagram; (b) 7490 pin layout.

The 7490 decade counter has four reset leads: $R0(1)$, $R0(2)$, $R9(1)$, and $R9(2)$. When the $R0$ leads are logic 1 together, the counter output resets to decimal 0 or $D_oC_oB_oA_o = 0000$. On the other hand, making the $R9$ leads logic 1 causes the output to reset to decimal 9 or $D_oC_oB_oA_o = 1001$. These leads can be used to change the count sequence.

Figure 16.12(a) gives the connection of a 7490 counter for normal decade operation. Its input/output waveforms are shown in Fig. 16.12(b).

EXAMPLE 16.7

What is the BCD count in the 7490 circuit of Fig. 16.12(a) after the eleventh clock pulse?

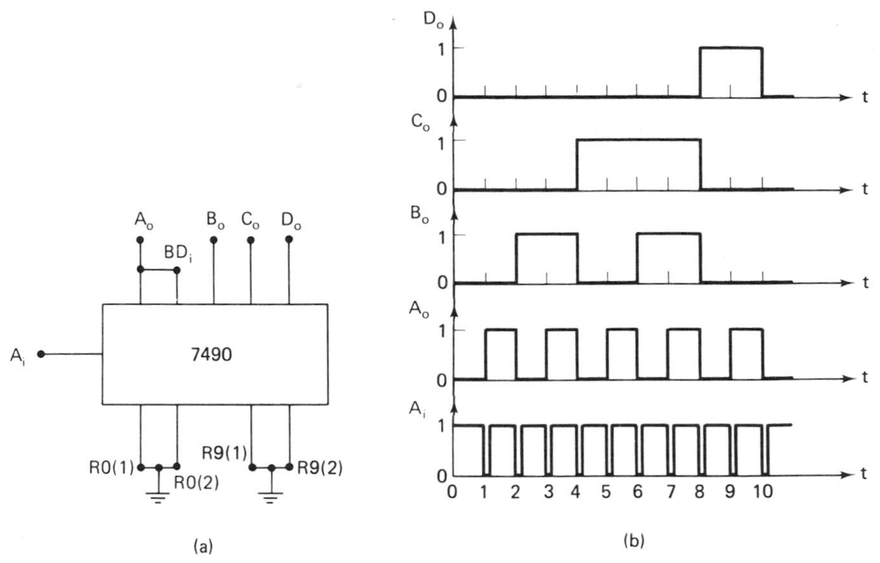

Figure 16.12 (a) 7490 counter connection; (b) Input/output waveforms.

SOLUTION:

The 7490 counter resets to $D_oC_oB_oA_o = 0000$ on the tenth pulse at A_i and sets to $D_oC_oB_oA_o = 0001$ as the eleventh pulse occurs:

$$D_oC_oB_oA_o = 0001$$

Cascaded Counters

To make a multidigit decimal counter circuit, integrated counter devices can be *cascaded*. For example, Fig. 16.13 shows a two-digit decimal counter. This circuit can count from 00 through 99 using an 8-bit BCD code.

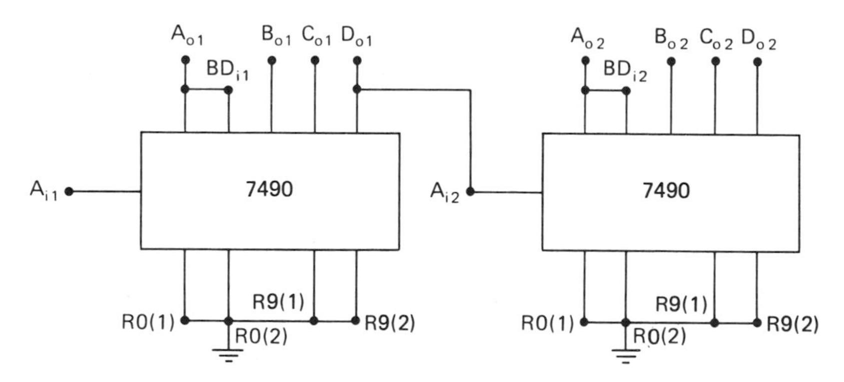

Figure 16.13 Cascaded decade counters.

Here two 7490 ICs are cascaded by wiring the D_{o1} output of the least significant digit device to the A_{i2} input on the most significant digit counter.

16.6 THE SYNCHRONOUS COUNTER

Another structure counter uses what is known as a *synchronous counter circuit*. A typical 4-bit synchronous binary counter circuit is shown in Fig. 16.14. In this circuit, the flip-flop connection differs from that in the ripple counter of Fig. 16.2. Here the flip-flops are used in both the toggle mode and set-reset mode of *J-K* flip-flop operation.

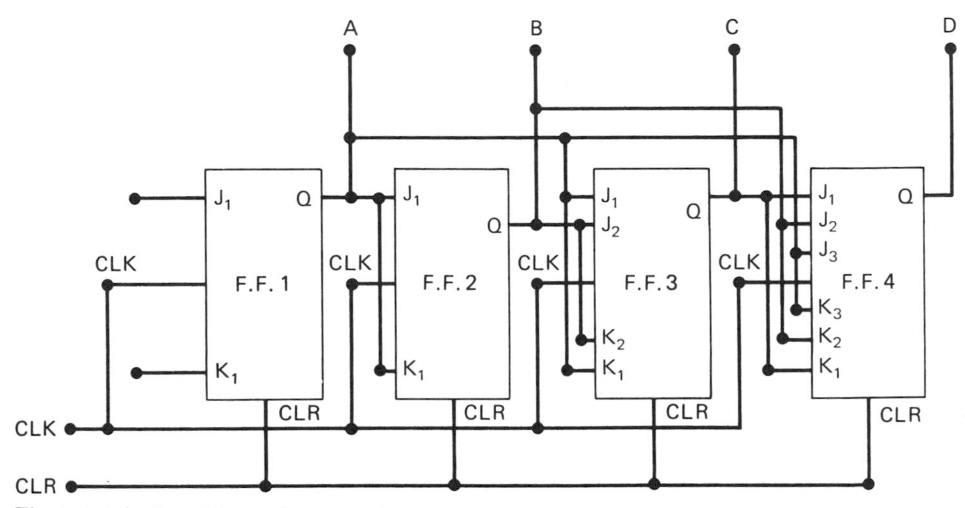

Figure 16.14 Four-bit synchronous binary counter.

The advantage in the use of this more complex circuit configuration is its high speed. An increase in speed is obtained because data are always at all *J-K* inputs and all flip-flops are clocked at the same time. Therefore, all outputs change logic states at the same time instead of data rippling down through the other flip-flops. In this way, propagation delay is decreased to that of one flip-flop.

74193 Synchronous Binary Counter

The 74193 integrated circuit contains a versatile *synchronous binary counter.* This circuit can be operated as an up-counter, down-counter, or *preset up/down-counter.*

Figure 16.15(a) is a block diagram of the 74193 counter circuit. The counter outputs are marked A_o, B_o, C_o, and D_o. There are two separate clock inputs CLK_U and CLK_D. The CLK_U lead is used during up-counter operation and CLK_D for a down-counter. The 74193 is positive edge triggered. Besides this, a clear input is provided, and it resets the output to $D_oC_oB_oA_o = 0000$ with a 1 logic level.

The 74193 has a set of data inputs labeled A_i, B_i, C_i, and D_i. These inputs are

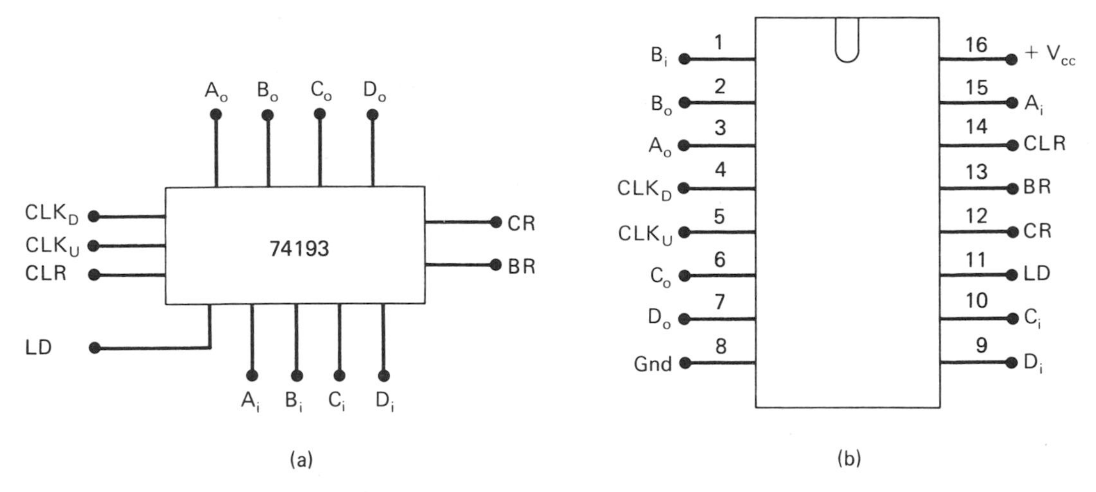

Figure 16.15 (a) 74193 block diagram; (b) 74193 pin layout.

used in conjunction with the load lead (*LD*) during preset counter operation. A count can be applied to the preset data inputs and entered into the counter output by pulsing *LD* to logic 0.

In addition, there is a *carry output* marked *CR* and a *borrow output BR*. The carry lead is wired to pulse the CLK_U input of a second 74193 device when connected in cascade. Moreover, borrow is tied to CLK_D when cascading.

A pin layout of the 74193 is given in Fig. 16.15(b). Its maximum operating frequency is 32 MHz, and typical up- or down-count propagation delays are 16 ns. Therefore, the 74193 is about four times faster than the 7493 ripple counter device.

Up-counter

To operate as an up-counter, the 74193 is connected as shown in Fig. 16.16(a). On this circuit, the load lead *LD* is made logic 1, and preset inputs all logic 0.

First, the count at output $D_oC_oB_oA_o$ is cleared to decimal 0 by pulsing the clear lead to logic 1. After this, the CLK_U clock input can be used to advance the count output from $D_oC_oB_oA_o = 0000$ toward 1111. As the output becomes 1111 and CLK_U is logic 0, the carry output switches to the 0 logic level.

Input/output waveforms for up-counter operation are given in Fig. 16.16(b).

Down-counter

A down-counter connection of the 74193 device is given in Fig. 16.17. For down-counter operation, the CLK_D input is used to make the output count from $D_oC_oB_oA_o = 1111$ down to 0000. When the outputs are all logic 0 and the clock returns to logic 0, the borrow output switches to logic 0.

EXAMPLE 16.8

The preset counter circuit in Fig. 16.18 has the preset count entered each time the down-counter resets. What is the count sequence?

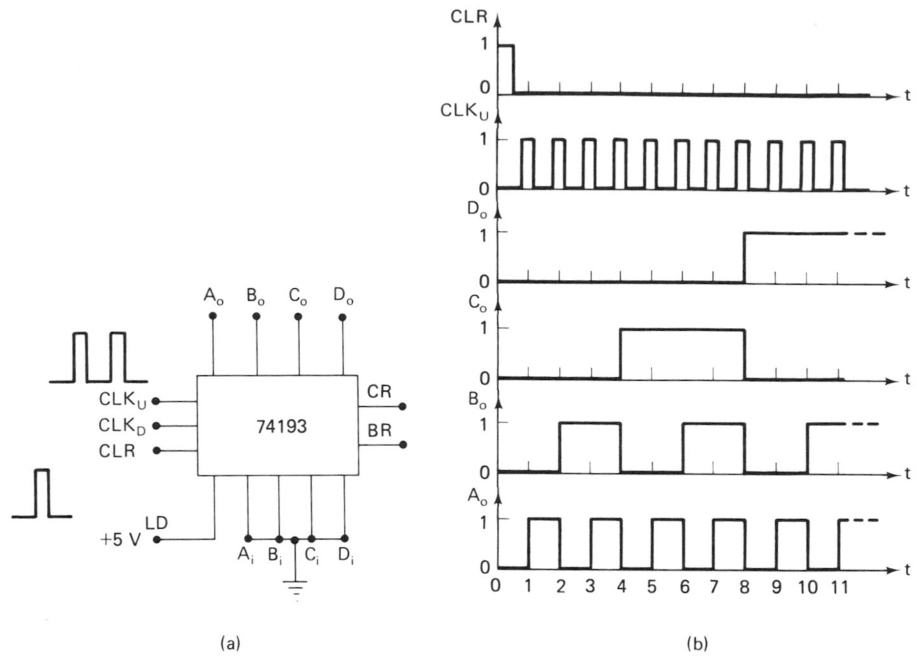

Figure 16.16 (a) Up-counter connection; (b) Input/output waveforms.

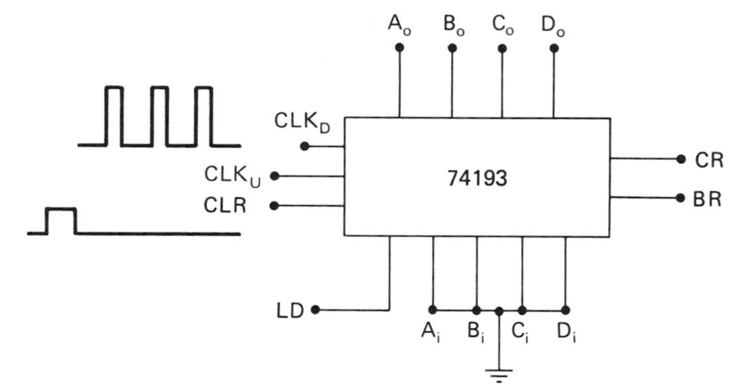

Figure 16.17 Down-counter connection.

SOLUTION:

The preset input count is

$$D_iC_iB_iA_i = 1010$$

Therefore, the output counts down from 10 equal to 1010 in binary form through 0 equal to 0000.

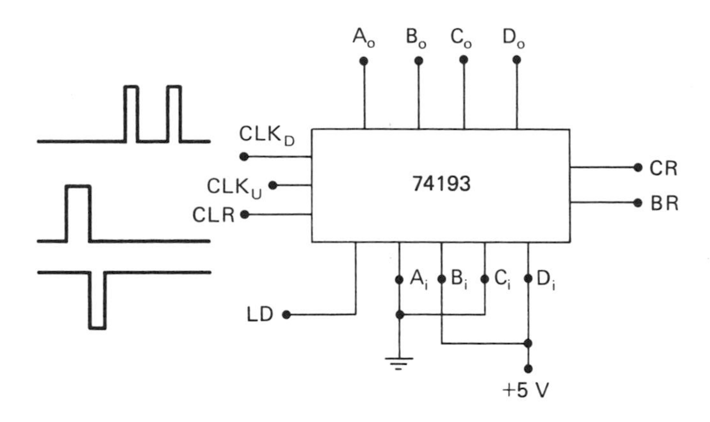

Figure 16.18 Preset counter connection.

ASSIGNMENTS

Section 16.2

1. What is the use of the *CLR* lead on a counter circuit? The *PST* lead?

2. Are set-reset or toggle mode flip-flops used to make a ripple counter circuit?

3. Is the *D* output the LSB or MSB of a ripple counter?

Section 16.3

4. What is thc binary output of the up-counter circuit in Fig. 16.3(b) just after the tenth clock pulse?

5. Make a drawing of the output waveforms for the decade counter in Fig. 16.4.

6. Draw a 4-bit binary up-counter using *D*-type flip-flops connected to work in the toggle mode.

7. Show how the counter in Problem 6 can be reset to have a count sequence from 0 through 6.

8. The input to a 4-bit binary up-counter is a 1-MHz square wave. Find the frequency of the waveforms at the *A*, *B*, *C*, and *D* outputs.

9. A 16-kHz symmetrical square wave is applied to the input of a 4-bit binary up-counter. What are the frequency, period, pulse width, and pulse interval of the *D* output?

Section 16.4

10. Give the binary output of the down-counter circuit of Fig. 16.6(b) after the fifteenth clock pulse.

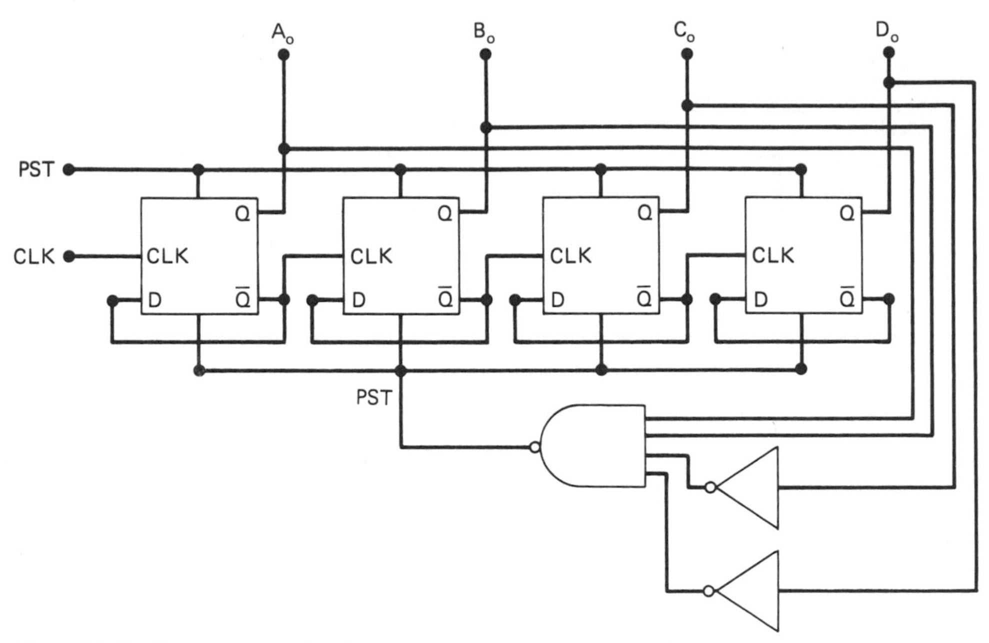

Figure 16.19 Down-counter circuit.

11. What is the count sequence of the down-counter circuit in Fig. 16.19? Show the outputs in the form of a truth table.

Section 16.5

12. Find the count held in the circuit of Fig. 16.9(a) just after the twentieth pulse at the clock input.

13. Which outputs on a 7493 IC should be returned to the *R*0(1) and *R*0(2) reset inputs to obtain decade counter operation?

14. What is the output count of the 7490 counter in Fig. 16.12(a) after the twentieth clock pulse?

15. At what count does the 7490 circuit in Fig. 16.20 reset?

16. Draw output waveforms for the counter circuit in Fig. 16.20.

Section 16.6

17. What is the binary output of the 74193 counter in Fig. 16.16(a) after the fifteenth pulse at the clock input?

18. Draw input/output waveforms for the 74193 down-counter connection of Fig. 16.17.

19. Find the binary output of the circuit in Fig. 16.18 caused by the third clock pulse after loading.

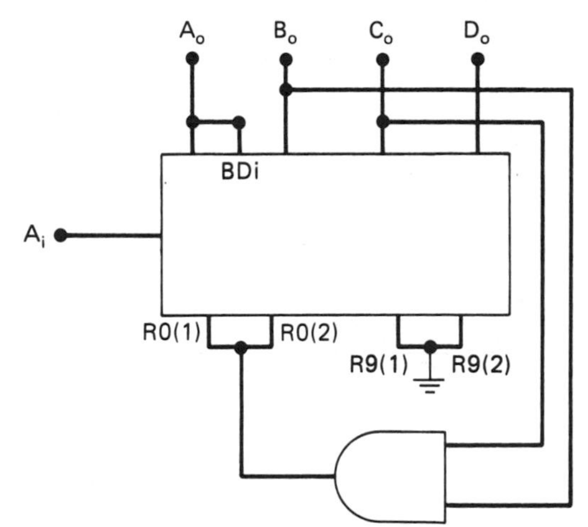

Figure 16.20 Reset 7490 counter.

Shift Registers 17

17.1 INTRODUCTION

The *shift register* is another important digital electronic circuit that uses the flip-flop as its building block. A shift register contains *S-R* flip-flop circuits and logic gate circuitry to control their operation.

In digital equipment, shift registers find a variety of applications. One common use is to delay a sequence of logic pulses by a set number of clock cycles. For this type of operation, a *serial-in, serial-out register* is needed. By serial-in, we mean that a series of logic pulses are inputted in time at a single input lead. Serial-out means the same except the series of pulses are outputted at a lead.

A second shift register application is in serial to parallel data conversion. In this use, serial data at a single input are changed to logic levels at a set of parallel outputs. To do this, we need a *serial-in, parallel-out register.* In a similar way, parallel data can be converted to serial form with a *parallel-in, serial-out register.*

In this chapter, the basic shift register configurations are introduced, and the operation of integrated register devices is described. The topics covered are as follows:

1. The serial-in, serial-out shift register
2. The parallel-in, serial-out shift register
3. Shift right, shift left register
4. Shift register counters

17.2 THE SERIAL-IN, SERIAL-OUT SHIFT REGISTER

As earlier mentioned, the serial-in, serial-out shift register can be used to delay data transfer between circuits. Figure 17.1(a) shows a block diagram of a 4-bit serial-in, serial-out shift register. On this block, the *serial input* lead is marked with the symbol S_i and the *serial output* S_o. The input and output of data are controlled by the signal at the clock input *CLK*.

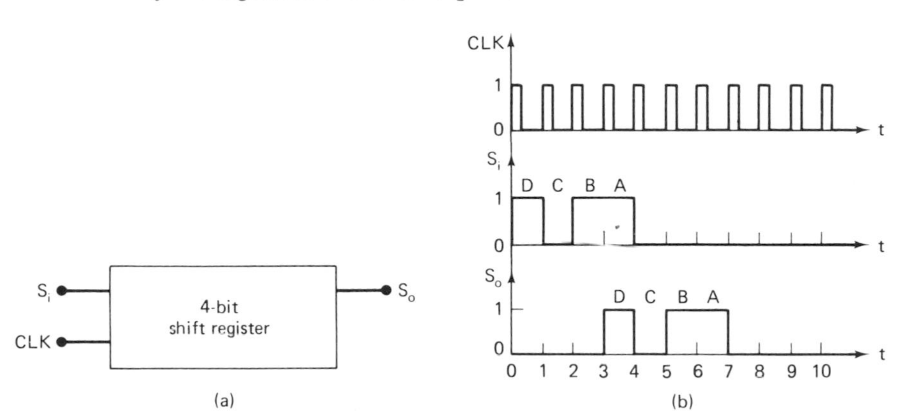

Figure 17.1 (a) Four-bit serial-in, serial-out shift register block; (b) Input/output waveforms.

Waveforms of the circuit's operation are given in Fig. 17.1(b). As each clock pulse occurs, another bit of data is entered into the register from the S_i input. At the same time, the data held in all flip-flops of the register shift 1 bit toward the S_o output. After the fourth clock pulse, the *D* bit of the input data appears at the output.

In this example, the serial data input is the binary code $DCBA = 1011$, and it is delayed by three clock cycles between input and output of the register. In this way, we see that a shift register delays the serial data by a number of clock cycles equal to one less than the number of flip-flops in the register.

7491 8-Bit Shift Register

The 7491 device is an MSI circuit that contains an 8-bit serial-in, serial-out shift register. In Fig. 17.2(a), the circuitry within the 7491 is shown. Here we see that the circuit has two serial inputs labeled *A* and *B*. The signal sent to the set input *S* of the first *S-R* flip-flop is the AND function of the logic signals *A* and *B*. On the other hand, the logic level of the reset input is the complement of the *S* data:

$$S = AB$$

$$R = \bar{S} = \overline{AB}$$

When both *A* and *B* inputs are logic 1, the set input of flip-flop 1 is logic 1 and reset logic 0. On the positive edge of the first clock pulse, the *Q* output on flip-flop

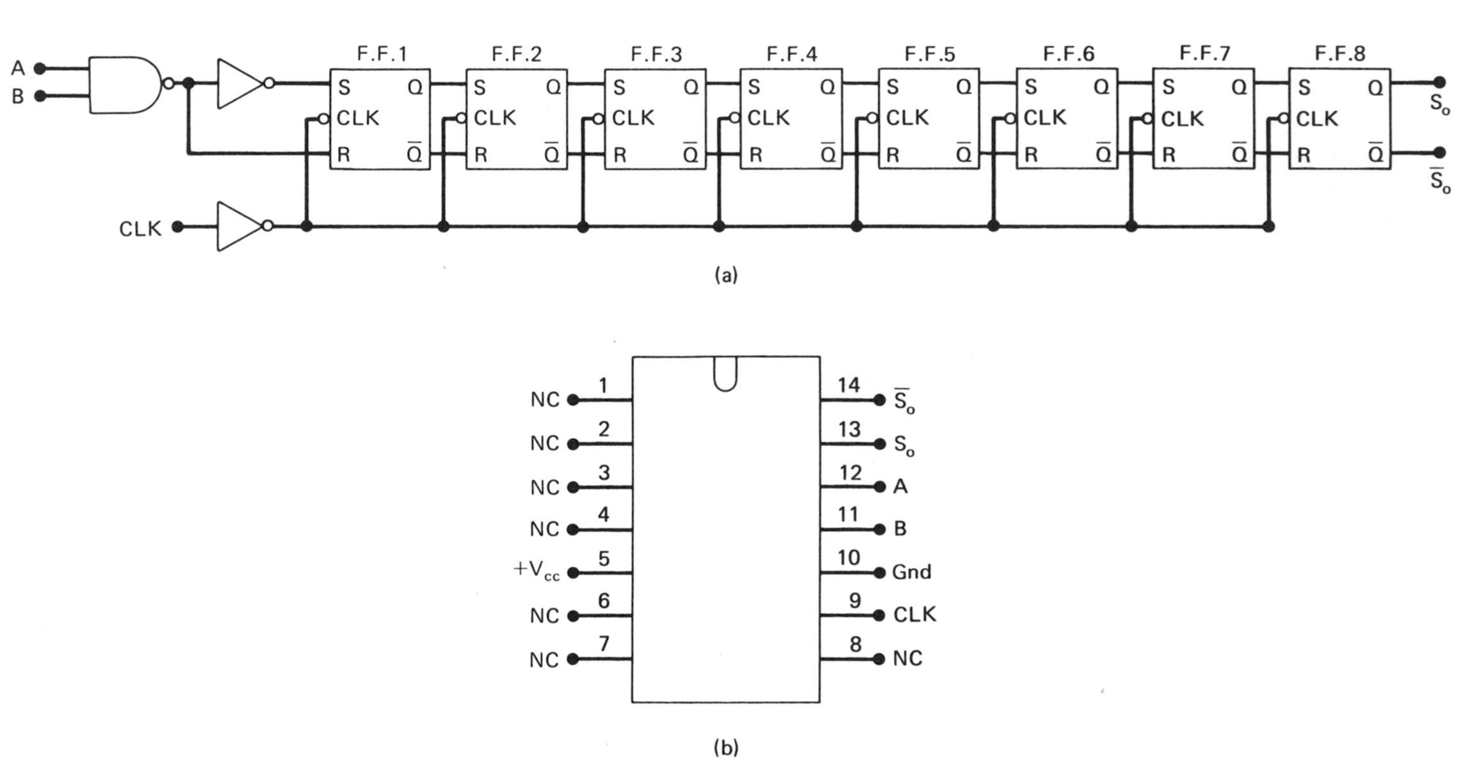

Figure 17.2 (a) 7491 8-bit shift register circuit; (b) 7491 pin layout.

1 switches to logic 1 and $\bar{Q}$ to logic 0. The outputs on flip-flop 1 are used as inputs to flip-flop 2. On the next clock pulse, the set input on flip-flop 2 is 1 and reset 0. These data cause its Q output to set to 1, and $\bar{Q}$ switches to 0. At the same time, a logic 0 or 1 due to a new A and B input is loaded into flip-flop 1.

As the clock input is pulsed one more time, data shift another flip-flop toward the output and a new data bit enters flip-flop 1. On the eighth clock pulse, the data bit entered at the serial input on the first clock pulse appears at the serial output S_o. The complement of the serial input is produced at the $\bar{S}_o$ output.

The period of the clock input can be found as the reciprocal of its frequency. For instance, for a 100-kHz clock the period of a clock cycle is found to be

$$T = \frac{1}{f} = \frac{1}{100\ \text{kHz}} = 10\ \mu\text{s}$$

For serial data to be passed from input to output, seven clock cycles must elapse. This gives a time delay t_d between serial input and serial output equal to 70 μs:

$$t_d = 7T = 7(10\ \mu\text{s}) = 70\ \mu\text{s}$$

A pin layout of the 7491 is given in Fig. 17.2(b).

EXAMPLE 17.1

Find the S_o output waveform for the 7491 shift register of Fig. 17.3(a). It has the A input and CLK signals shown in Fig. 17.3(b).

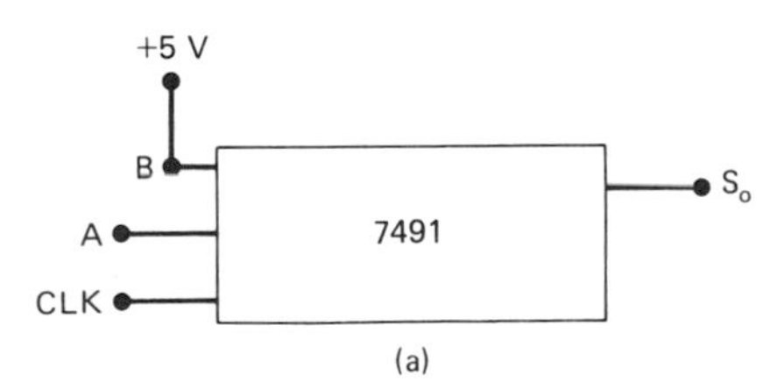

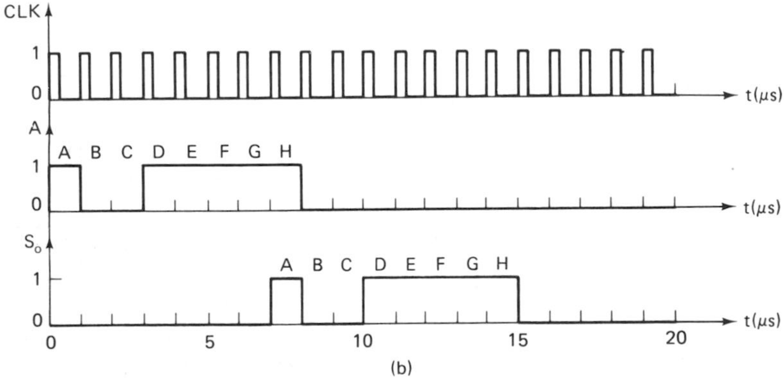

Figure 17.3 (a) 7491 serial-in, serial-out shift register; (b) Input/output waveforms.

SOLUTION:

When the B input is logic 1, the A data are entered to S-R flip-flop 1 in the register. On the eighth clock pulse, the A data bit appears at the S_o output. This occurs at time $t = 7\ \mu s$. For the ninth clock pulse and on, data bits B through H are provided at the output.

17.3 THE PARALLEL-IN, SERIAL-OUT SHIFT REGISTER

In many digital systems, it is necessary to convert parallel-coded information to serial form. For this, we use a parallel-in, serial-out shift register circuit.

A block diagram of this type of register is shown in Fig. 17.4(a). Here, we find a clock lead CLK, load control input LD, serial output S_o, and four parallel data inputs A_i, B_i, C_i, and D_i. By parallel inputs, it is meant that all 4 bits of input data are present at the same time. Of these, D_i is the most significant bit.

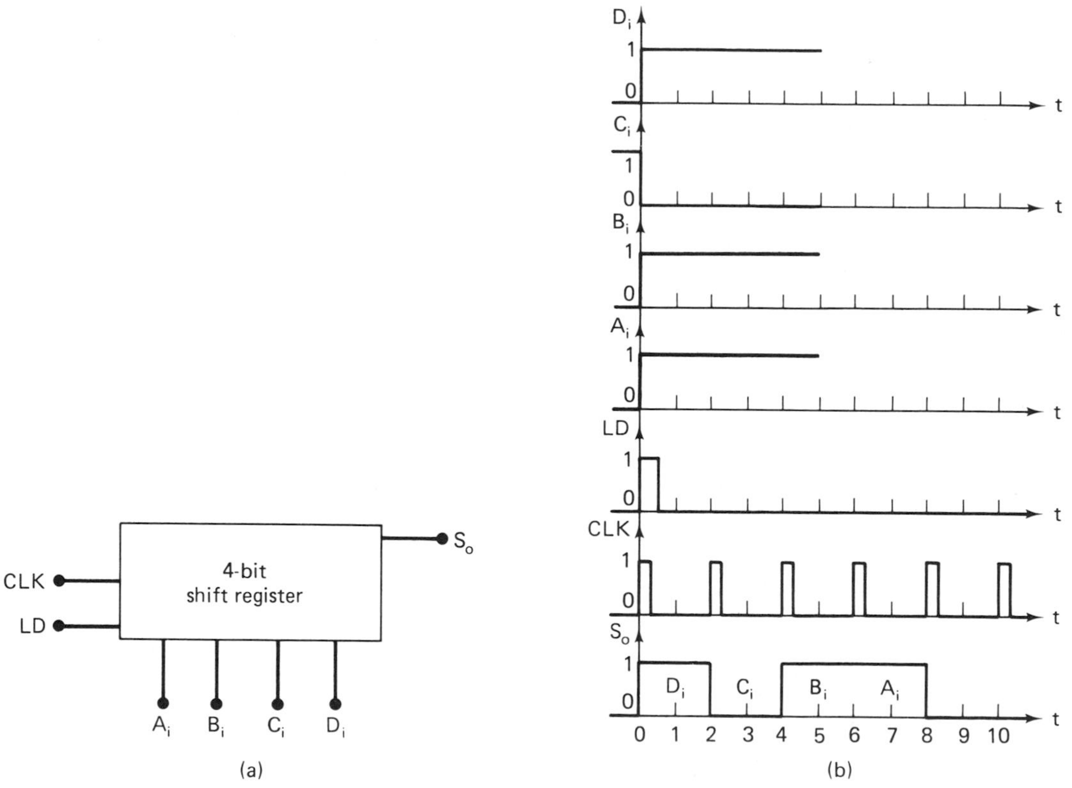

Figure 17.4 (a) Parallel-in, serial-out shift register block; (b) Input/output waveforms.

In normal operation, a 4-bit code at the parallel inputs is loaded by applying a pulse to the load (LD) control lead. After loading, the data word can be clocked out of the register in serial form at the S_o lead.

Figure 17.4(b) shows input/output waveforms for a parallel-in, serial-out shift register. Looking at these signals, we see that a pulse at time $t = 0$ at the load lead causes the data word $D_iC_iB_iA_i = 1011$ to enter the flip-flops of the register. The logic 1 loaded at the D_i parallel input makes the S_o output switch to the 1 level. On the clock pulse at time $t = 2$, the output switches to logic 0 for one clock cycle and then back to 1 for the next two cycles. If we compare the 4-bit input word 1011 to the waveform at the output lead, it is found that both are the same. However, the input is expressed in parallel form, and the output is serial.

7494 4-Bit Shift Register

A 7494 device is an integrated shift register which contains the circuitry for both serial-in, serial-out operation and parallel-in, serial-out operation. This circuit is shown in Fig. 17.5(a) and its pin layout in Fig. 17.5(b).

The parallel-in, serial-out operation of the 7494 shift register circuit is similar to that described for the block in Fig. 17.4(a). However, there are two sets of parallel inputs and a clear lead. Parallel inputs $D_1C_1B_1A_1$ are loaded by the logic level of control input PE_1 and $D_2C_2B_2A_2$ with parallel enable PE_2.

When operated as a parallel-in, serial-out shift register, the 7494 is connected as shown in Fig. 17.6. The enable inputs PE_1 and PE_2 load data into the register when made logic 1. In this circuit PE_2 is wired to logic 0; therefore, parallel inputs $D_2C_2B_2A_2$ are not in use. To load data from the $D_1C_1B_1A_1$ inputs, the register must first be cleared by pulsing the CLR lead to logic 1. Then PE_1 is pulsed to the 1 logic level to parallel load the register.

After loading, data are clocked out of the register in serial form at output S_o. The clock input is positive edge triggered.

EXAMPLE 17.2

The input $D_2C_2B_2A_2 = 1001$ is applied to the parallel inputs of a 7494 register. The CLR lead is first pulsed to logic 1 to clear out the register, and then PE_2 is pulsed to logic 1 to enter the data. Draw the serial output waveform clocked to the S_o output.

SOLUTION:

When the CLR input goes to logic 1, flip-flops A through D are reset. As PE_2 is pulsed to 1, the A and D flip-flops are set, while the other two stay reset. This makes the register content

$$Q_4Q_3Q_2Q_1 = 1001$$

Since Q_4 is logic 1, the S_o output is logic 1:

$$S_o = 1$$

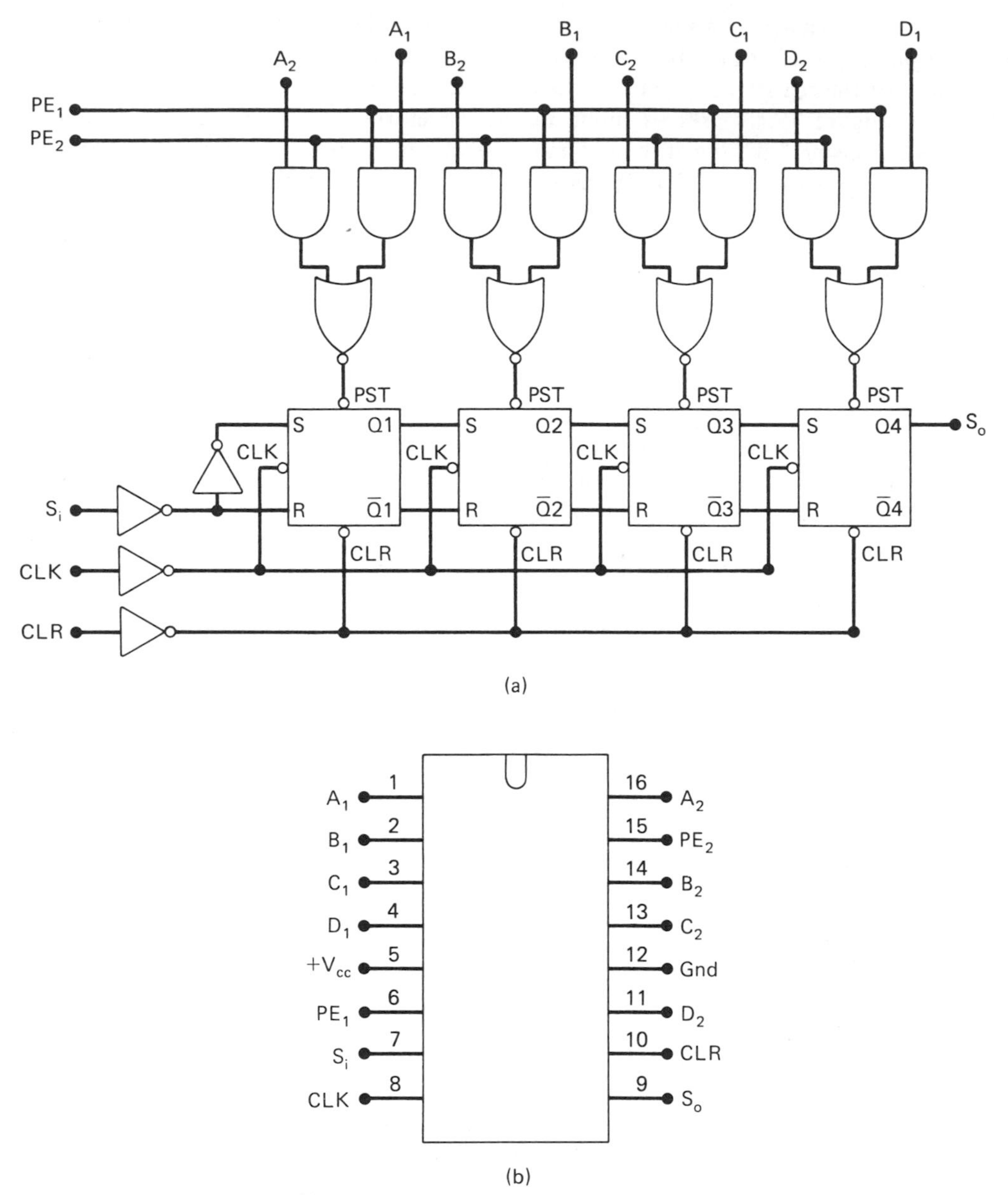

Figure 17.5 (a) 7494 4-bit shift register circuit; (b) 7494 pin layout.

The output waveform is shown in Fig. 17.7. On the first clock pulse, the data content of all flip-flops shifts one bit toward the output. This causes the serial output to switch to logic 0. For the second pulse at *CLK*, output S_o remains at logic 0, and on the third clock pulse it switches back to logic 1.

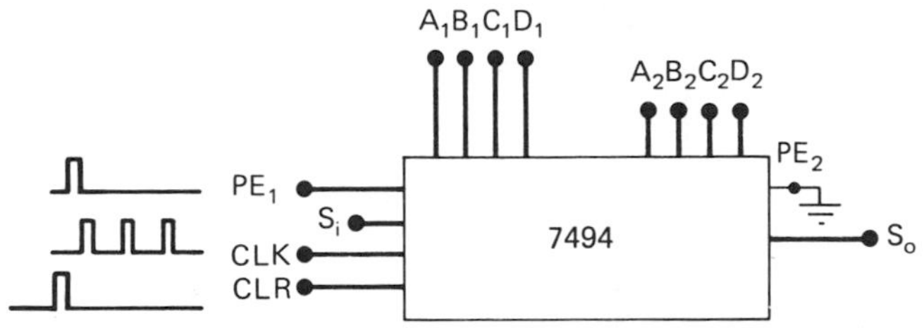

Figure 17.6 7494 parallel-in, serial-out connection.

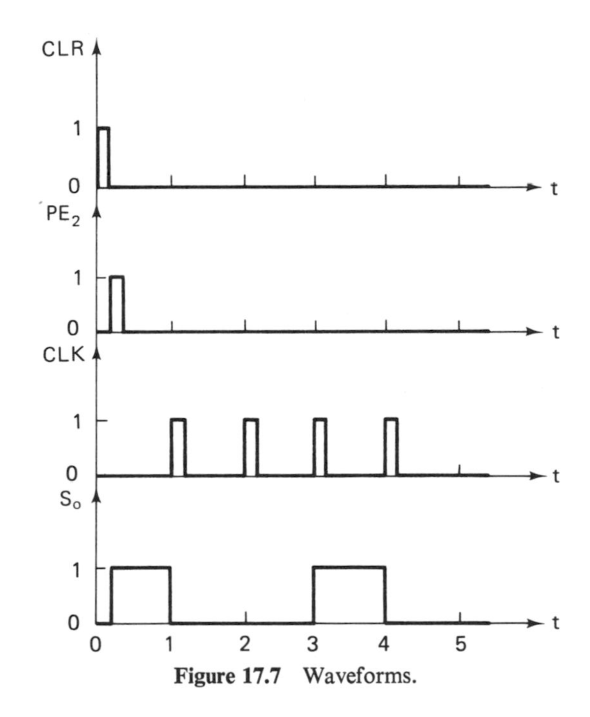

Figure 17.7 Waveforms.

Cascaded Shift Registers

Shift register devices can be cascaded to form large-capacity registers. For instance, Fig. 17.8 shows two 7494 devices connected to make an 8-bit parallel-in, serial-out register circuit. Here, group 1 parallel inputs on both devices are used. The PE_1, CLR, and CLK leads are paralleled, and the serial output of the least significant bit register is inputted to the serial input of the MSB register.

With this circuit, an 8-bit data word can be loaded at the H_1 through A_1 inputs by clearing and then pulsing the PE_1 lead. Serial data are clocked out of the register at the S_o lead on the MSB register.

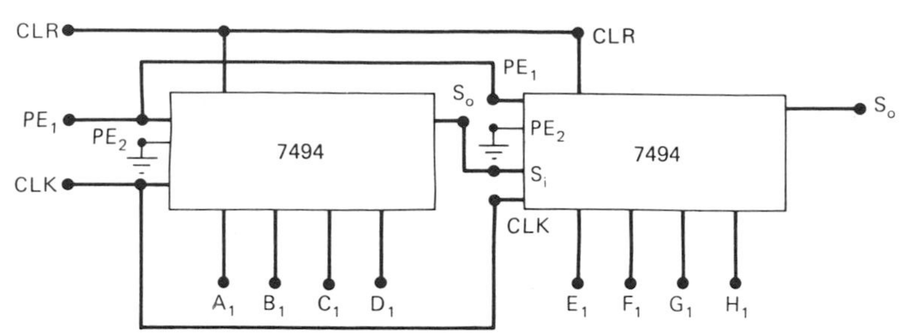

Figure 17.8 Eight-bit parallel-in, serial-out shift register.

17.4 SHIFT RIGHT, SHIFT LEFT REGISTER

In each of the shift register circuits we have considered up to this point, data are shifted from the A flip-flop toward the D flip-flop. When data shift in this direction, the circuit is known as a *shift right register*. A variation on the standard shift right circuit can be made to transfer data in both directions. This is called a *shift right, shift left register.*

Looking at the shift right, shift left register in Fig. 17.9, we find that the register has a serial input S_i and 4-bit parallel output $D_oC_oB_oA_o$. Moreover, there is a control lead marked $MODE$. The logic level at the $MODE$ lead selects shift right or shift left operation. Besides this, there is a separate shift right clock input CLK_R and shift left clock CLK_L.

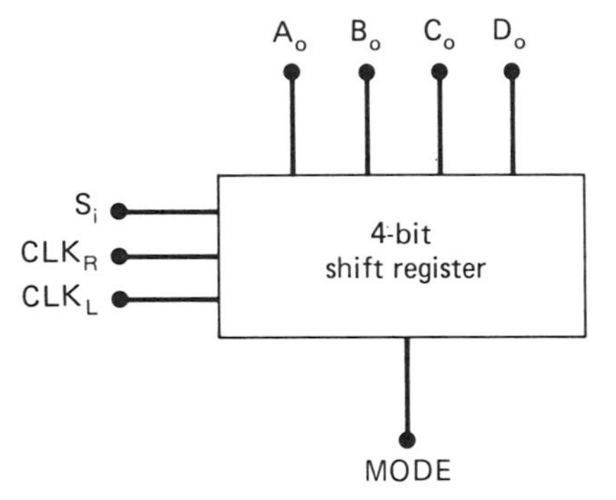

Figure 17.9 Shift right, shift left register block.

Shift Right Operation

For shift right operation, data are entered at the serial input and shifted toward the D parallel output. In this case, $MODE$ must first be set to the correct logic level for shift right operation. After this, data are clocked into the register by pulses applied to the CLK_R lead.

For example, the binary word $DCBA = 0110$ could be entered into the register with four clock pulses. Now the output of the register is $D_oC_oB_oA_o = 0110$.

On the next pulse at CLK_R, the data word shifts one more bit toward the D_o flip-flop. This makes a new output of $D_oC_oB_oA_o = 1100$.

The first output word corresponds to the decimal number 6, and after it is shifted 1 bit to the right the number is 12. In this way, we notice that a shift right of 1 bit causes the register content to be multiplied by 2.

Shift Left Operation

In shift left operation, the CLK_L input is pulsed, and data within the register are shifted through the flip-flops toward the A output. Let us assume the content of the register is again $D_oC_oB_oA_o = 0110$ or decimal 6. As the CLK_L input is pulsed once, data move 1 bit toward the A flip-flop. This gives the output $D_oC_oB_oA_o = 0011$ or 3. This shows that shift left operation divides the register content by 2 instead of multiplying by 2.

7495 Shift Right, Shift Left Register

The 7495 MSI device is a shift right, shift left register like the block given in Fig. 17.9. However, it also has parallel input leads A_i through D_i, as shown in Fig. 17.10(a). A pin layout of the 7495 register is given in Fig. 17.10(b).

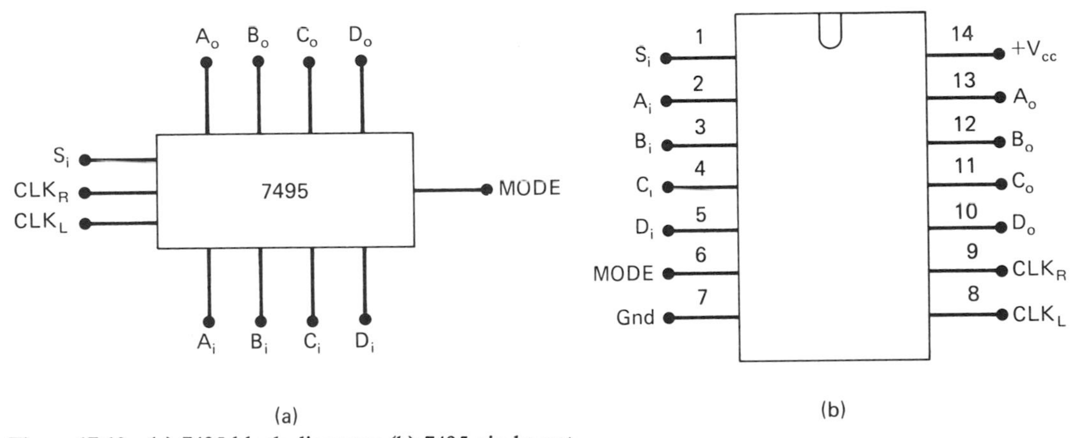

Figure 17.10 (a) 7495 block diagram; (b) 7495 pin layout.

Operation of the register begins by loading a 4-bit word at data inputs $D_iC_iB_iA_i$. To do this, the $MODE$ control must be made logic 1 and the logic level of the CLK_L lead switched from 1 to 0. As data are parallel-loaded, they immediately appear at the $D_oC_oB_oA_o$ parallel outputs.

For shift right operation, the $MODE$ control must be logic 0, and the shift of data between flip-flops occurs on the change of the CLK_R clock signal from 1 to 0. A shift right connection of the 7495 is shown in Fig. 17.11(a).

When the $MODE$ control is made logic 1, the register is set for shift left operation. But for shift left operation, the B_o output must be wired to the A_i input, C_o

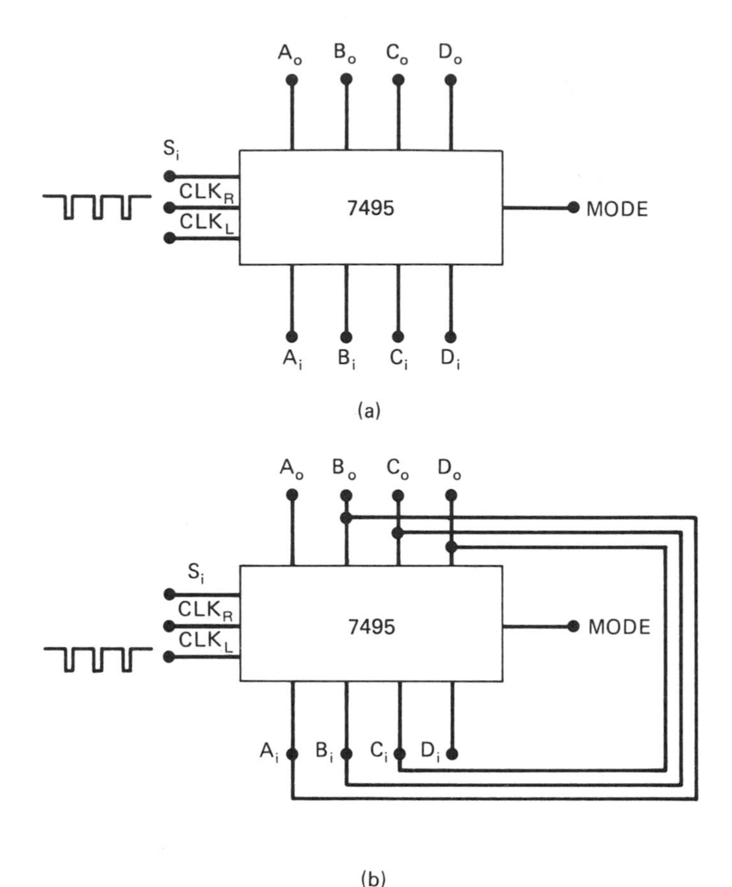

Figure 17.11 (a) 7495 shift right register connection; (b) 7495 shift left register connection.

output to input B_i, and D_o output to C_i, as shown in Fig. 17.11(b). In this circuit, serial data are entered at the D_i input, and data shift in the direction of the A flip-flop output A_o as the CLK_L input is pulsed.

Actually, the shift register connection in Fig. 17.11(b) can be switched between the shift left and shift right modes of operation. This is done by changing the logic level of $MODE$.

EXAMPLE 17.3

The data $D_iC_iB_iA_i = 0001$ are parallel-loaded into a 7495 register set for shift right operation. What is the output of the register after the second pulse at CLK_R?

SOLUTION:

The register is loaded with the binary number for decimal 1:

$$1 = 0001$$

On the first clock pulse at CLK_R, data shift one flip-flop toward the output. This results in the output 0010 or decimal 2:

$$2 = 0010$$

As CLK_R is pulsed once more, the 1 logic level is shifted into the C flip-flop, and the output is made

$$4 = 0100$$

From these results, we find that a shift of one digit to the right multiplies by 2 and a two-digit shift multiplies by 2 again, for a total of 4.

EXAMPLE 17.4

The 7495 shift register in Fig. 17.11(b) has been loaded with the word $D_iC_iB_iA_i = 1000$. Find the serial output at the A_o lead as the shift left clock input CLK_L is pulsed.

SOLUTION:

As the register is loaded, the parallel output is

$$D_oC_oB_oA_o = 1000$$

This makes the serial output at A_o start at logic 0. For the first and second clock pulse the outputs are

$$D_oC_oB_oA_o = 0100$$

and

$$D_oC_oB_oA_o = 0010$$

So we see that the serial output at the A_o lead stays at the 0 logic level. However, on the third clock pulse at CLK_L the logic 1 is loaded into the A flip-flop, and the A_o or serial output changes to the 1 logic level:

$$D_oC_oB_oA_o = 0001$$

The clock and serial output waveforms are shown in Fig. 17.12.

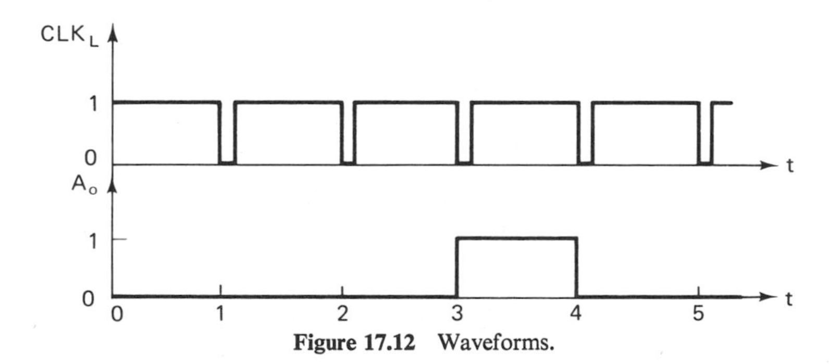

Figure 17.12 Waveforms.

17.5 SHIFT REGISTER COUNTERS

The shift register can be used to make a number of special counter circuits. Two of the most widely used shift register counters are known as the *ring counter* and *Johnson counter circuits.* Here we shall cover these circuits and their operation as well as introduce their input/output waveforms.

The Ring Counter Circuit

A truth table of ring counter operation is shown in Fig. 17.13(a). In this diagram, we see that the circuit is used to shift a logic 1 level from the *A* output toward

Clock pulse	D_o	C_o	B_o	A_o
0	0	0	0	1
1	0	0	1	0
2	0	1	0	0
3	1	0	0	0
4	0	0	0	1
5	0	0	1	0
6	0	1	0	0
7	1	0	0	0

(a)

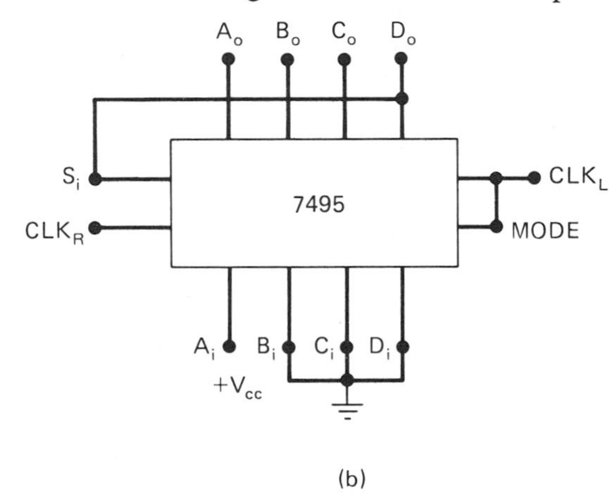

(b)

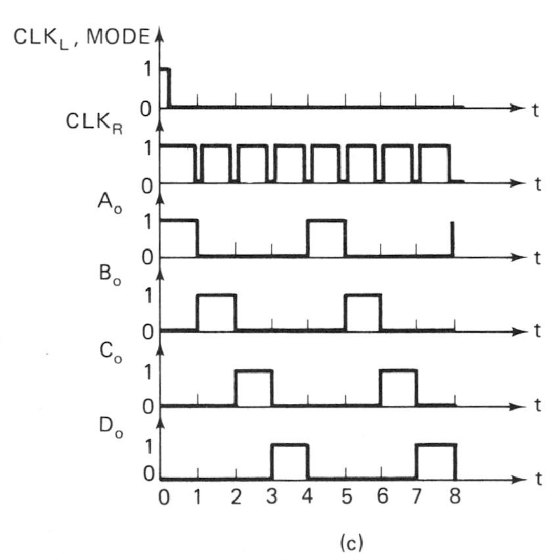

(c)

Figure 17.13 (a) Ring counter truth table; (b) 7495 connected as a ring counter; (c) Input/output waveforms.

the D output as the clock input is pulsed. On the fourth clock pulse, the logic 1 at the D output is returned to the A output, and the count sequence starts all over.

Figure 17.13(b) gives the connection of a 7495 shift right, shift left register as a ring counter circuit. Looking at the circuit, we find that the parallel inputs are $D_iC_iB_iA_i = 0001$. Moreover, the $MODE$ control lead is wired to the clock left input CLK_L. As shown in the input/output waveforms of Fig. 17.13(c), the CLK_L and $MODE$ leads are pulsed to logic 1 to parallel-load the output. This initializes the count at the output to start at $D_oC_oB_oA_o = 0001$, and the serial input S_i is logic 0.

On the first clock pulse at CLK_R, we see that the logic 1 at the A_o output is shifted one flip-flop toward the D_o output. The result is the count $D_oC_oB_oA_o = 0010$. Pulsing CLK_R two more times shifts the 1 to the D_o output. Up to now, the S_i input was at logic 0. But as D_o switches to 1 the logic level of the serial input is changed to logic 1. As the CLK_R lead is pulsed one more time, the 1 at S_i is reloaded into the A_o output, and the count sequence repeats.

In this way, we see that the ring counter is first loaded with a logic 1 into the LSB flip-flop A_o. After this, the logic 1 is just clocked through the register and returned to the serial input for reloading. For this reason, the ring counter is also known as a *recirculating shift register.*

EXAMPLE 17.5

What is the output of the ring counter in Fig. 17.13(b) after the tenth pulse at CLK_R?

SOLUTION:

After the tenth clock pulse at CLK_R, the output of the 7495 register is

$$D_oC_oB_oA_o = 0100$$

The Johnson Counter

A second type of shift register counter is described by the truth table in Fig. 17.14(a). It is known as a Johnson counter circuit. Looking at this table, we notice that as the register is clocked the outputs are loaded 1 bit at a time starting at the A_o output until all outputs are logic 1. At that time, the flip-flop outputs are reset to logic 0 one at a time until all register outputs are again logic 0. After this, the count sequence repeats.

The connection of a 7495 shift right, shift left register for Johnson counter operation is shown in Fig. 17.14(b). This circuit is similar to the ring counter in Fig. 17.13(b) except the D_o output is inverted before it is returned to the S_i input of the register. Input/output waveforms are given in Fig. 17.14(c).

From the waveforms, we see that the register output is first loaded with the count $D_oC_oB_oA_o = 0000$ by a pulse at the CLK_L and $MODE$ leads. This makes the serial input $S_i = \bar{D}_o$ equal to logic 1. As CLK_R is pulsed, the register outputs

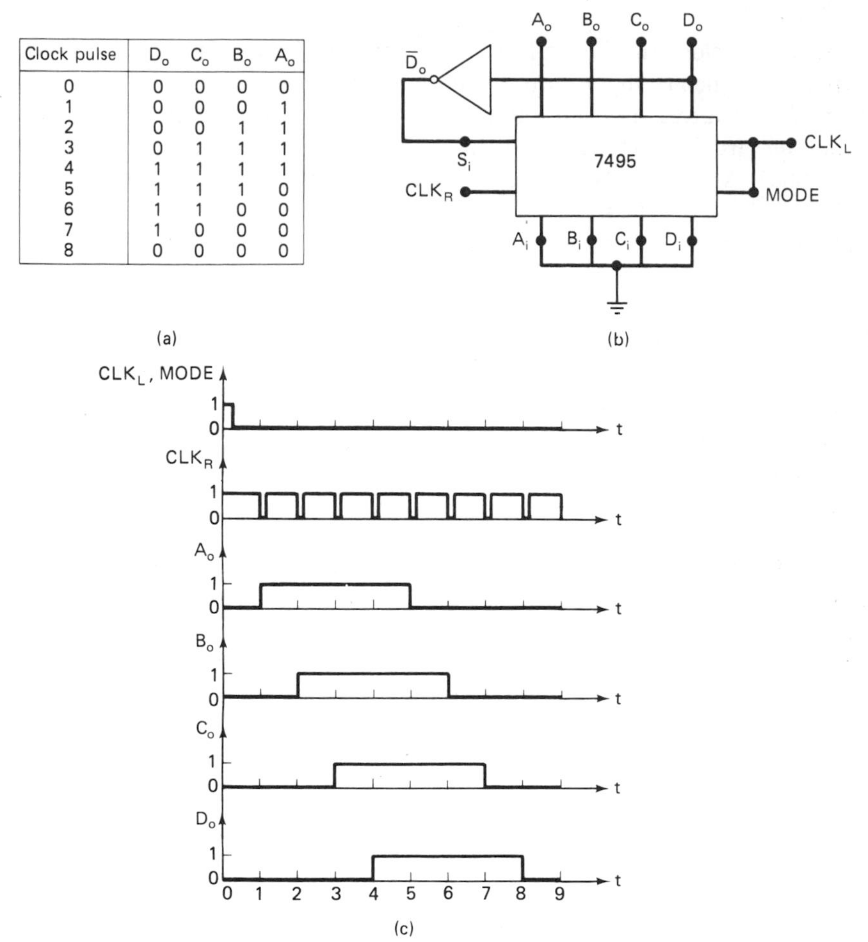

Clock pulse	D_o	C_o	B_o	A_o
0	0	0	0	0
1	0	0	0	1
2	0	0	1	1
3	0	1	1	1
4	1	1	1	1
5	1	1	1	0
6	1	1	0	0
7	1	0	0	0
8	0	0	0	0

Figure 17.14 (a) Johnson counter truth table; (b) 7495 connected as a Johnson counter; (c) Input/output waveforms.

are loaded one after the other with a logic 1. During this time, the D_o output stays at logic 0 and the serial input logic 1.

As D_o switches to logic 1, the S_i input goes to logic 0. On the next four clock pulses, the A_o through D_o outputs are reset one at a time to logic 0.

EXAMPLE 17.6

Find the output of the Johnson counter in Fig. 17.14(b) after the tenth clock pulse at the CLK_R input.

SOLUTION:

After the tenth clock pulse at CLK_R, the output of the Johnson counter is

$$D_oC_oB_oA_o = 0011$$

ASSIGNMENTS

Section 17.2

1. Give a general description of operation for a serial-in, serial-out shift register.

2. Draw the BCD code 0101 in serial form.

3. The input to a 7491 shift register is clocked through the register by a 1-MHz signal. How long is the time delay between serial input and serial output?

4. Figure 17.15 shows the B input and clock waveforms of a 7491 shift register. The A input is held at the 1 logic level. If the flip-flops are all initially reset, draw the waveforms at outputs S_o and $\bar{S}_o$.

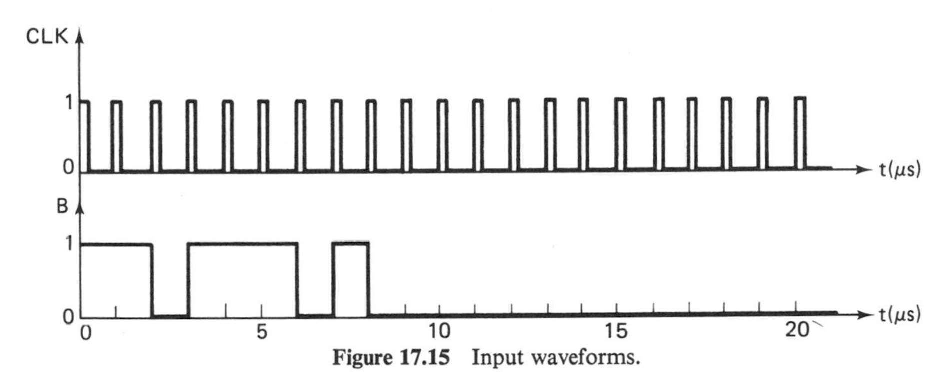

Figure 17.15 Input waveforms.

Section 17.3

5. Describe the general operation of a parallel-in, serial-out shift register.

6. What is the 8-bit parallel binary code for the first 8 bits in the B waveform of Fig. 17.15?

7. The binary data word 1101 is parallel-loaded into a 4-bit serial output shift register. After this, it is clocked through the register with a 100-kHz square wave. Draw the serial output waveform.

8. What must be done just before parallel data can be entered into a 7494 shift register?

9. The 8-bit word $HGFEDCBA = 10101110$ is parallel-loaded into the cascaded 7494 register of Fig. 17.8. Draw the serial output waveform if data are clocked through the register with a 200-kHz signal at the CLK lead.

Section 17.4

10. Describe shift right register operation and shift left operation.

11. A 7495 register is parallel-loaded with the code $DCBA = 1110$. When the *MODE* control is set to logic 0, draw the serial output. The clock is a 1-MHz square wave.

12. The data word $D_iC_iB_iA_i = 1110$ is parallel-loaded into a 7495 register set for shift left operation. Find the parallel output of the register after the first clock pulse, second clock pulse, and third clock pulse.

Section 17.5

13. Make a drawing of how a 7495 shift right, shift left register can be connected as a 4-bit ring counter where the output pulse is to logic 0.

14. Find the output of the ring counter in Problem 13 after the seventh pulse at the CLK_R lead.

15. Draw the input/output waveforms of the circuit drawn in Problem 13 for the first eight clock pulses.

16. What is the output of the 7495 Johnson counter circuit in Fig. 17.14(b) after the twentieth clock pulse?

CMOS Integrated Circuits 18

18.1 INTRODUCTION

Another important group of digital integrated circuits is the CMOS family of devices. The term CMOS stands for *complementary metal oxide semiconductor* logic. This name corresponds to the type of devices and circuitry used within the integrated circuit.

The same basic TTL combinational logic and sequential circuits we studied in earlier chapters are available in standard CMOS integrated circuits. In this chapter, we begin with an introduction to the switching devices and circuits used to make CMOS circuitry. This is followed by a survey of CMOS integrated circuit devices.

The topics covered in this chapter are as follows:

1. CMOS devices and circuitry
2. The CMOS NAND gate and electrical characteristics
3. Other CMOS gates
4. CMOS combinational logic networks
5. Integrated CMOS flip-flops
6. CMOS sequential circuitry

18.2 CMOS DEVICES AND CIRCUITRY

Our study of CMOS or complementary metal oxide semiconductor logic begins with an introduction to its building block, the *field effect transistor*, and its use as an electronic switch.

Field Effect Transistor

The electronic switching device used to make CMOS integrated circuitry is the field effect transistor or FET. The FET device in CMOS ICs is made with a metal oxide process. *Metal-oxide-type FET* or MOSFET devices can be made with what is known as an *N-channel* structure or *P-channel* structure. The schematic symbol of an *N*-channel MOSFET is shown in Fig. 18.1(a) and that of a *P*-channel MOSFET in Fig. 18.1(b).

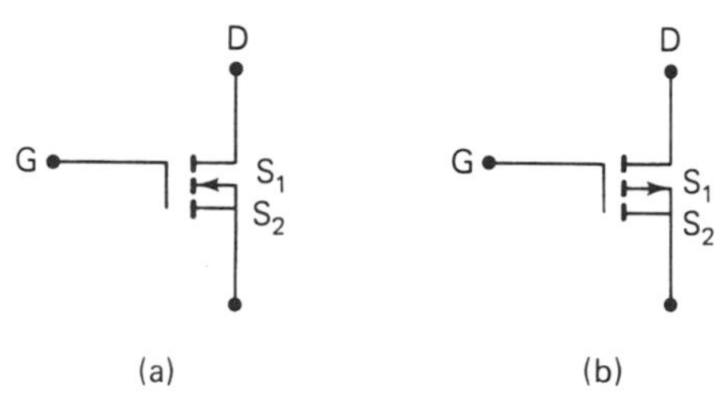

Figure 18.1 (a) *N*-channel enhancement mode FET; (b) *P*-channel enhancement mode FET.

Looking at the symbols in Fig. 18.1, we notice that the FET device has four leads. The lead marked *D* is the *drain* of the FET; *G* stands for the *gate* of the device, S_1 the *substrate* lead, and S_2 the *source* lead. Normally, the substrate and source leads are tied together. For this reason, they are often labeled with a single *S* at the source location.

MOSFET devices can be made with either *enhancement mode* electrical characteristics or *depletion mode* characteristics. However, the enhancement mode device can be turned ON and OFF with the same polarity voltage and is exclusively used in CMOS digital integrated circuits.

MOSFET Switch Circuit

The circuit in Fig. 18.2(a) is a *MOSFET switch circuit*. In this circuit, an *N*-channel, enhancement mode FET Q_1 is used as the electronic switch. The drain of the FET is wired through a resistor to the $+V_{cc}$ power supply. On the other hand, the source is directly connected to ground.

Input signal V_i is applied to the gate lead of the FET and an output V_o is produced at the drain. The logic levels of V_i and V_o are measured with respect to ground.

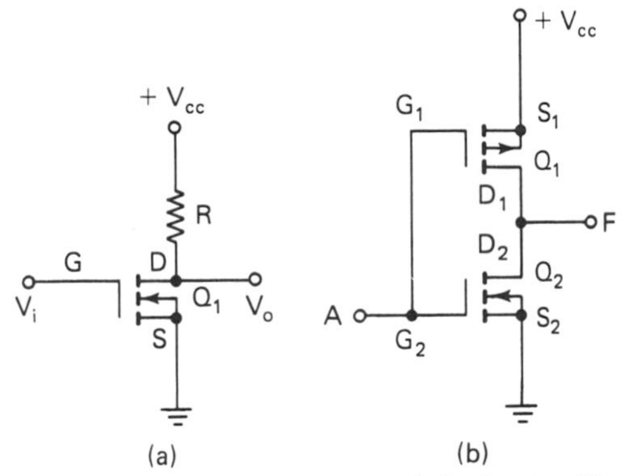

Figure 18.2 (a) MOSFET switch circuit; (b) CMOS switch circuit.

When the input V_i is 0 V, the gate lead of the FET is electrically at ground. This makes the FET turn OFF and act like an open circuit from drain to source. In turn, the output voltage V_o switches to $+V_{cc}$ volts:

$$V_i = 0 \text{ V}$$
$$V_o = +V_{cc}$$

Using positive logic, the input is logic 0 and output binary 1.

If the input V_i is set to voltage $+V_{cc}$, the FET switches ON. Now it operates like a short circuit from drain to source, and output V_o is held at 0 V:

$$V_i = +V_{cc}$$
$$V_o = 0 \text{ V}$$

This makes the input binary 1 and output logic 0.

In this way, we see that the input and output are always opposite logic levels. Therefore, the FET switch circuit in Fig. 18.2(a) operates like an inverter gate.

EXAMPLE 18.1

What is the value of the output voltage for the MOSFET switch circuit in Fig. 18.2(a) when the input voltage level is 0 V? The circuit operates off a +5-V dc power supply.

SOLUTION:

An input of 0 V to the gate lead of Q_1 switches the FET to the OFF state, and the output goes to +5 V:

$$V_o = +5 \text{ V}$$

CMOS Switch Circuit

A second switch circuit made with MOSFET devices is given in Fig. 18.2(b). In this circuit configuration, a *P*-channel FET Q_1 is used to source the output *F* to $+V_{cc}$ when it turns ON and an N-channel FET Q_2 to sink the output to ground. The P-channel FET replaces the load resistor R of the MOSFET switch circuit in Fig. 18.2(a).

This circuit structure which has both P-channel and N-channel FET devices is known as a *complementary circuit* and leads to the name CMOS for the complementary metal oxide logic family.

In the CMOS switch circuit, a logic 0 at the A input causes Q_1 to turn ON and Q_2 to switch OFF. This sources the F output to the $+V_{cc}$ supply through Q_1 and gives a logic 1 output:

$$A = 0$$

$$F = 1$$

As the A input is changed to logic 1, Q_1 goes OFF and Q_2 ON. In this way, the F output is sinked to ground through Q_2, and the output is held at the 0 logic level:

$$A = 1$$

$$F = 0$$

18.3 CMOS NAND GATE

The complementary connection of N-channel and P-channel MOSFET devices can be used to make each of the basic logic gate circuits. In this section, we begin our study of the CMOS integrated circuit family with the NAND gate.

74C00 Two-Input NAND Gate

Figure 18.3(a) shows the basic CMOS NAND gate circuit configuration. Here parallel-connected MOSFETs Q_1 and Q_2 are used to source the output F

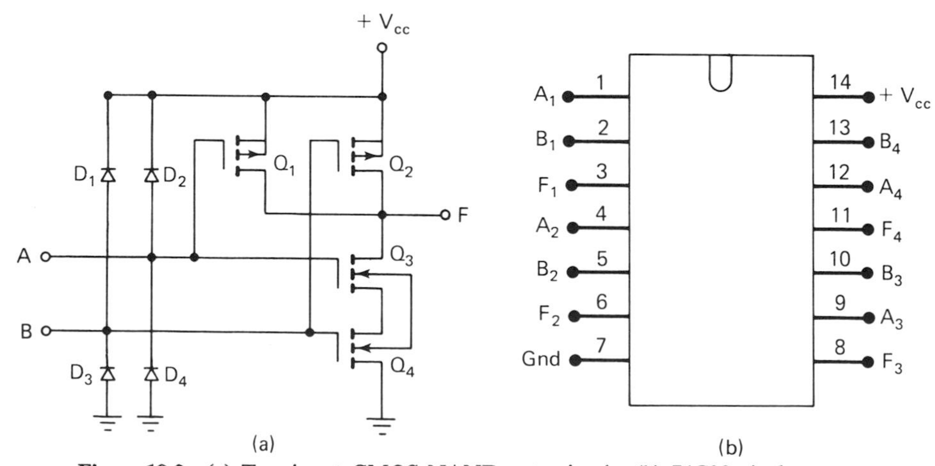

Figure 18.3 (a) Two-input CMOS NAND gate circuit; (b) 74C00 pin layout.

to the $+V_{cc}$ supply. If either MOSFET is ON, output F is logic 1. This condition occurs if the A or B input is at logic 0.

On the other hand, F is returned to ground through series-connected MOSFETs Q_3 and Q_4. For the output to be logic 0, both devices must switch ON. This condition happens when A and B are both logic 1.

In this way, we see that the output is logic 1 if at least one input is logic 0, and F switches to 0 when both A and B are made logic 1. Operation of this circuit is consistent with that of a NAND gate.

This NAND gate circuit is available in the 74C00 CMOS integrated circuit. The 74C00 device is a quad two-input NAND gate with the pin layout of Fig. 18.3(b). Looking at this diagram, we find that the pin layout is the same as for the 7400 TTL NAND gate.

The 74C00 CMOS NAND gate can be operated off a +5-V dc power supply, and typical power consumption is 10 nW/device. Comparing this value to the tens of milliwatts needed by a TTL gate, we find that CMOS devices draw about one-millionth the power needed to operate an equivalent TTL device. Low power consumption is the major advantage of CMOS integrated circuits.

Electrical Switching Characteristics

The electrical switching characteristics of a 74C00 integrated circuit are given in the table of Fig. 18.4. The table includes the input/output logic levels, propagation delays, and noise immunity ratings.

Characteristic	Rating	Value	Condition
Input and output logic levels	V_{OH}	4.5 V	Minimum
	V_{OL}	0.5 V	Maximum
	V_{IH}	3.5 V	Minimum
	V_{IL}	1.5 V	Maximum
Propagation delays	t_{PLH}	90 ns	Maximum
	t_{PHL}	90 ns	Maximum
Noise immunity	NI	1 V	Minimum

Figure 18.4 74C00 switching characteristics.

Here we see that the minimum high input and output logic levels are $V_{IH} = 3.5$ V and $V_{OH} = 4.5$ V. These values are closer to the +5-V supply $+V_{cc}$ than for TTL circuits. Moreover, the maximum low logic levels $V_{IL} = 1.5$ V and $V_{OL} = .5$ V are higher than on TTL devices.

Along with this, the propagation delay of a signal through a CMOS gate is much longer than for TTL devices. For instance, the switch of the output from the 0 level to 1 is delayed by a maximum of 90 ns after the input logic levels are set. So TTL gates which take about 20 ns are four times faster than the similar CMOS device.

The minimum noise immunity of a CMOS device is 1 V. Compared to the .4-V noise immunity of a TTL gate, we find a second improvement or advantage of CMOS circuitry. That is, CMOS devices are less sensitive to noise than those made with TTL circuitry.

EXAMPLE 18.2

Both the A and B inputs of a NAND gate on a 74C00 device are 4.7 V; what is the maximum value of output voltage?

SOLUTION:

A value of 4.7 V at the inputs of the gate make A and B logic 1. For this reason, the output F is logic 0, and the maximum value of output voltage is .5 V:

$$V_o = .5 \text{ V}$$

18.4 OTHER CMOS LOGIC GATE DEVICES

As with TTL ICs, other gate circuits are available in standard CMOS devices. Some examples are the 74C02 quad two-input NOR gate, 74C04 hex-inverter, and 74C10 triple three-input NAND gate integrated circuits. The pin layouts of these three devices are identical to their equivalent 7400 TTL IC.

Each of these devices has the same electrical switching characteristics as the 74C00 NAND gate device. These switching characteristics are given in the table of Fig. 18.4.

EXAMPLE 18.3

The A input of a 74C04 inverter is 4.7 V. What is the maximum value of output voltage?

SOLUTION:

An input of 4.7 V on a CMOS inverter is a logic 1, and the output switches to logic 0. The maximum output value for logic 0 is .5 V.

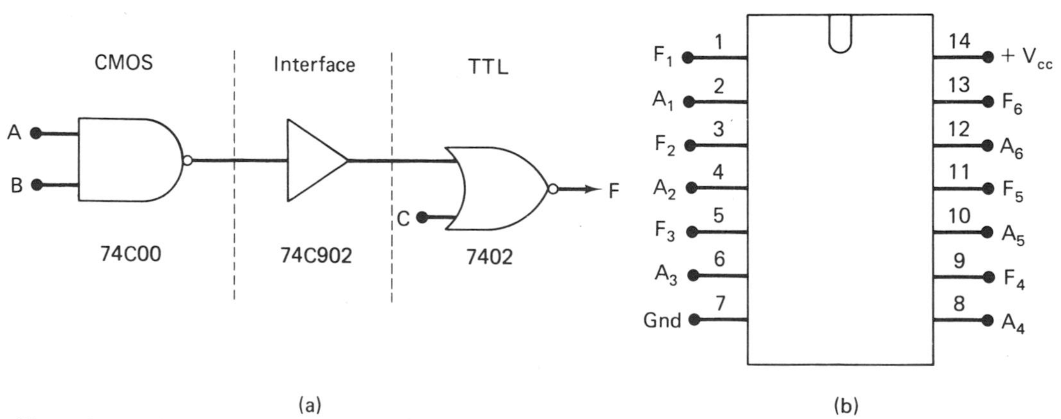

Figure 18.5 (a) CMOS to TTL interface; (b) 74C902 pin layout.

74C902 CMOS to TTL Buffer

In many electronic systems, CMOS and TTL devices are used in different sections of circuitry. However, the outputs of a standard CMOS device cannot drive an input of a TTL device. For this reason, buffers must be used to interface the two types of circuitry.

Special interface devices are available in the 74C00 CMOS family of integrated circuits. Figure 18.5(a) shows the interface connection of a 74C00 CMOS NAND gate to a standard 7402 TTL NOR gate. The interface gate used is a 74C902 CMOS to TTL buffer. This gate has a fan out of two TTL loads. A pin layout of the 74C902 device is given in Fig. 18.5(b).

18.5 CMOS COMBINATIONAL LOGIC CIRCUITS

In our study of CMOS logic gates, we found that the same basic circuits introduced as TTL integrated circuits are available in the 74C00 CMOS family. The same is true for medium-scale combinational logic networks.

For instance, the 74C86 quad exclusive-OR gate of Fig. 18.6(a) has the same

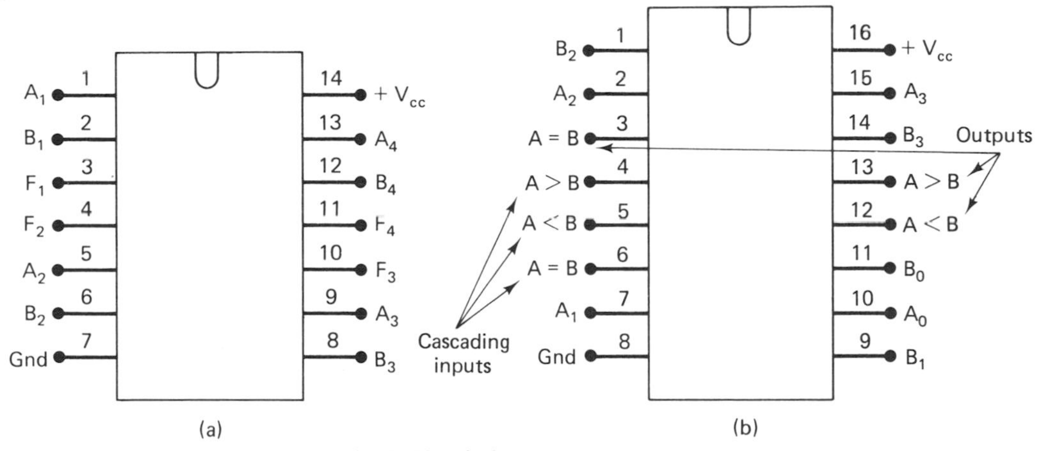

Figure 18.6 (a) 74C86 pin layout; (b) 74C85 pin layout.

operation as the 7486 TTL device. Another closely related device is the 4-bit magnitude comparator of Fig. 18.6(b). This device provides the same data comparison capability as the 7485 TTL integrated circuit. Moreover, the 74C85 device can also be cascaded to compare larger data words. The major advantage of a CMOS device over its equivalent TTL device is the low power consumption. The pin layout of both these devices are different from the corresponding TTL ICs.

Two other CMOS devices containing MSI combinational logic networks are the 74C42 and 74C83 IC. The 74C42 is a low-power BCD to decimal decoder and the 74C83 a 4-bit binary adder.

18.6 INTEGRATED CMOS FLIP-FLOPS

The standard *D*-type and *J-K* flip-flop circuits are available in CMOS integrated circuits. The 74C73 and 74C74 are two such devices. A pin layout of the 74C73 dual *J-K* flip-flop is shown in Fig. 18.7(a).

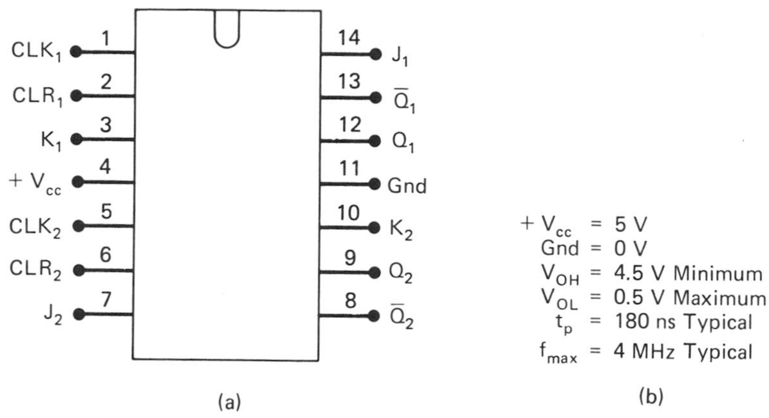

Figure 18.7 (a) 74C73 pin layout; (b) Electrical characteristics.

The 74C73 *J-K* flip-flop operates the same as its equivalent TTL device, the 7473. However, its electrical switching properties are somewhat different. In Fig. 18.7(b), a list of these characteristics is given. Looking at these values, we see that the power supply and output logic levels are consistent with other devices in the 74C00 CMOS family.

For the operation of a flip-flop, two important switching properties are its propagation delay t_p and maximum operating frequency f_{max}. The 74C73 has a 180-ns propagation delay of signals transferred between input and output. This value is much longer than the 16-ns delay found in the 7473 TTL device. Moreover, the 4-MHz maximum operating frequency of the 74C73 is five times lower than the 20-MHz rating of the 7473 TTL circuit.

74C173 Quad Three-State *D*-Type Flip-flop

Another flip-flop available in the 74C00 CMOS family is the 74C173 device shown in Fig. 18.8(a). This integrated circuit contains a quad *D*-type flip-flop circuit. In this device, all flip-flops are clocked from a single lead marked *CLK* and cleared with the *CLR* lead. The clear lead operates off a 1 logic level.

The 4-bit data inputs are labeled *A*, *B*, *C*, and *D*. On the positive edge of the clock pulse, the input data word *DCBA* is loaded into the flip-flops and produces a 4-bit output word $Q_D Q_C Q_B Q_A$.

However, to work in this way both the input and output circuitry must be enabled. The operation of the input circuitry is controlled by the *input disable leads* ID_1 and ID_2. To enable the inputs, both *ID* leads must be logic 0. In many

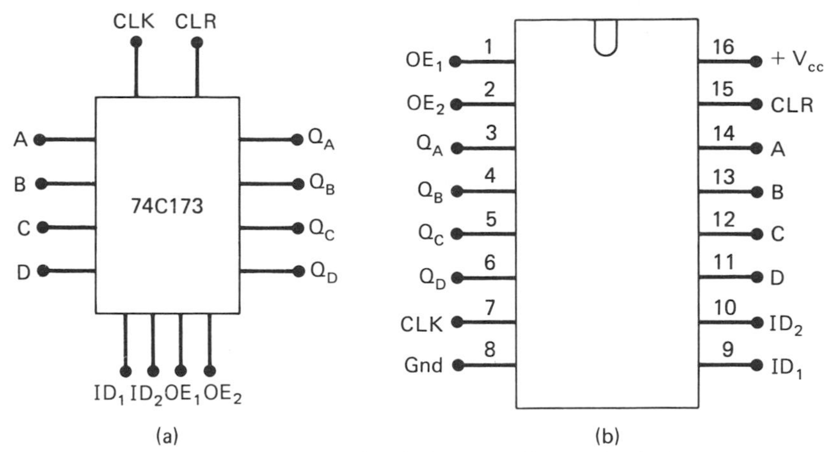

Figure 18.8 (a) 74C173 block diagram; (b) 74C173 pin layout.

applications, the input disable leads are directly wired to ground to permanently enable the inputs.

The Q_D, Q_C, Q_B, and Q_A outputs of the 74C173 are three-state circuits. To set the outputs into the hi-Z state, either of the two *output enable leads*, OE_1 and OE_2, can be set to logic 1. When this condition occurs, the four outputs act like open circuits. A pin layout of the 74C173 is shown in Fig. 18.8(b).

The three-state output operation is important in systems where data are to be parallel-fed into a processing unit. In this case, the outputs of several devices can be directly wired in parallel and sent to the processor.

18.7 CMOS SEQUENTIAL CIRCUITS

Many medium-scale sequential circuits are provided in the 74C00 CMOS family. Two examples, the 74C90 decade counter and 74C95 4-bit shift left, shift right register operate similar to the 7490 and 7495 TTL devices, respectively.

74C161 Presettable Binary Counter

A versatile 74C00 CMOS family counter circuit is the 74C161 device shown in Fig. 18.9. This integrated circuit contains a 4-bit presettable binary counter circuit.

The 74C161 counter works a lot like the standard 4-bit binary counter; however, its output can be preset to any binary count. From this value, the count can be incremented by pulsing the *CLK* lead.

As an example, let us assume the binary input is $D_iC_iB_iA_i = 0011$. This count can be loaded into the circuit by pulsing the load lead (*LD*) to logic 0. In this way, the count sequence at output $D_oC_oB_oA_o$ starts at the binary number for 3 instead of 0.

Now the counter is preset and ready for normal up-counter operation. On the

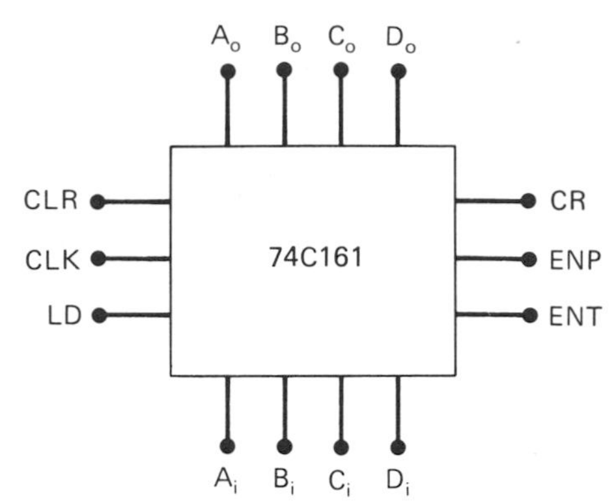

Figure 18.9 74C161 block diagram.

positive edge of the next clock pulse, the output count changes to $D_oC_oB_oA_o = 0100$ or decimal 4. As the *CLK* input is further pulsed, the circuit counts up through binary 15. At the next clock pulse, the counter output resets to decimal 0, or the preset count of 3 can be reloaded.

At any point in the sequence, the output count can be cleared to $D_oC_oB_oA_o = 0000$ by pulsing the *CLR* lead to logic 0. For normal counter operation, the *ENP* and *ENT* leads must be set to logic 1.

ASSIGNMENTS

Section 18.2

1. What are the two structure MOSFET devices manufactured?

2. Are FET devices with enhancement mode or depletion mode characteristics used in CMOS digital ICs?

3. What is the output voltage V_o of the switch circuit in Fig. 18.2(a) if the power supply $+V_{cc}$ is $+5$ V and the input voltage V_i is also $+5$ V?

4. The CMOS switch circuit in Fig. 18.2(b) is powered by a +5-V dc supply. Which FETs are ON and OFF when the input at *A* is 0 V? What is the voltage at output *F*?

Section 18.3

5. The inputs of a 74C00 NAND gate are both logic 1. What is the binary output?

6. The inputs of a 74C00 NAND gate are both logic 1, and then one input changes to the 0 logic level. What is the maximum amount of time that can elapse before the output switches to its new logic state?

7. The input voltages on a 74C00 gate are .15 and 4.8 V. Find the minimum voltage value that can be measured at the output.

8. Figure 18.10 shows the input waveforms to a 74C00 NAND gate. Draw the output waveform.

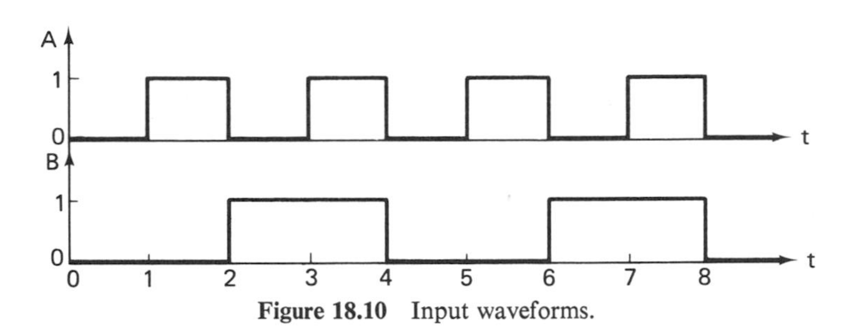

Figure 18.10 Input waveforms.

Section 18.4

9. The inputs on a 74C10 NAND gate are measured to be .1, 4.85, and 4.9 V. Is the output binary 1 or 0?

10. The signal at the input of a 74C04 inverter changes from logic 0 to 1. How long is the maximum time it can take the output to change logic levels?

11. The waveforms of Fig. 18.10 are applied to the inputs on a 74C02 NOR gate. Make a drawing of the output waveform.

Section 18.5

12. The inputs to one gate on a 74C86 device are given by the waveforms in Fig. 18.10. Construct the waveform at the output of the gate.

13. A 74C42 decoder has an input of $DCBA = 0111$. Which of its outputs is at the 0 logic level?

Section 18.6

14. Both the *J* and *K* inputs on a 74C73 *J-K* flip-flop are wired to +5 V. How long is the delay between the clock input and the toggle mode change of output logic levels?

15. Make a block drawing of the 74C173 *D*-type flip-flop set for 4-bit data latch operation. The input disable leads are to be permanently enabled. However, the outputs are to be enabled by a 1 logic level of signal *DS*.

Section 18.7

16. A 74C161 counter is loaded with the binary code $D_iC_iB_iA_i = 0100$. What is the output count after four pulses at the clock input?

The Microcomputer 19

19.1 INTRODUCTION

The microcomputer is a recent, important development in digital equipment design. A microcomputer is a miniature digital information processing system made with large-scale integrated circuit devices. An example of a digital system based on the use of a microcomputer is the electronic calculator.

In this chapter, we shall introduce the PPS-4 microcomputer, its operation and the large-scale integrated (LSI) devices used to construct a microcomputer circuit. A list of the topics covered in this chapter is as follows:

1. Microcomputer block diagram
2. The clock generator
3. Central processing unit
4. Read only memory
5. Random access memory
6. General-purpose input/output
7. The program
8. Operating cycles

19.2 THE MICROCOMPUTER SYSTEM

A microcomputer is a small digital computer data processing system. It contains all the elements of a large computer, but each is provided with a smaller capacity.

The block diagram of Fig. 19.1 shows the structure of a Rockwell International, Inc. PPS-4 microcomputer circuit. Looking at this diagram, we find that the basic building blocks of a microcomputer are the *clock*, *central processing unit*, *read only memory*, *random access memory*, and *input/output section*.

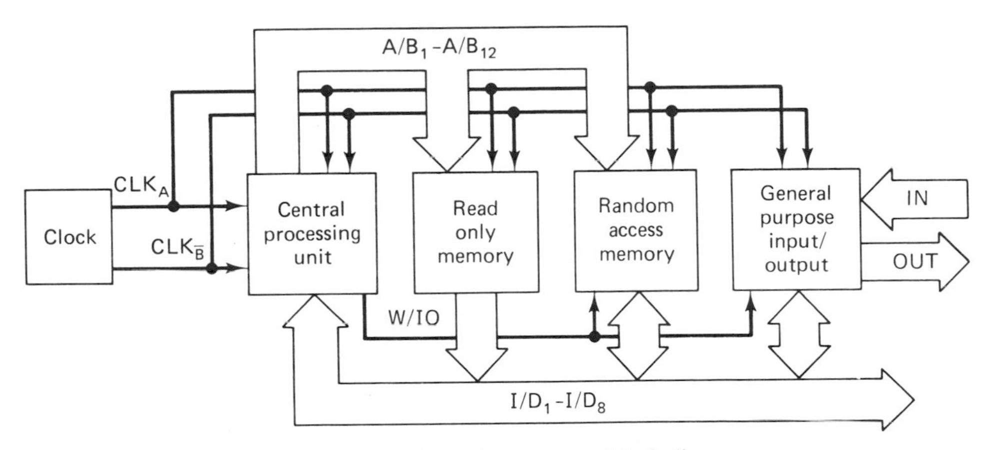

Figure 19.1 PPS-4 microcomputer block diagram.

Each of these sections can be implemented with a separate LSI chip. However, sections like the read only memory and random access memory can use many devices. Moreover, a general purpose input/output device can be used for each peripheral device.

Block Functions

For a microcomputer circuit, each block provides a separate function of its operation. We shall begin our study with an overview of the use of each block.

In the central processing unit, data operations and logic decisions are performed. For example, it has the ability to add binary numbers and make OR, AND, and exclusive-OR decisions. Besides the basic arithmetic and logic operations, the central processing unit can perform many data transfer operations, control transfer operations, and input/output operations.

The sequence of operations performed by the central processing unit is controlled by a set of instructions called a *microprogram*. This program is permanently stored in the read only memory part of the microcomputer and instructions are read into the central processing unit for execution. The program held in read only memory is known as the *firmware* of the system. A typical microprogram can contain several thousand *microinstructions*.

Random access memory is used to store data and results for use by the microcomputer. For this reason, it is sometimes called a *scratch pad memory*. In a microcomputer, we can *write* data from the central processing unit into a storage location of the random access memory or *read* data held at a memory location back to the central processing unit.

Data are transferred between the microcomputer and peripheral devices through an interface section called the input/output. For instance, a keyboard can be used to enter data to the microcomputer through one input/output device, and data can be output to a seven-segment display unit with a similar device.

The clock circuit produces timing signals that determine the operating cycles of the microcomputer. These clock signals synchronize the operation of the different sections in the microcomputer to the central processing unit.

Buses and Control Signals

In Fig. 19.1, we have indicated the electrical connection of the different circuit blocks with *bus notation*. A bus is a set of parallel-connected wires. Instead of drawing each wire separately, an arrow notation is used to indicate the direction of information transfer over the lines, and it is labeled to give the number and meaning of the wires.

As an example, let us take the *address bus A/B* of the microcomputer. This bus contains 12 parallel lines A/B_1 through A/B_{12}, and it is used to carry a 12-bit address word from the central processing unit to the read only memory or random access memory. The address code put on this bus is used to select a storage location in one of these two memories. The time at which the address occurs in the microcomputer cycle determines if the code belongs to a location in the read only memory or random access memory.

The *instruction/data bus I/D* is used to carry program instructions from the read only memory to the central processing unit. It is also used to transfer data between the central processing unit and random access memory. This bus consists of eight lines marked I/D_1 through I/D_8. Since data can go either way on these wires, this bus is marked with an arrow on each end. This notation indicates that it is a *bidirectional bus*. Another use of the *I/D* bus is to carry instructions and data to the input/output section.

The line marked *W/IO* supplies a control signal important to PPS-4 microcomputer operation. This is the *write/input-output enable* signal. Depending on the time in the operating cycle, the logic level of *W/IO* indicates that a random access memory or input/output operation is in progress and if the operation is a read or write to memory.

The signals CLK_A and CLK_B are the two clock signals of the microcomputer. These timing signals are used to synchronize the operation of all devices to the central processing unit.

Parallel Processing

For a PPS-4 microcomputer, 12-bit address words, 8-bit instruction words, and 4-bit data words are transferred between blocks over bus lines. In each case, all bits of the code or data are produced at the same time, and each bit is carried over a separate wire to its destination. For instance, a 12-bit address word is output by the central processing unit on bus lines A/B_1 through A/B_{12} and inputted to the read only memory.

This type of operation is known as *parallel data transfer.* For this reason, the PPS-4 is said to be a *parallel processing microcomputer.*

Logic Levels

The PPS-4 microcomputer works off a -12-V dc supply V_{DD} and $+5$-V supply V_{SS}. Besides this, it uses negative logic for input/output logic levels. The ideal levels are $1 = -12$ V and $0 = +5$ V. However, the minimum 1 and 0 output logic levels are specified to be -2.5 and $+4$ V, respectively.

19.3 THE CLOCK GENERATOR

The operation of the LSI circuits in a PPS-4 microcomputer is synchronized by the timing signals CLK_A and $CLK_{\bar{B}}$. These two clock waveforms are produced by a 10706 crystal-controlled oscillator device. A pin layout of the 10706 is shown in Fig. 19.2(a).

In Fig. 19.2(b), a block diagram of the clock generator is shown. Here we see that a 3.58-MHz crystal must be externally connected at the *XTAL* leads of the clock circuit to set its operating frequency. Besides this, power supply voltages $V_{SS} = +5$ V and $V_{DD} = -12$ V must be applied to the IC.

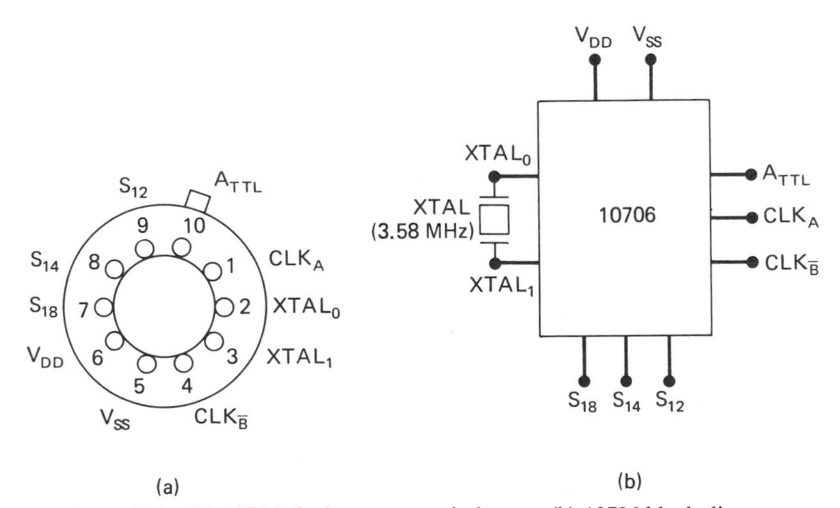

Figure 19.2 (a) 10706 clock generator pin layout; (b) 10706 block diagram.

The output waveforms at the CLK_A and $CLK_{\bar{B}}$ leads of the clock generator circuit are given in Fig. 19.3(a). Looking at this diagram, we find that the A clock signal is a 200-kHz square wave with period equal to 5 μs. Moreover, output $\bar{B}$ produces a pulse waveform corresponding to each of the four phases in the 5-μs clock cycle. These phases are labeled ϕ_1, ϕ_2, ϕ_3, and ϕ_4.

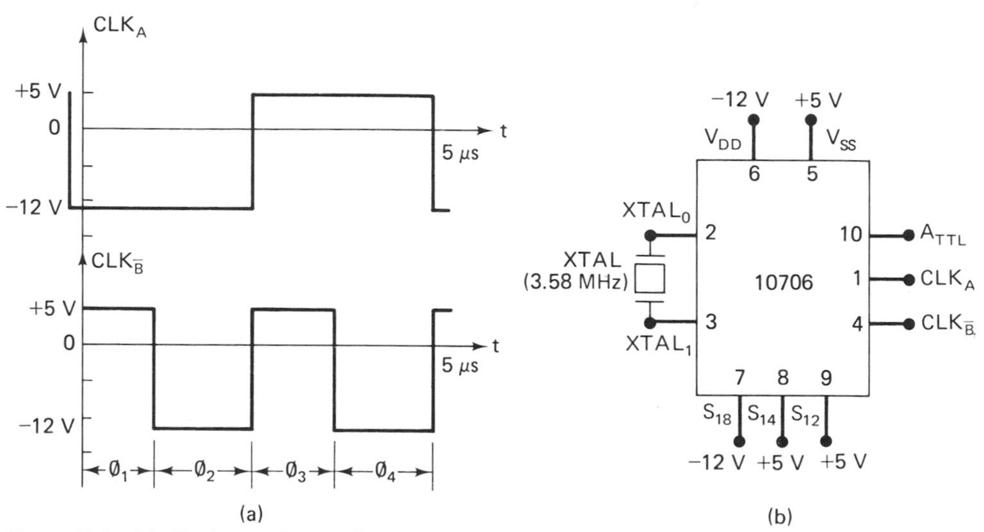

Figure 19.3 (a) Clock waveforms; (b) 200-kHz clock connection.

Both CLK_A and $CLK_{\bar{B}}$ are shown to switch between ideal negative logic levels of $1 = -12$ V and $0 = +5$ V. The minimum levels of these signals are slightly different from those for the bus lines. Here, we get a minimum of -5 V for logic 1 and $+4.5$ V for the 0 level. The CLK_A timing signal is produced with TTL logic levels at the A_{TTL} output lead.

The frequency of the A clock output can be changed to one of three values by setting the logic levels at frequency select straps S_{12}, S_{14}, and S_{18}. The notation used to mark these leads indicates the 3.58-MHz natural frequency of the crystal will be divided by a factor of 12, 14, or 18, respectively, within the clock generator. When a division factor is selected, the corresponding strap is made logic 1. This is done by wiring the lead to the V_{DD} or -12-V power supply.

For example, to select normal 200-kHz output the crystal frequency must be divided by 18. To do this, S_{18} is connected to -12 V as shown in Fig. 19.3(b).

19.4 THE CENTRAL PROCESSING UNIT

The central processing unit or CPU in a PPS-4 microcomputer is the 10660 large-scale integrated circuit device. In a microcomputer, the CPU performs all arithmetic and logic operations. Moreover, it controls the operation of the other devices in the microcomputer system. The CPU in a microcomputer is also known as a microprocessor.

Address and Instruction/Data Buses

A pin layout and block diagram of the 10660 device are shown in Figs. 19.4(a) and (b), respectively. From the block diagram, we see that the CPU works off a 12-bit address word, 8-bit instruction word, and 4-bit data word. The address bus leads are labeled A/B_1 through A/B_{12}. Of these, A/B_{12} is the MSB and A/B_1 the LSB. Address words are sent to both the read only memory and random access memory in parallel over these bus lines.

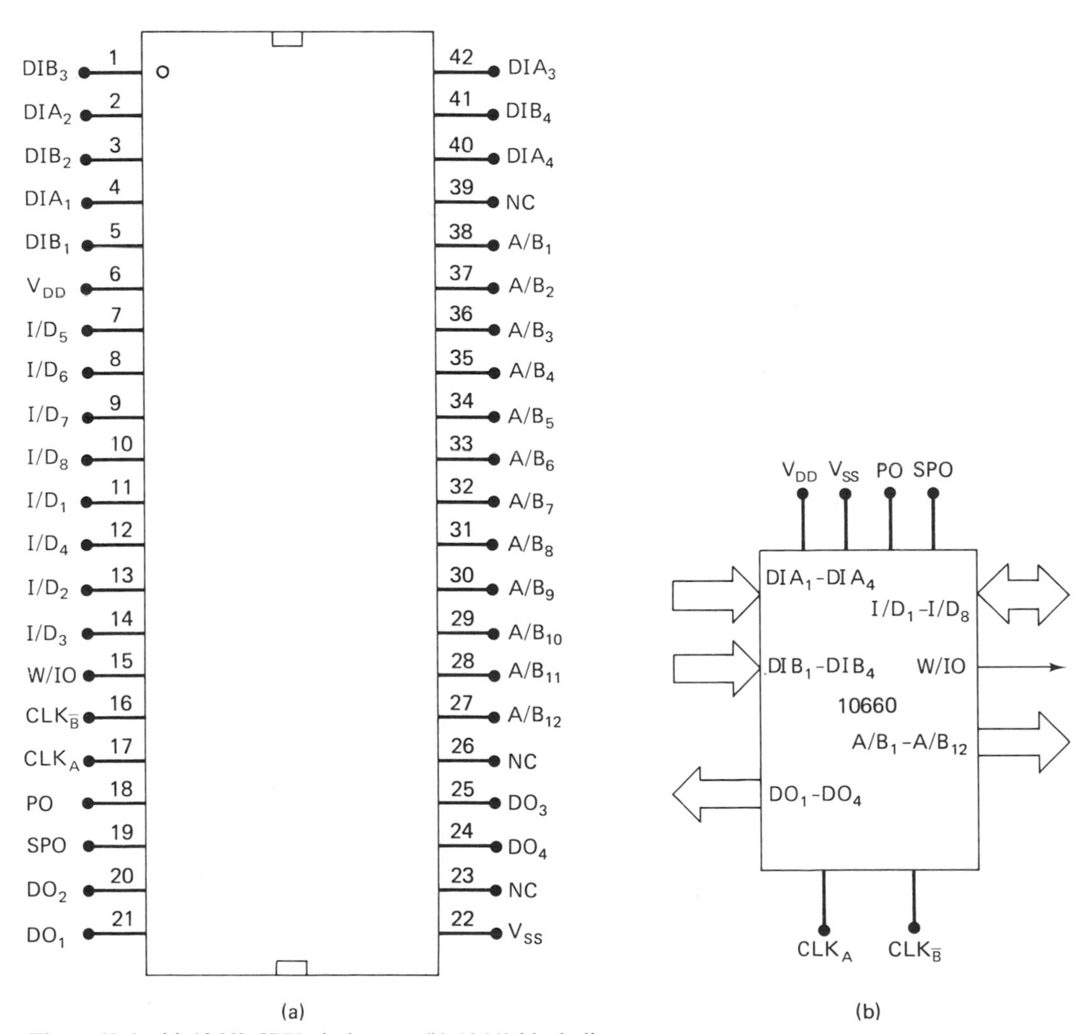

Figure 19.4 (a) 10660 CPU pin layout; (b) 10660 block diagram.

Instructions and data are carried on instruction/data bus lines I/D_1 through I/D_8. For this bus, the MSB is I/D_8 and LSB I/D_1. During normal microcomputer operation, the CPU sends a 12-bit address to the read only memory over the A/B bus, and the memory returns an 8-bit instruction to the CPU over the I/D lines. This instruction is loaded into the instruction decoder within the CPU. It is the output of this decoder that controls the operation of the microprocessor as it performs the operation indicated by the instruction.

On the other hand, the instruction/data bus is divided in half to perform 4-bit read or write operations to the random access memory. When the CPU addresses the random access memory section over the A/B bus, a 4-bit data word can be written from the CPU into the memory. In this case, data are sent to memory on bus lines I/D_5 through I/D_8.

But if a 4-bit data word stored in random access memory is to be returned to the CPU, a read operation must be performed instead of a write operation. For this type of operation, data transfer is performed over lines I/D_1 through I/D_4.

Write/Input-Output Enable

The logic level at the write/input-output enable line W/IO is used to signal the random access memory that a read or write operation is to be performed. When W/IO is logic 0, a read operation is done between the CPU and random access memory. However, a logic 1 at W/IO initiates a write operation. The logic level of this lead is also used to indicate that an input/output operation is to be performed to a peripheral device.

Discrete Inputs and Outputs

In addition to the address bus and instruction/data bus lines, a set of eight *discrete inputs* and four *discrete output* lines are provided on the 10660 CPU. The discrete inputs are divided into two independent 4-bit groups marked DIA_1 through DIA_4 and DIB_1 through DIB_4.

The CPU can execute an instruction from the read only memory that causes the 4 bits of data at the group A or group B lines to be inputted to the CPU. Moreover, the CPU can be instructed to output a 4-bit data word at discrete output group DO_1 through DO_4. Data are latched into the discrete outputs until another instruction is processed by the CPU to cause it to clear or reload the discrete outputs.

19.5 THE READ ONLY MEMORY

A read only memory, or ROM, as it is known, is one type of electronic memory device. Information stored in an ROM is permanently programmed at address locations during the manufacture of the device. For this reason, data can only be read out of this type of memory element. Moreover, stored information is not lost when power supply voltage is removed from the device, and it is said to be a *nonvolatile memory*.

Memory Capacity

In a PPS-4 microcomputer, ROM devices are programmed with the instructions used to control the operation of the central processing unit. For instance, the A05 LSI ROM device in Fig. 19.5(a) could be used as the ROM memory in a microcomputer circuit. This device is what is known as a $1k \times 8$ ROM. That is, it can store up to 1024 8-bit instruction words.

Using just one A05 device, we can store a program containing up to 1024 CPU instructions. However, in most microcomputer systems larger programs are required. For this reason, the 10660 CPU is made so it can address up to 16 A05 devices. This extends the ROM memory capacity to $16k \times 8$ and program capability to 16,384 instructions.

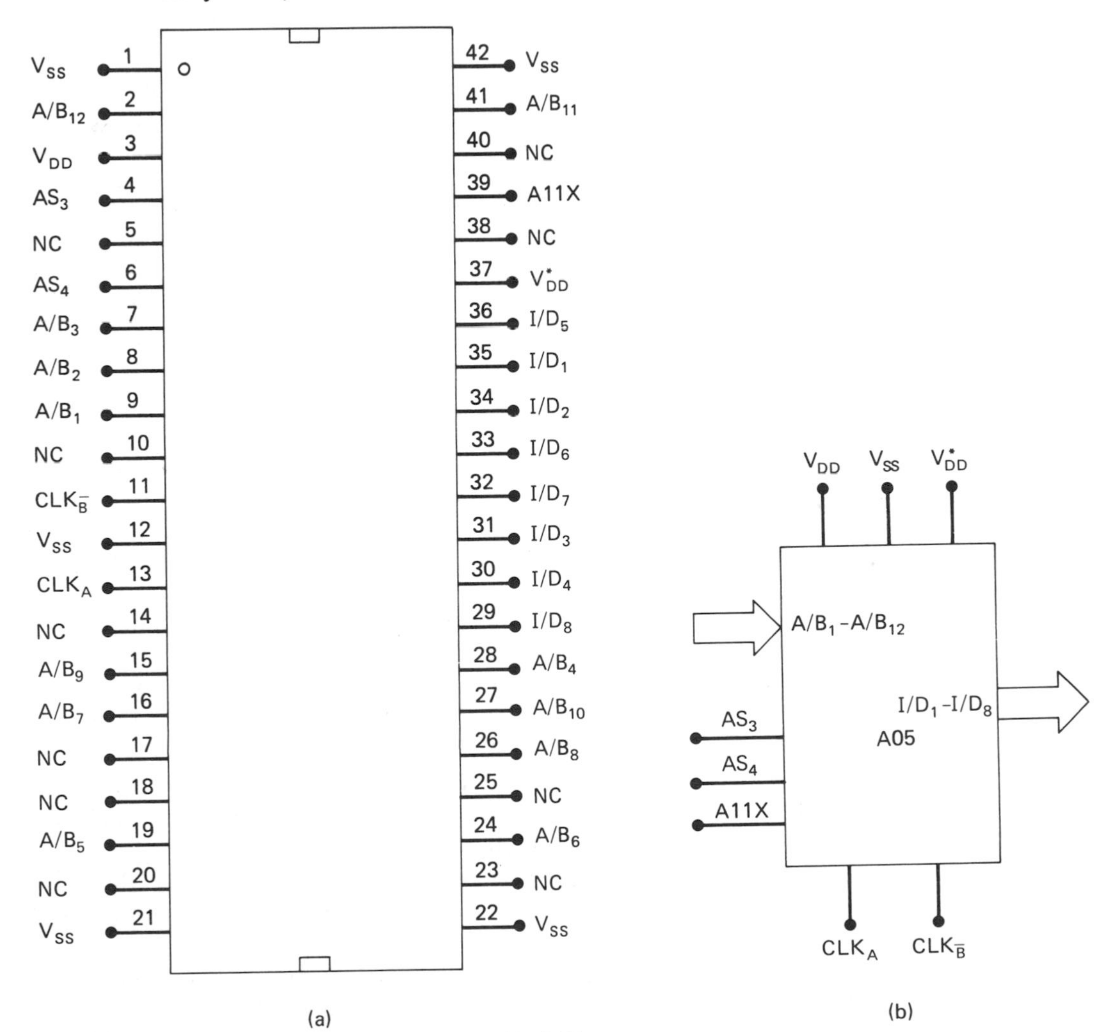

Figure 19.5 (a) A05 ROM pin layout; (b) A05 block diagram.

Address and Instruction/Data Buses

The block diagram in Fig. 19.5(b) shows the structure of an A05 ROM memory. During normal operation, the CPU outputs an address on the A/B bus to the ROM devices. In turn, the ROM memory puts the single instruction stored at the addressed location on the I/D bus for return to the CPU.

The AO5 device has a 12-bit address bus made from lines A/B_1 through A/B_{12}. Of these, A/B_{12} is again the MSB and A/B_1 the LSB. Using a 12-bit address bus, codes 000000000000 through 111111111111 can be sent to ROM in order to select locations within memory. This binary range corresponds to the hexadecimal range 000_{16} to FFF_{16}. This gives a total of 4096 different addresses and shows that four parallel-connected A05 devices can be directly addressed off the A/B bus.

The $A11X$ lead can be used to switch the storage organization of the A05 ROM from $1k \times 8$ to a $2k \times 4$ structure.

The 8-bit instruction data bus I/D_1 through I/D_8 is used to return the instruction stored at the addressed ROM location to the CPU. On this bus the I/D_8 line corresponds to the MSB.

Address Format

The address supplied to the ROM is divided into two parts as shown in Fig. 19.6. Here we see that the 10 bits A/B_1 through A/B_{10} are needed to select one of the $1k$ instructions stored in ROM. This segment of the address is called the *instruction select code.*

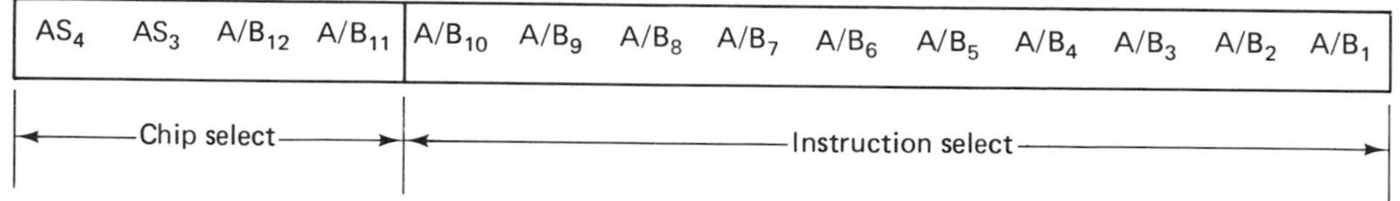

Figure 19.6 Address word format.

The upper two bits A/B_{11} and A/B_{12} are used to select one of four ROM devices in a $4k \times 8$ ROM section. However, when $16k$ of ROM memory is needed the ROM address straps AS_4, and AS_3 can be used along with A/B_{12} and A/B_{11} to give a 4-bit *chip select code.* In this way, we get 16 different 4-bit binary codes and one for each of the ROM devices in the memory.

During manufacture, each A05 ROM memory device has a different chip select code built in. To be selected, bits AS_4, AS_3, A/B_{12}, and A/B_{11} of the address are first compared to the internal ROM select code, and the device with the matching code is selected. This enables the instruction select code on lines A/B_{10} through A/B_1 to enter the ROM, and the addressed instruction is outputted on I/D_1 through I/D_8.

As an example, let us take the address 110000000000 applied to a $4k$ memory of A05 ROMs. The ROM devices are numbered 0 through 3 and correspond to the decimal value of the chip select code. In this case, the chip select bits of the address are $A/B_{12}A/B_{11} = 11$ and ROM 3 is selected for operation. The remaining bits give the instruction select code 0000000000. This corresponds to the first instruction in ROM 3. This address can be written in hexadecimal form as $C00_{16}$.

EXAMPLE 19.1

Find the chip select code and instruction select code for the ROM address 502_{16} applied to a $4k$ ROM memory made with A05 devices.

SOLUTION:

Converting the address to binary form, we get

$$502_{16} = 010100000010$$

The upper 2 bits give a chip select code of 01:

$$01 = \text{ROM } 1$$

The instruction select code is 0100000010 and gives the instruction at location 258 in ROM 1:

$$0100000010 = 258_{10}$$

19.6 THE RANDOM ACCESS MEMORY

A second type of electronic memory needed in a microcomputer system is random access memory or RAM. This type is also sometimes called a *read/write memory*. In a PPS-4 microcomputer, the CPU can write 4-bit data words into an addressed location of the RAM memory. Here the data are stored for additional processing at a later time. When new data is written to a RAM location, it replaces the old data; and the old data is lost. This is called *destructive write-in*. To return data held in RAM to the CPU, a read operation must be performed. During a read operation of RAM, the content of the memory location is retained. For this reason, it is said that a RAM has a *non-destructive read-out*. If voltage is removed from a RAM device, its data content is lost. For this reason, it is said to be a *volatile memory*.

Memory Capacity

Figure 19.7(a) is a pin layout of a 10432 RAM device. This device contains 256 4-bit storage locations. With the 12-bit address of the 10660 CPU, 16 of these RAM devices can be used in parallel to give a total storage capacity of 4096 4-bit words or $4k$ of RAM memory.

Buses and Control Leads

A block diagram of the 10432 RAM is given in Fig. 19.7(b). From this block, we see that its leads are essentially the same as those found on the A05 ROM device. However, here we have a control line marked W/IO and five chip select straps SC_1 through SC_5.

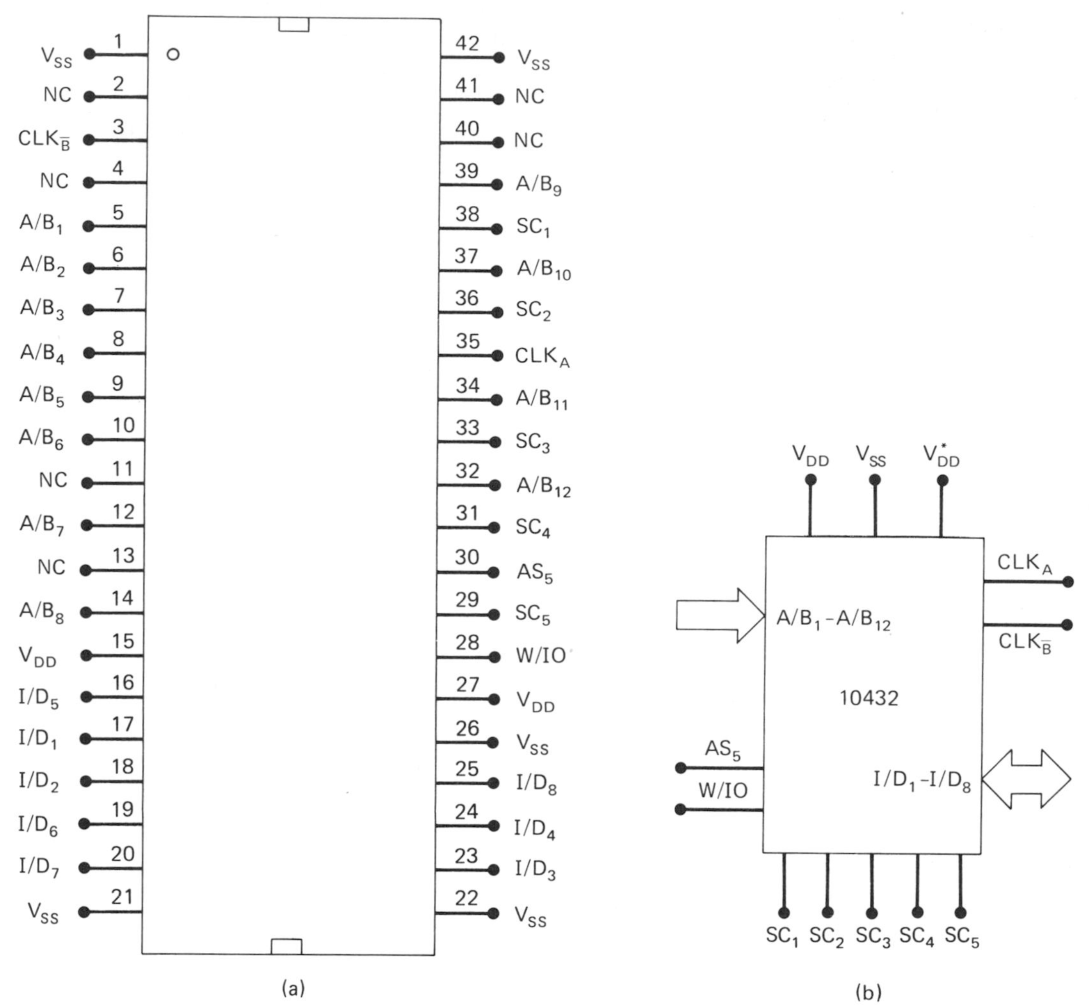

Figure 19.7 (a) 10432 RAM pin layout; (b) 10432 block diagram.

The write/input-output enable control line (*W/IO*) is used to signal that the RAM is to perform a read or write operation. When *W/IO* equals 1 in phase ϕ_2 of a RAM cycle, a write operation of data to the RAM memory will follow. For this case, a 4-bit data word is sent from the CPU into the addressed location of memory for storage. On the other hand, a 0 on *W/IO* at this time indicates that a read operation is to be performed. During a read operation of the RAM memory, 4-bit data content at the addressed memory location is sent back to the CPU.

Chip Selection

On the A05 ROM memory device, a chip select code which was part of the address word was compared to a built-in code to select one of the many parallel-connected devices. RAM devices in the random access memory section of the microcomputer are selected in a similar way except on the 10432 device the select code must be wired to chip select straps SC_1, SC_2, SC_3, and SC_4.

For instance, strapping all chip select leads on a device to −12 V makes the chip select code 1111. This corresponds to RAM 15 if the 16 devices are labeled RAM 0 through RAM 15. If the code on address lines A/B_9 through A/B_{12} matches this strapped code, the RAM device is selected and enabled to perform a read or write operation.

The remaining 8 bits A/B_1 through A/B_8 select one of the 256 memory locations. When W/IO is logic 1 in phase ϕ_2, a write operation is performed to the selected device and data storage location. So a 4-bit data word is sent out from the CPU on lines I/D_5 through I/D_8 of the instruction/data bus and into the storage location. If W/IO is at the 0 level in ϕ_2, 4 bits of data are read from the selected RAM location over bus lines I/D_1 through I/D_4 to the CPU. Formats of the RAM address and data words are shown in Fig. 19.8.

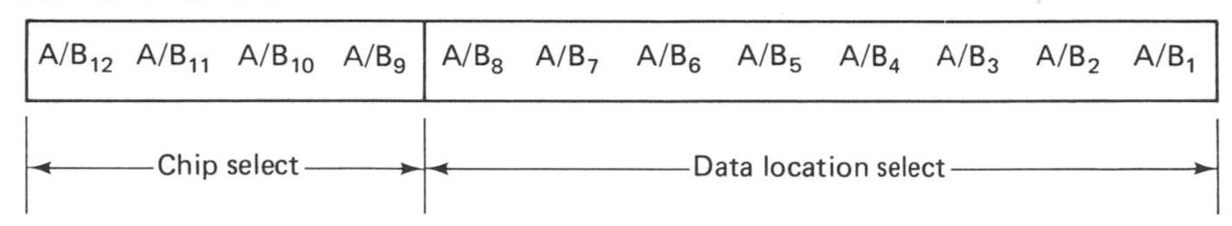

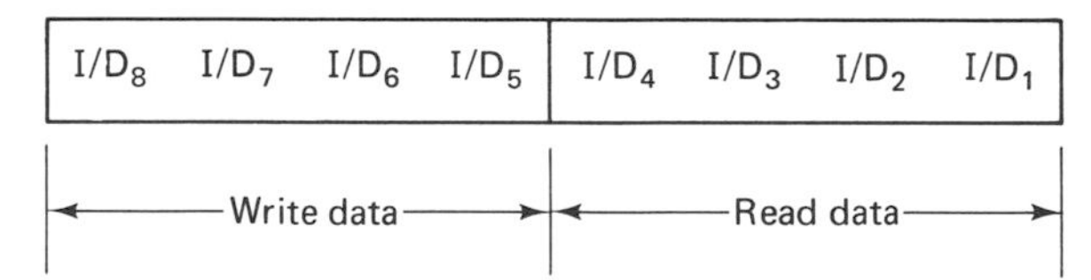

Figure 19.8 RAM address and data word formats.

The SC_5 chip select strap can be used to extend the RAM memory to a second section of 16 RAM devices. In this way, the memory capacity can be extended to $8k \times 4$ bits.

EXAMPLE 19.2

A PPS-4 microcomputer uses 16 256×4 RAM devices. These devices are strapped with codes 0000 to 1111 for devices RAM 0 through RAM 15. From which device and storage location are data read when the CPU outputs the hexadecimal address $A05_{16}$?

SOLUTION:

First, we must express the address in binary form:

$$A05_{16} = 101000000101$$

From the binary address, we find that the chip select code is 1010, and RAM 10 is enabled for operation:

$$1010 = \text{RAM } 10$$

The addressed location in RAM 10 is given by the code 00000101. Converting to decimal form, we get

$$00000101 = 5_{10}$$

Therefore, we read the data stored at location 5 in RAM 10 out onto bus lines I/D_1 through I/D_4.

19.7 THE GENERAL-PURPOSE INPUT/OUTPUT DEVICE

The last part essential to a microcomputer system is some type of *input/output controller device*. It is needed to input or output data for peripheral devices such as keyboards, displays, and printers.

In a PPS-4 system, the 10696 GPI/O or general-purpose input/output integrated circuit can be used to interface the microcomputer with the peripheral device. Furthermore, special-purpose input/output controllers are available as standard LSI circuits. Two examples are the 10788 keyboard/display controller (KDC) and the 10815 keyboard/printer controller (KPC).

Buses and Controls

In Fig. 19.9(a), a pin layout of the 10696 GPI/O integrated circuit is shown, and its block diagram is given in Fig. 19.9(b). Here, we see that the device has 12 input bus lines labeled IN_1 through IN_{12}. In addition, there are also 12 outputs OUT_1 through OUT_{12}. Both inputs and outputs are divided into three 4-bit sections called group A, group B, and group C starting from the LSBs IN_1 and OUT_1, respectively.

Four chip select straps SC_1, SC_2, SC_3, and SC_4 are provided for selection of the chip. By using different strapping codes, up to 16 different GPI/O devices can be operated off a PPS-4 microcomputer. In most applications, one or several input/output devices are needed.

When an input/output operation is initiated, the CPU uses the W/IO control line to signal that the operation is in progress. This is indicated by a 1 logic level in phase ϕ_4 of the CPU clock cycle.

GPI/O Operation

The CPU in a microcomputer can command a GPI/O to perform either an input or output operation. The time needed to execute an input/output operation between the CPU and GPI/O is two clock cycles or 10 μs, because one clock cycle is required to select the I/O device and command it to perform a specific input/output operation, but the 4 bits of data are not transferred until the next cycle.

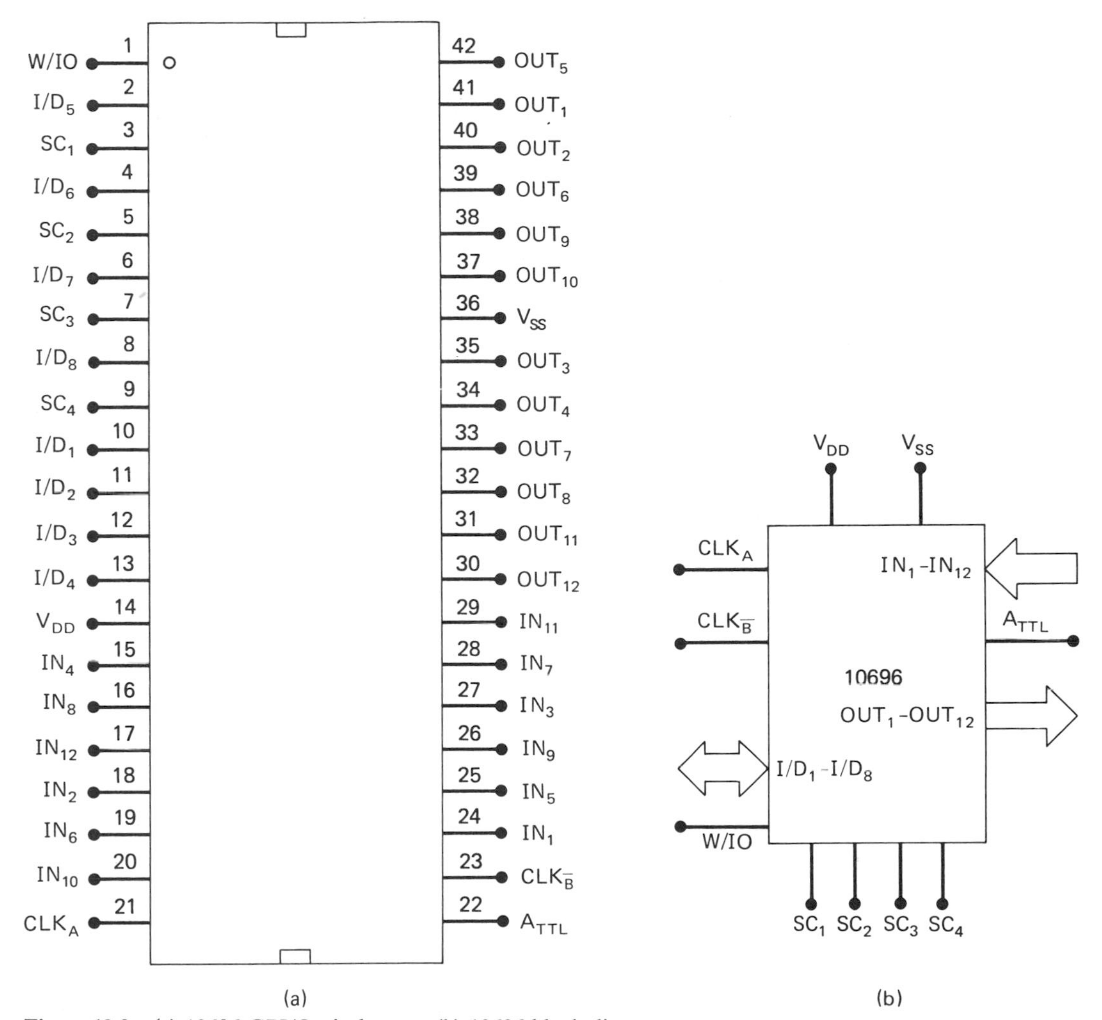

Figure 19.9 (a) 10696 GPI/O pin layout; (b) 10696 block diagram.

The command sent out to the GPI/O can make the device load a 4-bit data word from the *I/D* bus into the group *A*, group *B*, or group *C* outputs. For this output operation, data are sent from the CPU to GPI/O on lines I/D_5 through I/D_8. These data are latched at the outputs of the GPI/O until they are replaced by new data with another input/output operation. To load the group *A*, group *B*, and group *C* outputs with a total of 12 bits of data, three separate input/output operations must be performed to the GPI/O.

On the other hand, a 4-bit data word at the group *A*, group *B*, or group *C* data inputs can be sent to the CPU with an input command. These input data are returned to the CPU on bus lines I/D_1 through I/D_4.

Instruction and Data Word Formats

Formats for the first cycle instruction word and second cycle data word of a GPI/O operation are given in Fig. 19.10(a). Here we find that the four MSBs of the instruction word carried on the *I/D* bus are the chip select code. This code is com-

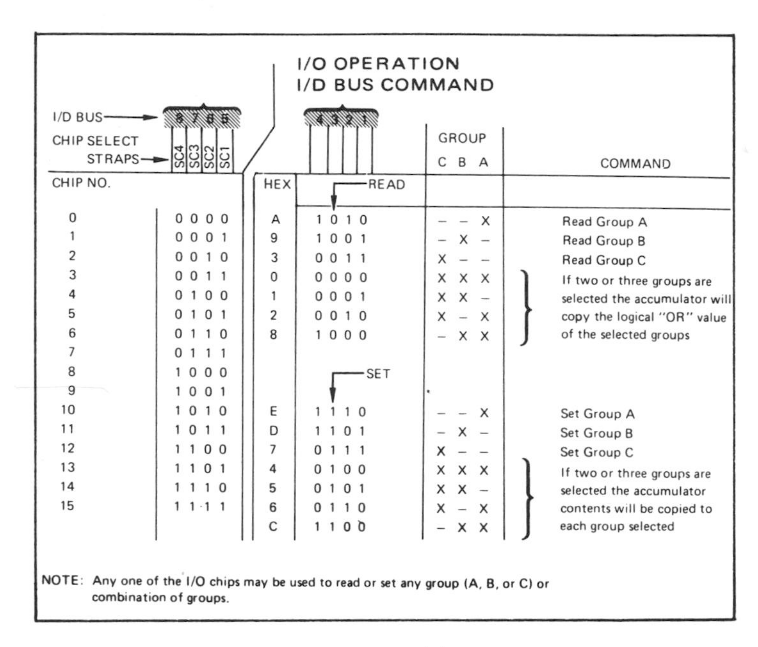

(a)

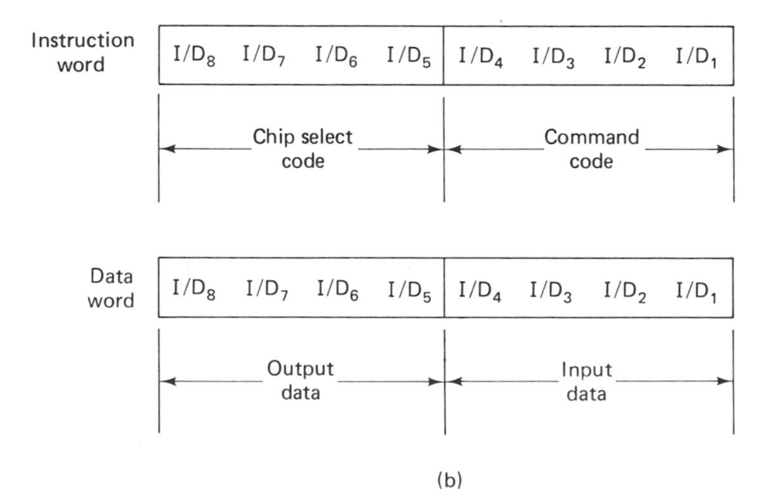

(b)

Figure 19.10 (a) GPI/O instruction and data format; (b) Command codes (Rockwell International, Inc.).

pared to the strapped code at $SC_4SC_3SC_2SC_1$ to determine if the GPI/O device is selected.

At the same time, a command is entered into the selected GPI/O over lines I/D_1 through I/D_4. This code sets up the internal circuitry of the GPI/O for the data transfer operation to be performed.

During the second cycle, data are passed from GPI/O to CPU on bus lines I/D_1 through I/D_4, or data are outputted on lines I/D_5 through I/D_8 to a peripheral device connected on the input/output side of the GPI/O.

Figure 19.10(b) lists the 14 command codes that can be executed by a GPI/O. Looking at this table, we see that there are 7 commands for input operations and 7 output commands. Along with each code, the operation performed is described.

EXAMPLE 19.3

The data at the input lines of a GPI/O are IN_{12} through IN_1 = 000000100011. If the device is selected by the instruction word I/D_8 through I/D_1 = 00011001, what is the chip select code strapped to the device? Find the input/output command to be performed. Give the data input to the CPU on bus lines I/D_1 through I/D_4.

SOLUTION:

From the instruction word bits I/D_8 through I/D_5, we find that the chip select code is 0001 for GPI/O 1:

$$0001 = \text{GPI/O } 1$$

The lower 4 bits identify the command to be executed. Looking at the table in Fig. 19.10(b), we see that the code 1001 corresponds to an input of group B data:

$$1001 = \text{input group } B$$

From the data at the GPI/O inputs, we find that the group B lines IN_8 through IN_5 are 0010 or 2_{16}. These data are transferred onto bus lines I/D_4 through I/D_1 for return to the CPU:

$$I/D_4 I/D_3 I/D_2 I/D_1 = 0010 = 2_{16}$$

19.8 THE PROGRAM

The CPU in a PPS-4 microcomputer system can execute 50 different instructions. This group of microinstructions is called its *instruction set*.[1] The instruction set can be divided into six categories of instructions:

1. Arithmetic instructions
2. Logic instructions

[1]A complete listing of the PPS-4 instruction set is provided in Appendix 2.

3. Data transfer instructions
4. Control transfer instructions
5. Input/output instructions
6. Special instructions

Of the 50 PPS-4 instructions, 44 are executed in a single CPU cycle of 5 μs. On the other hand, the remaining 6 instructions take two cycles or 10 μs. Using these instructions, a program is written to control the operation of the CPU. This sequence of instructions makes the CPU execute the operations to be performed by the microcomputer system.

For instance, a program can be written to add binary numbers. A simple adder program is shown in Fig. 19.11(a), and Fig. 19.11(b) gives a list of the instructions used in the program.

Instruction	Binary code	Hexadecimal code
RC	00100100	24
LD(6)	00111001	39
ADI2	01101101	6D
ADCSK	00001000	08
DC	01100101	65

(a)

Mnemonic	Name
RC	Reset carry flip-flop
LD	Load accumulator from memory
ADI	Add immediate and skip on carry out
ADCSK	Add with carry in and skip on carry out
DC	Decimal correction

(b)

Figure 19.11 (a) Binary addition program for $6 + 2 = ?$; (b) Instructions.

Here, we notice that each instruction has a mnemonic name and an 8-bit binary code. The binary code is normally expressed in hexadecimal form when programming. The complete program of a practical microcomputer system can contain thousands of instructions.

The binary combination of the instructions in the program are built into ROM devices at consecutive address locations. The programmed memory is sometimes called the firmware of the microcomputer system, because it contains a fixed program that controls the operation of the CPU.

19.9 MICROCOMPUTER OPERATING CYCLES

At this point, we have introduced the separate devices used in a PPS-4 microcomputer system. Figure 19.12 shows an actual interconnection of these devices. Furthermore, we described the operation of the ROM, RAM, and input/output devices with respect to the CPU.

However, ROM, RAM, and input/output operations are synchronized to the four phases ϕ_1, ϕ_2, ϕ_3, and ϕ_4 of clock signals CLK_A and CLK_B. In this section we shall describe the basic ROM cycle, RAM cycle, and input/output cycle as they occur in the CPU timing cycle.

Figure 19.12 PPS-4 microcomputer system (Rockwell International, Inc.).

ROM Memory Cycle

The block diagram in Fig. 19.13(a) shows the devices and buses involved in the execution of a ROM memory cycle of the microcomputer. The sequence of operations that occur during the cycle are summarized in Fig. 19.13(b).

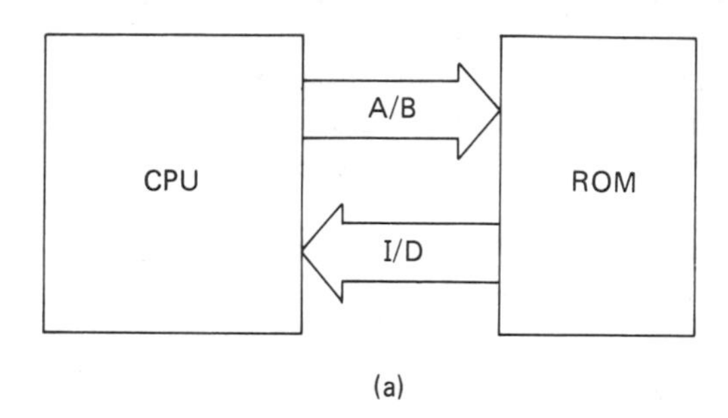

(a)

Clock phase	ϕ_1	ϕ_2	ϕ_3	ϕ_4
A/B bus	Clear	ROM address	Clear	–
I/D bus	Clear	–	Clear	Instruction

(b)

Figure 19.13 (a) ROM cycle block diagram; (b) ROM cycle timing.

A 12-bit address code is generated within the CPU and held in its program counter. In phase ϕ_1 of the clock cycle, all address bus lines A/B_1 through A/B_{12} and instruction/data bus lines I/D_1 through I/D_8 are cleared to logic 0.

During phase ϕ_2 of the cycle, the address held in the program counter of the CPU is put on the A/B bus and sent to the ROM memory. After this, the program counter is incremented by 1 to give the next address location. As phase ϕ_3 of the timing cycle begins, the A/B and I/D buses are once again cleared to logic 0.

In phase ϕ_4 the 8-bit instruction held at the addressed location in ROM is put on the I/D bus for return to the CPU. This completes one ROM or instruction cycle. After this, the CPU performs the operation indicated by the instruction. For instance, it can execute an addition of two numbers.

RAM Cycle

The timing of a RAM memory cycle is opposite that of a ROM cycle. Figure 19.14(a) shows a block diagram of the circuitry involved in a RAM cycle, and Fig. 19.14(b) gives a timing diagram for the operation.

In these diagrams, we see that an address is sent out of the CPU and over the

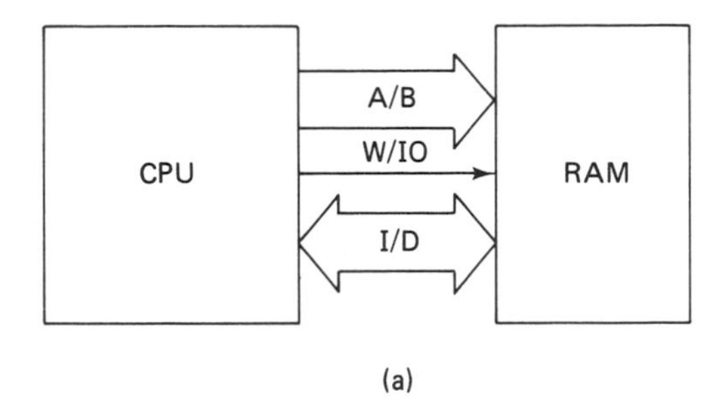

(a)

Clock phase	$\emptyset_4$	$\emptyset_1$	$\emptyset_2$	$\emptyset_3$
A/B bus	RAM address	Clear	—	Clear
I/D bus	—	Clear	Read/write data	Clear
W/IO	0 = RAM operation	Clear	0 = Read 1 = Write	Clear

(b)

Figure 19.14 (a) RAM cycle block diagram; (b) RAM cycle timing.

A/B bus to the RAM memory during phase ϕ_4 of the CPU cycle. During the phase ϕ_1 that follows, the bus and control lines are cleared to logic 0.

As the next phase ϕ_2 begins, a 4-bit read or write operation is performed between the CPU and RAM memory section. The data transfer occurs over the I/D bus. The logic level of control signal W/IO is set by the CPU in phase ϕ_2 to signal the RAM memory that a read or write operation is in progress.

When W/IO is logic 0, a read operation is performed, and a 4-bit data word is transferred from the addressed location in RAM back to the CPU on bus lines I/D_1 through I/D_4 in phase ϕ_2 of the clock cycle. On the other hand, a logic 1 at W/IO causes a write operation to be performed to RAM instead of a read. In this case, a 4-bit data word is sent into the addressed storage location on lines I/D_5 through I/D_8.

When the phase ϕ_2 data transfer operation is complete, the A/B and I/D buses are again cleared for phase ϕ_3 of the CPU cycle.

Combined ROM/RAM Cycle

In actual operation, the PPS-4 microcomputer can perform a ROM and RAM operation in the same cycle. This timing is illustrated in Fig. 19.15. Here we see that an overlapping timing configuration is used where the A/B and I/D buses

	1st cycle				2nd cycle	
Clock phase	$\emptyset_1$	$\emptyset_2$	$\emptyset_3$	$\emptyset_4$	$\emptyset_1$	$\emptyset_2$
A/B bus	Clear	ROM address	Clear	RAM address	Clear	–
I/D bus	Clear	–	Clear	Instruction	Clear	Read/write data
W/IO	Clear	–	Clear	0 = RAM operation 1 = I/O operation	Clear	0 = Read 1 = Write

Figure 19.15 Overlapping ROM/RAM cycle timing (Rockwell International, Inc.).

are both carrying information during phases ϕ_2 and ϕ_4. However, one is carrying information for a ROM operation and the other for a RAM operation.

Let us take an example to show how ROM and RAM cycles are involved in microcomputer operation. It is common to use a specific section of address locations in RAM memory to store a total. In the operation of the system, the value in this storage location must be updated by adding to values held in the CPU. To do this, a RAM cycle must be executed to read the number stored in the memory location back to the CPU. After this, a ROM cycle is performed to make the CPU add the value from memory to the value held within the CPU. The new total is returned to its storage location by a write operation to RAM.

Input/Output Cycle

When an input/output cycle is to be performed to a device such as the GPI/O, the CPU must perform what is called an input/output long instruction. The mnemonic for this instruction is *IOL*, and it takes two complete CPU cycles.

The circuitry involved in an input/output operation is shown in Fig. 19.16(a) and the *IOL* cycle timing in Fig. 19.16(b). During phase ϕ_2 of the CPU cycle, an address is sent out to the ROM memory over lines A/B_1 through A/B_{12}. In phase ϕ_4 of the same cycle. an instruction is sent from ROM to CPU on bus lines I/D_1 through I/D_8. The instruction is $1C_{16}$ or 00011100 and indicates to the CPU that an *IOL* operation is to be performed. This instruction is held in the instruction decoder of the CPU for execution.

As the second cycle begins, another address is sent to ROM in phase ϕ_2. The instruction output by ROM on I/D_1 through I/D_8 in ϕ_4 goes to the GPI/O instead of the CPU. This instruction selects a GPI/O device with the code at bits I/D_5

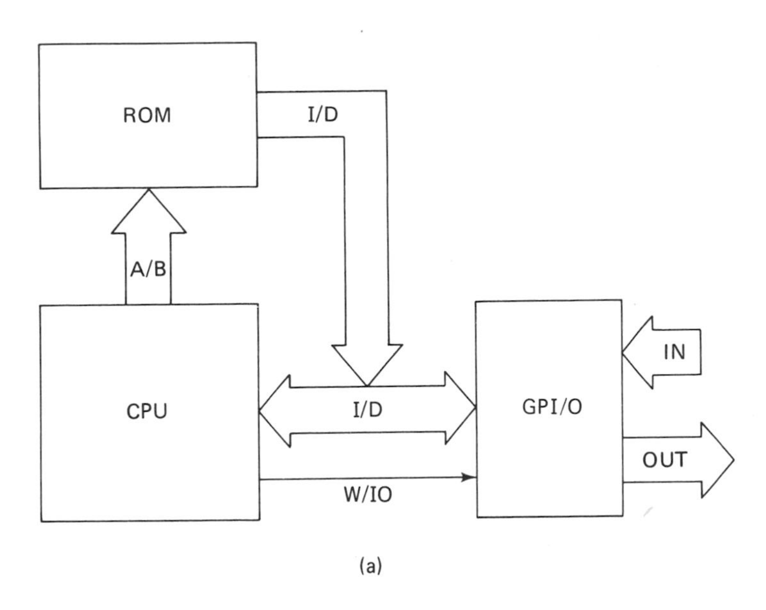

	1st cycle				2nd cycle				3rd cycle	
Clock phase	$\emptyset_1$	$\emptyset_2$	$\emptyset_3$	$\emptyset_4$	$\emptyset_1$	$\emptyset_2$	$\emptyset_3$	$\emptyset_4$	$\emptyset_1$	$\emptyset_2$
A/B bus	Clear	ROM address	Clear	—	Clear	ROM address	Clear	—	Clear	—
I/D bus	Clear	—	Clear	IOL instruction $1C_{16}$	Clear	—	Clear	Select code (I/D_5–I/D_8) Command code (I/D_1–I/D_4)	Clear	I/O data output (I/D_5–I/D_8) I/O data input (I/D_1–I/D_4)
W/IO	Clear	—	Clear	—	Clear	—	Clear	1 = I/O operation	Clear	—

(b)

Figure 19.16 (a) Input/output cycle block diagram; (b) Input/output cycle timing (Rockwell International, Inc.).

through I/D_8 and commands a certain input or output operation with bits I/D_1 through I/D_4.

The input/output data operation performed between GPI/O and CPU happens in phase ϕ_2 of a third cycle. Input data are carried from the GPI/O to CPU on bus lines I/D_1 through I/D_4. On the other hand, output data go between the CPU and GPI/O on lines I/D_5 through I/D_8.

An example of a use that requires an input/output cycle is when the results of a calculation held in the CPU are to be printed or displayed instead of put in memory. In these cases, the data held in the CPU must be output through a GPI/O or special LSI peripheral controller device to the circuitry that controls the printer or display.

ASSIGNMENTS

Section 19.2

1. What are the basic blocks of a PPS-4 microcomputer circuit?

2. Are SSI, MSI, or LSI devices used to provide the basic blocks in a microcomputer circuit?

3. Briefly describe the function of each block of the PPS-4 microcomputer system in Fig. 19.1.

4. List the important buses and control signals of the PPS-4 microcomputer.

5. Is the PPS-4 microcomputer a serial or parallel processing system?

6. Does the PPS-4 microcomputer use positive or negative logic?

Section 19.3

7. What is used to set the frequency of the 10706 clock generator output?

8. Give the frequency and period of the CLK_A timing signal.

9. How are the phases of the CPU timing cycle labeled?

10. Describe the use of the frequency select straps S_{12}, S_{14}, and S_{18}. Find the clock frequency if S_{12} is wired to -12 V and both S_{14} and S_{18} are $+5$ V.

Section 19.4

11. How many bits are the address word, instruction word, and data word for the PPS-4 microcomputer?

12. Over which parts of the instruction/data bus are read and write data words carried?

13. What does the logic level of W/IO indicate?

14. Which CPU bus lines other than the I/D bus can be used to output a 4-bit data word?

Section 19.5

15. Is ROM volatile or nonvolatile?

16. What capacity of A05 ROM can be directly addressed with the 12-bit A/B address bus?

17. Which bits of a 12-bit ROM address select the device, and which bits select the instruction?

18. The address applied to a $4k$ ROM made from A05 devices is $CF4_{16}$. Which ROM device from ROM 0 through ROM 3 is selected, and what instruction is selected?

Section 19.6

19. Is RAM volatile or nonvolatile?

20. Give the capacity of 10432 RAM that can be directly addressed off the 12-bit A/B bus.

21. What type of RAM operation is indicated by $W/IO = 0$ and $W/IO = 1$ in phase ϕ_2 of the CPU clock cycle?

22. If all chip select straps on a 10432 RAM are wired to +5 V, what is the chip select code?

23. A $4k$ random access memory consists of 16 10432 RAM devices labeled RAM 0 through RAM 15. Into which device and storage location is a write operation performed when the CPU outputs the hexadecimal address 0D9?

Section 19.7

24. What does the notation GPI/O stand for?

25. During which phase of the CPU cycle does a 1 on the W/IO control line indicate an input/output operation?

26. Which bits of the instruction word to a GPI/O device are used to select the device and input/output operation command? What is transferred between the CPU and GPI/O during the data word part of the cycle?

27. A GPI/O device is selected by the instruction word I/D_8 through I/D_1 = 00001110, and data are put on bus line I/D_8 through I/D_5 in the phase ϕ_2 that follows. Find the GPI/O chip selected and type of operation performed.

Section 19.8

28. What is an instruction set?

29. When writing a program, are instructions expressed in binary, octal, or hexadecimal form?

30. Is a program built into ROM using binary, octal, or hexadecimal code to make the system firmware?

Section 19.9

31. In which phase of the CPU timing cycle is a ROM address outputted by the CPU?

32. During which phase of the timing cycle does the ROM send an 8-bit instruction to the CPU?

33. For which phases of the clock cycle is a RAM address on the A/B bus and RAM data on the I/D bus?

34. What are the phase ϕ_4, ϕ_1, ϕ_2, and ϕ_3 levels of the signal *W/IO* during a write operation to the RAM?

35. Give the hexadecimal code of the instruction sent to the CPU on the *I/D* bus in phase ϕ_4 of the first cycle of an *IOL* operation.

36. To which device is the ROM instruction in the second cycle of an *IOL* operation sent?

37. During which phase and cycle of an *IOL* operation is data transferred through a GPI/O device?

Answers to Selected Odd Numbered Problems

CHAPTER 1

Section 1.2

3. Input unit, central processing unit, memory unit, and output unit; **5.** Minicomputer.

Section 1.4

7. Discrete circuitry; **9.** 14, 16, and 24 pins.

Section 1.5

11. Diode-transistor logic; **13.** Inverter, AND gate, OR gate, NOR gate, and NAND gate.

Section 1.6

17. Monolithic technology; **19.** Silicon.

CHAPTER 2

Section 2.2

1. 10; **3.** Weight; **5.** 9 and 10^3, 8 and 10^{-3}.

Section 2.3

7. Bit; **9.** 2^{-1}.

Section 2.4

13. 71; **15.** 111110100.

Section 2.5

19. 10.

Section 2.6

21. (a) 14, (b) 56, (c) 702, (d) 7257; **23.** (a) 11010, (b) 100111, (c) 10001000, (d) 110111110000.

Section 2.7

27. 16^1.

Section 2.8

29. (a) 39, (b) E2, (c) 9A, (d) 3A0; **31.** (a) 1101011, (b) 11110011, (c) 1000101, (d) 1010110000.

Section 2.9

35. 70; **37.** (a) -21, (b) $+96$.

CHAPTER 3

Section 3.2

5. Voltage mode, positive logic.

Section 3.3

7. Left-hand side of the truth table.

Section 3.4

13. $F = A + B$; **17.** F equals NOT C; **19.** $F = \overline{A \cdot B \cdot C \cdot D} = \overline{ABCD}$.

Section 3.5

21.

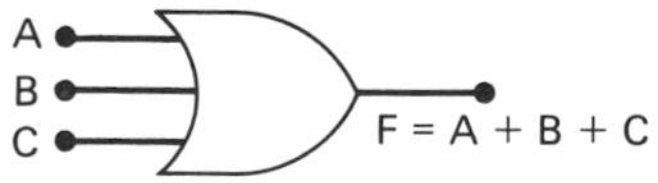

A
B
C
F = A + B + C

Section 3.6

27. $F = 0$; **29.** $F = 1$.

Section 3.7

31. $F = 0$; **33.** $F = 1$.

Section 3.8

35. $F = 1$.

Section 3.9

39. $F = 1$.

Section 3.10

43. $F = 0$; **45.**

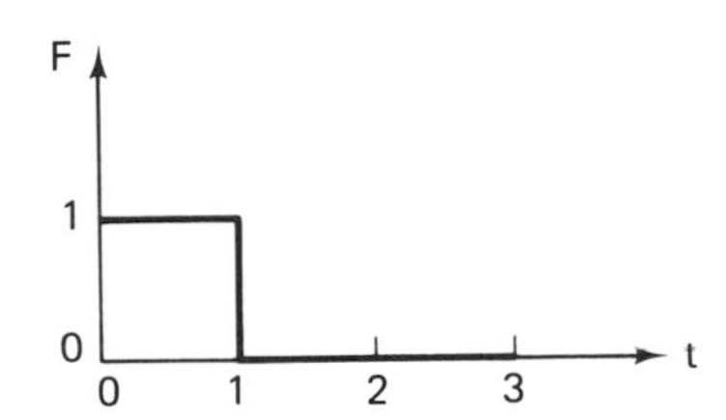

CHAPTER 4

Section 4.2

3. 5 V dc.

Section 4.3

7. $A_2B_2 = 11$, $F_2 = 0$.

Section 4.4

9. $F = 1$, $V_F = 2.4$ V; **11.** 10 TTL loads; **13.** .7 V.

Section 4.5

15. $F = 0$, $V_F = .2$ V.

Section 4.6

17. Q_6; **19.** $F = 0$, $V_F = .2$ V.

Section 4.7

21. Q_3.

Section 4.8

23. $F = 1$; **25.** $F_1 = 1$.

Section 4.9

27. $+5$ V; **29.** $F = 0$.

Section 4.10

31. 30 V; **33.** High current (fan out of 30).

CHAPTER 5

Section 5.3

5.

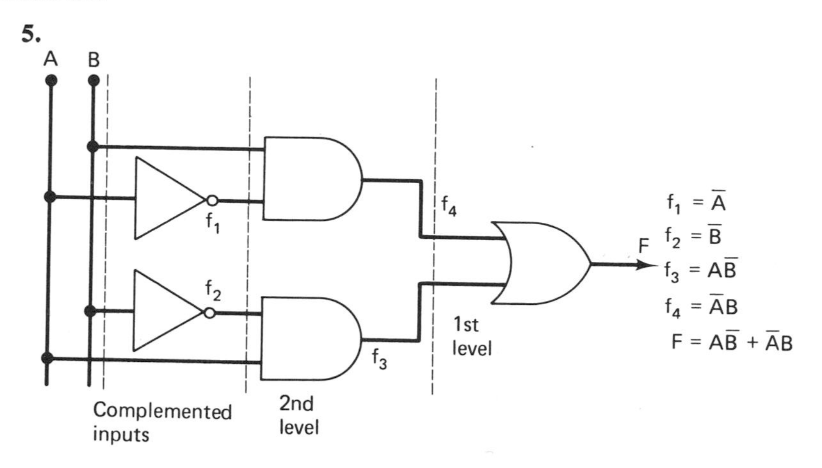

Sections 5.4 and 5.5

7.

A	B	C	f_1	f_2	f_3	F
0	0	0	1	0	0	0
0	0	1	1	0	0	0
0	1	0	0	0	0	0
0	1	1	0	1	0	1
1	0	0	1	0	1	1
1	0	1	1	0	1	1
1	1	0	0	0	0	0
1	1	1	0	1	0	1

Section 5.6

9.

A	B	C	f_1	f_2	f_3	F
0	0	0	1	1	1	0
0	0	1	1	1	1	0
0	1	0	0	1	1	0
0	1	1	0	1	0	1
1	0	0	1	0	1	1
1	0	1	1	0	1	1
1	1	0	0	1	1	0
1	1	1	0	1	0	1

Section 5.7

13.

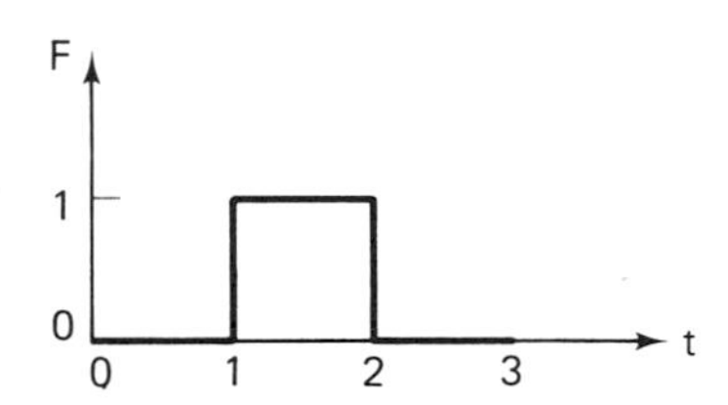

Section 5.9

19.

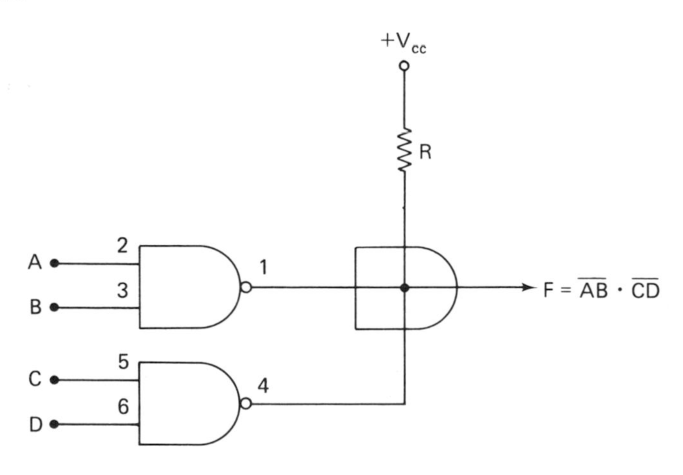

CHAPTER 6

Section 6.2

1.

A	B	F
0	0	1
0	1	1
1	0	0
1	1	0

Section 6.3

5. $F = \bar{A}\bar{B}\bar{C} + \bar{A}B\bar{C} + \bar{A}BC.$

Section 6.4

9. $F = \bar{A}\bar{B}\bar{C} + A\bar{B}\bar{C}.$

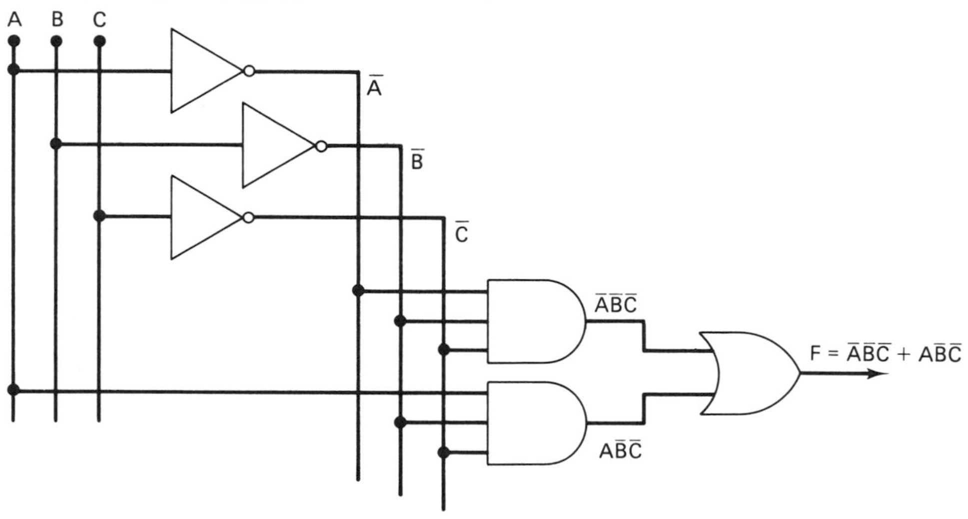

Section 6.5

11.

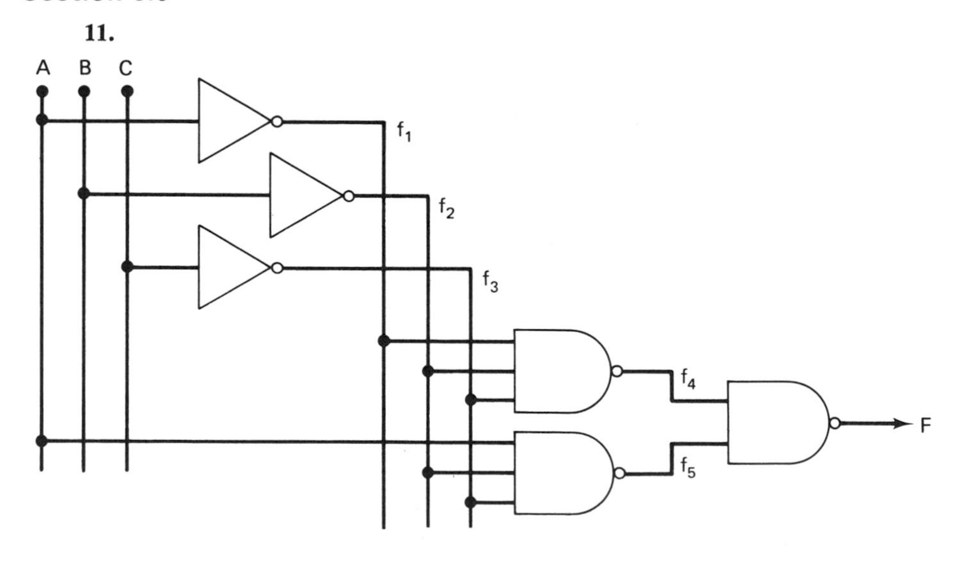

Section 6.6

15. (a)

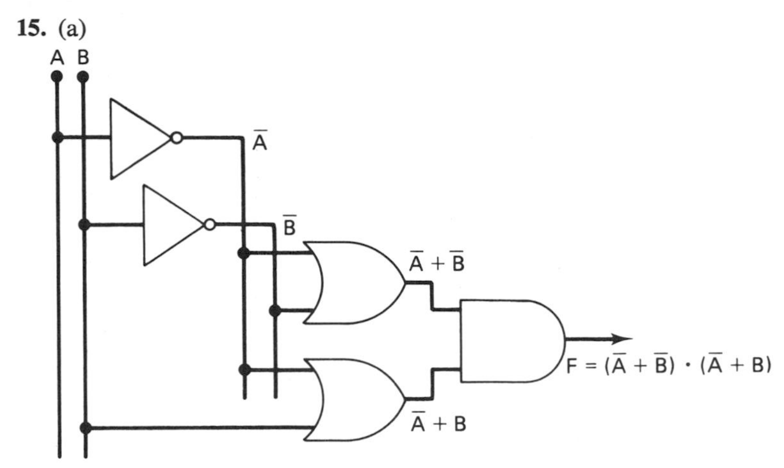

Section 6.7

17. (a)

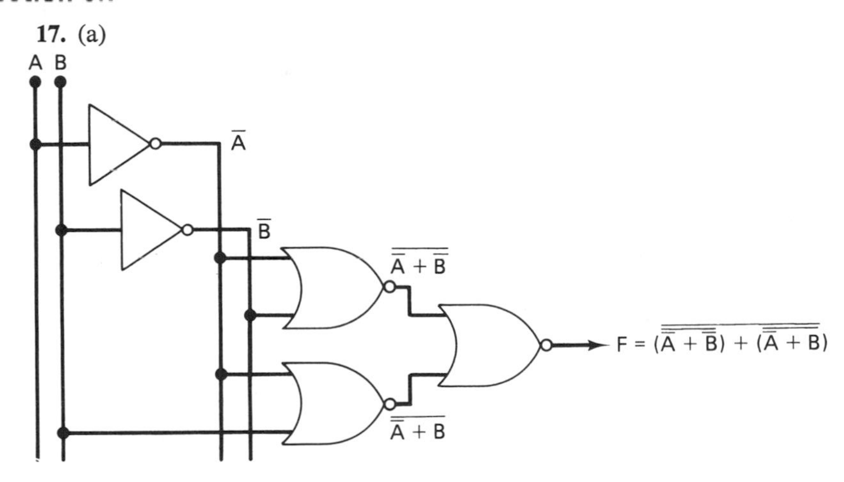

Section 6.8

19.

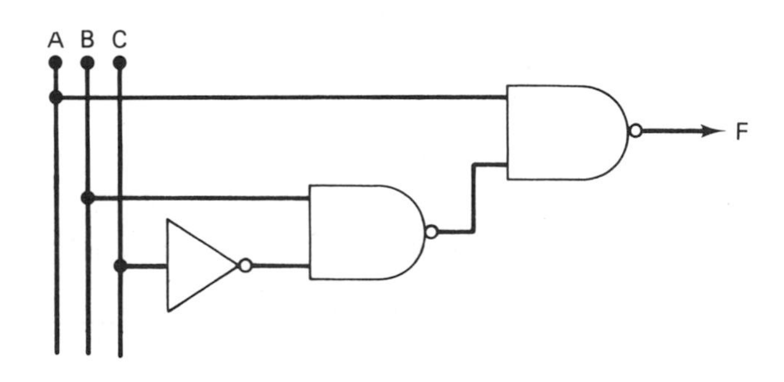

Section 6.9

21. $F = \bar{A}B + AB$.

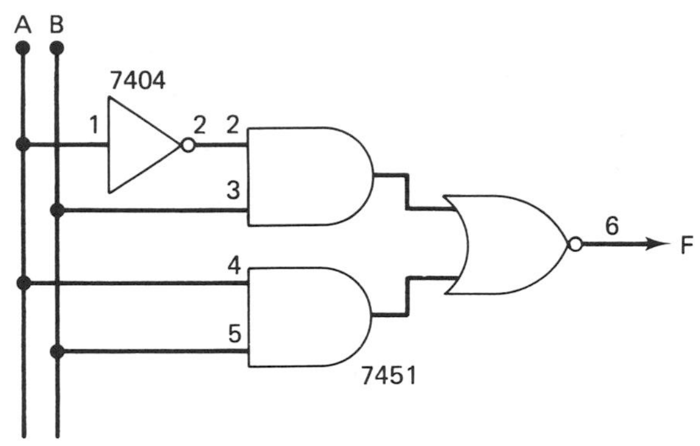

CHAPTER 7

Section 7.2

1. *P4b*; **3.** When $X = 0$, both inputs of the OR gate are logic 0, and the output switches to logic 0; if $X = 1$, the input to the OR gate is 10, and its output switches to logic 1; so the output is identical to the X input.

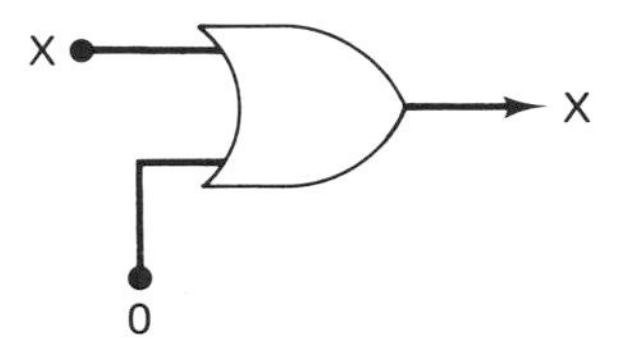

Section 7.3

7.

X	Y	Z	XY	XZ	XY + XZ
0	0	0	0	0	0
0	0	1	0	0	0
0	1	0	0	0	0
0	1	1	0	0	0
1	0	0	0	0	0
1	0	1	0	1	1
1	1	0	1	0	1
1	1	1	1	1	1

X	Y	Z	Y + Z	X(Y + Z)
0	0	0	0	0
0	0	1	1	0
0	1	0	1	0
0	1	1	1	0
1	0	0	0	0
1	0	1	1	1
1	1	0	1	1
1	1	1	1	1

13. $X(\bar{X} + Y) = XY,\ X\bar{X} + XY = XY,\ XY = XY;$
15.

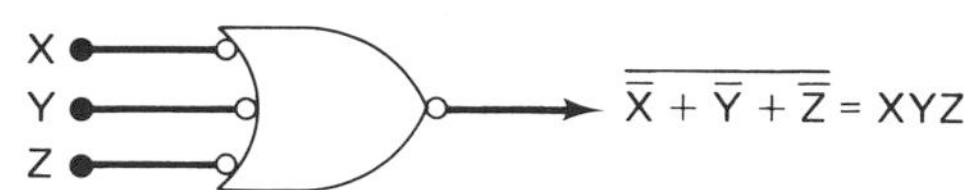

Section 7.4

17. (a) $F = A$, (b) $F = \bar{A}$; **19.** (a) $F = AC + A\bar{B}$

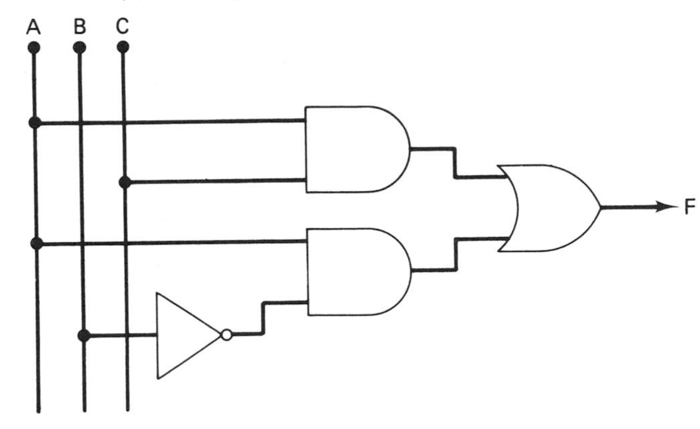

Section 7.5

21. $F = A\bar{B}C + AB\bar{C} + ABC$, $F = AB + AC$

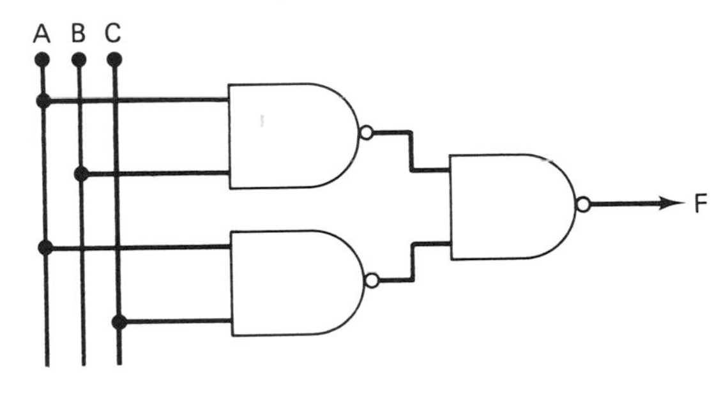

CHAPTER 8

Section 8.2

3. $F = \bar{B}C + B\bar{C}$, $F = B \oplus C$.

Section 8.3

7.

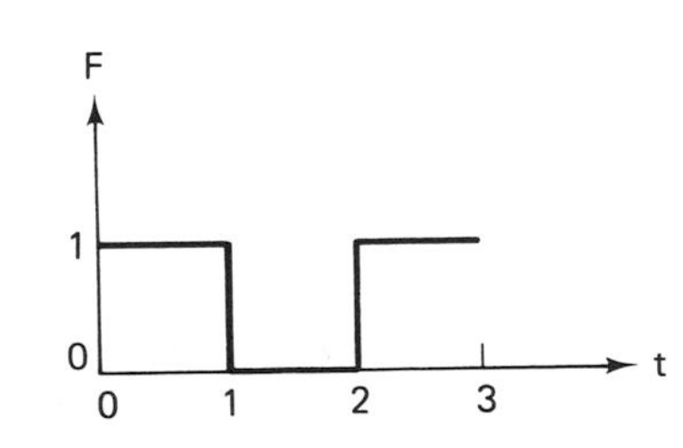

Section 8.4

9. $F = 0$; **11.** $f_1 = 1, f_2 = 1, f_3 = 1, f_4 = 0$, and $F = 1$.

Section 8.6

17. $f_1 = \overline{A_0 \oplus B_0}, f_2 = \overline{A_1 \oplus B_1}, F = (\overline{A_0 \oplus B_0})(\overline{A_1 \oplus B_1})$.

Section 8.7

19. $A > B = 0$, $A = B = 1$, and $A < B = 0$.

CHAPTER 9

Section 9.3

5. (a) 9, (b) 39, (c) 870; **7.** (a) 214 = 001000010100.

Section 9.4

9. (a) 103, (b) 99; **11.** (a) 73 = 10100110, (b) 817 = 101101001010.

Section 9.5

13. 34; **15.**

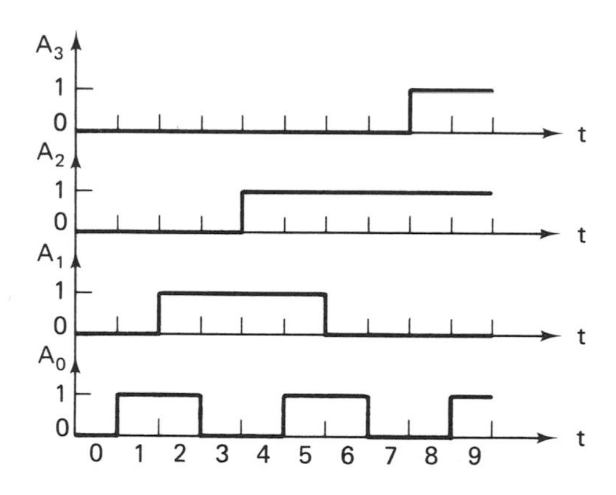

Section 9.6

19.

P	A	B	C	D
1	0	0	1	1
0	0	1	0	0
1	0	1	0	1
1	0	1	1	0
0	0	1	1	1
0	1	0	0	0
1	1	0	0	1
1	1	0	1	0
0	1	0	1	1
1	1	1	0	0

Section 9.7

21. The output is logic 1 when the input has odd parity.

A	B	C	D	f_1	f_2	P
0	0	1	1	0	0	0
0	1	0	0	1	0	1
0	1	0	1	1	1	0
0	1	1	0	1	1	0
0	1	1	1	1	0	1
1	0	0	0	1	0	1
1	0	0	1	1	1	0
1	0	1	0	1	1	0
1	0	1	1	1	0	1
1	1	0	0	0	0	0

Section 9.9

25. $\sum_{\text{even}} = 1, \sum_{\text{odd}} = 0.$

CHAPTER 10

Section 10.3

5.

A	B	C	D	$\bar{A}$	$\bar{B}$	$\bar{C}$	$\bar{D}$	0	1	2	3	4	5	6	7	8	9
0	0	0	0	1	1	1	1	1	0	0	0	0	0	0	0	0	0
0	0	0	1	1	1	1	0	0	1	0	0	0	0	0	0	0	0
0	0	1	1	1	1	0	0	0	0	1	0	0	0	0	0	0	0
0	0	1	0	1	1	0	1	0	0	0	1	0	0	0	0	0	0
0	1	1	0	1	0	0	1	0	0	0	0	1	0	0	0	0	0
0	1	1	1	1	0	0	0	0	0	0	0	0	1	0	0	0	0
0	1	0	1	1	0	1	0	0	0	0	0	0	0	1	0	0	0
0	1	0	0	1	0	1	1	0	0	0	0	0	0	0	1	0	0
1	1	0	0	0	0	1	1	0	0	0	0	0	0	0	0	1	0
1	1	0	1	0	0	1	0	0	0	0	0	0	0	0	0	0	1

Section 10.4

11. MSB; **13.** 16 mA/output; **15.**

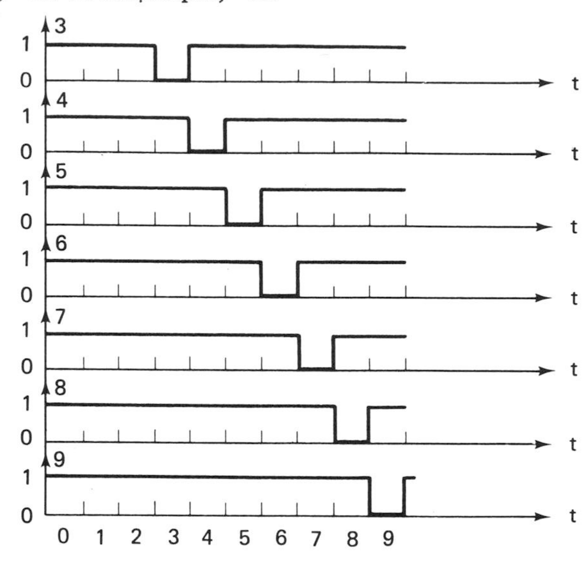

CHAPTER 11

Section 11.2

1. Nixie tube.

Section 11.3

5. Sink; **7.** $f_0 = 1, f_1 = 1, f_2 = 1, f_3 = 1, f_4 = 1, f_5 = 1, f_6 = 1, f_7 = 1, f_8 = 1$, and $f_9 = 0$; the number 9 lights on the display.

Section 11.4

9. Clockwise; **11.** 140 mA; **15.** 560 nA.

Section 11.5

19. *abcdefg* = 1001100; the number 4 lights on the display.

Section 11.6

21. $RBI = 1, LT = 1$; **23.** $RBI = 0$.

CHAPTER 12

Section 12.3

3.

B	A	ST	D_0	D_1	D_2	D_3	$\overline{ST}$	$\overline{B}$	$\overline{A}$	f_1	f_2	f_3	f_4	F
0	0	0	0	1	0	1	1	1	1	0	0	0	0	0
0	1	0	0	1	0	1	1	1	0	0	1	0	0	1
1	0	0	0	1	0	1	1	0	1	0	0	0	0	0
1	1	0	0	1	0	1	1	0	0	0	0	0	1	1

Section 12.4

7. $F = D_3$

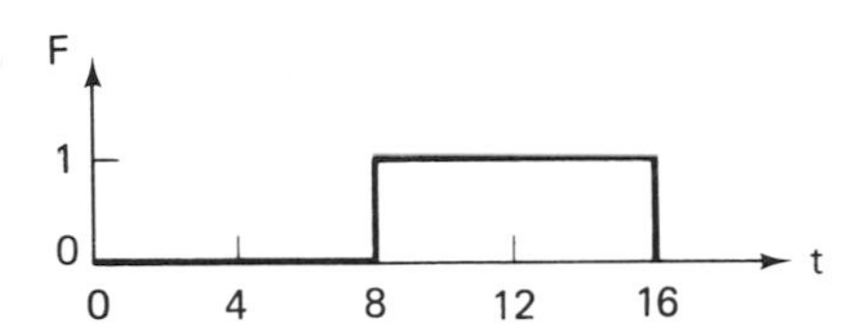

Section 12.6

11.

B	A	ST	D_1	$\overline{D}_1$	$\overline{B}$	$\overline{A}$	$\overline{ST + \overline{D}_1}$	$\overline{F}_1$	$\overline{F}_2$	$\overline{F}_3$	$\overline{F}_4$
0	0	0	1	0	1	1	1	0	1	1	1
0	1	0	1	0	1	0	1	1	0	1	1
1	0	0	1	0	0	1	1	1	1	0	1
1	1	0	1	0	0	0	1	1	1	1	0

Section 12.7

15. $F_3F_2F_1F_0 = B_3B_2B_1B_0$.

CHAPTER 13

Section 13.2

1. $S = 0$, $C = 1$; **3.**

A	B	$\overline{A}$	$\overline{B}$	$\overline{\overline{A}B}$	$\overline{A\overline{B}}$	$\overline{AB}$	S	C
0	0	1	1	1	1	1	0	0
0	1	1	0	0	1	1	1	0
1	0	0	1	1	0	1	1	0
1	1	0	0	1	1	0	0	1

Section 13.3

7. 110011, 51, 01010001; **9.** $f_1 = 0, f_2 = 0, f_3 = 1, S = 1$, and $C_o = 1$.

Section 13.4

13. $\Sigma = 0, \overline{\Sigma} = 1$, and $\bar{C}_o = 0$.

Section 13.5

15. $D = 0, Br = 0$; **17.** $f_1 = 0, D = 0$, and $Br = 0$.

Section 13.6

19. (a) 00011, (b) 10001101.

CHAPTER 14

Section 14.2

3. $T_1 = 10\ \mu s, T_2 = 10\ \mu s$; **5.** $T = 20\ \mu s, f = 50$ kHz.

Section 14.3

9. $T_1 = 10.4\ \mu s, T_2 = 10.4\ \mu s, T = 20.8\ \mu s$, and $f = 48.1$ kHz.

Section 14.5

17.

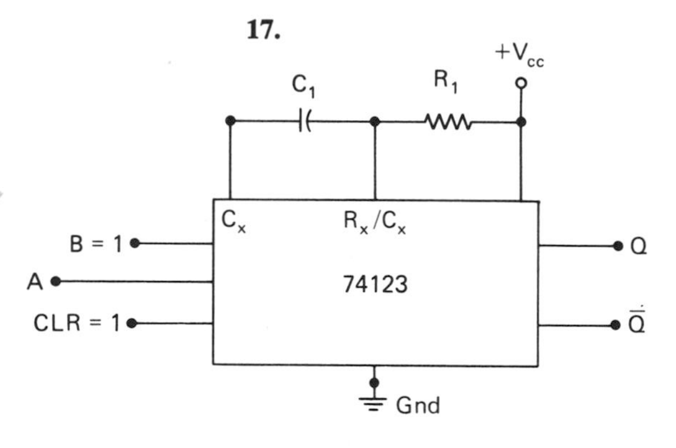

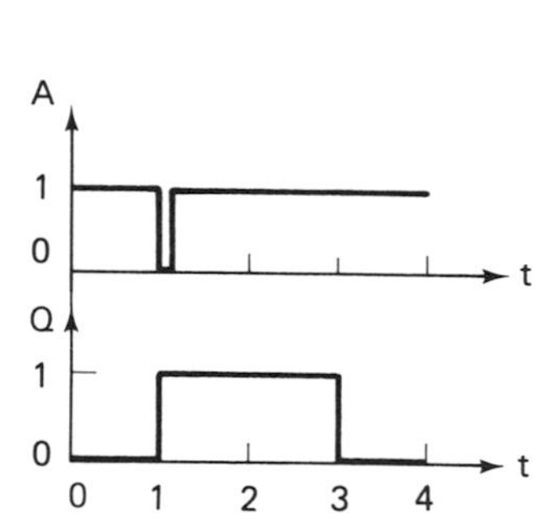

19. The retriggerable feature of the 74123 means that the duration of its output pulse can be lengthened by retriggering the input before the output pulse is complete.

Section 14.6

23. The value of R_1 or C_1 can be used to adjust the duration of the output pulse of a 555 monostable circuit.

CHAPTER 15

Section 15.2

3.

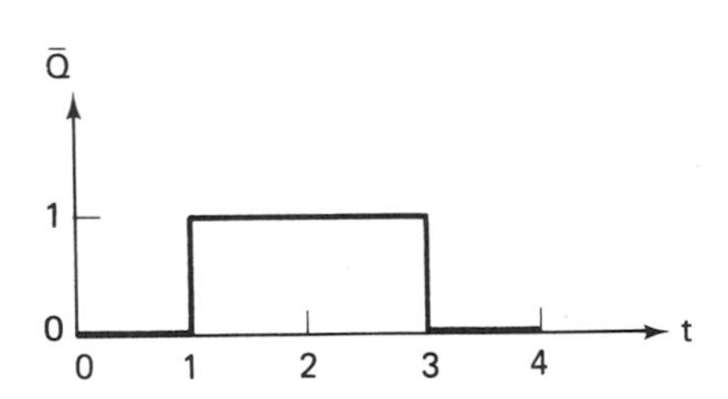

Section 15.3

7. $Q = 0$, $\bar{Q} = 1$.

Section 15.4

9. When the *CLK* input enables the latch, the *Q* output of the data latch becomes equal to the *D* or data input.

Section 15.6

15. $Q_1 = 0$, $\bar{Q}_1 = 1$; **17.**

Section 15.7

21. Input: $T = 10$ ms, $T_1 = 5$ ms, and $T_2 = 5$ ms; output: $T = 20$ ms, $T_1 = 10$ ms, and $T_2 = 10$ ms.

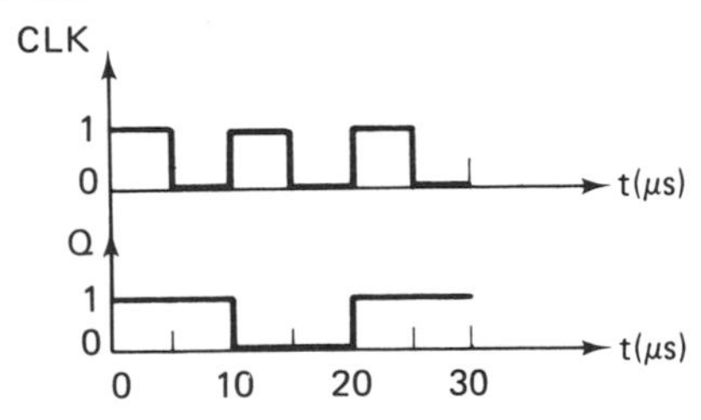

CHAPTER 16

Section 16.2

1. The CLR lead is used to clear the output count to $DCBA = 0000$; on the other hand, PST is used to preset the count to $DCBA = 1111$.

Section 16.3

7.

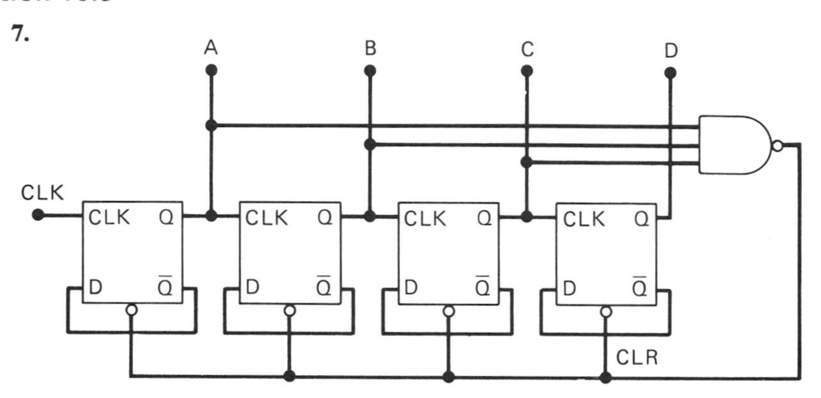

9. $f_D = 1$ kHz, $T_D = 1$ ms, $T_{D1} = .5$ ms, and $T_{D2} = .5$ ms.

Section 16.5

13. $D_o = R0(1)$ and $B_o = R0(2)$; **15.** $D_oC_oB_oA_o = 0110$.

Section 16.6

17. $D_oC_oB_oA_o = 1111$; **19.** $D_oC_oB_oA_o = 0111$.

CHAPTER 17

Section 17.2

3. $t_d = 7\ \mu s$.

Section 17.3

5. The parallel-in, serial-out shift register is loaded with a parallel code. This code is outputted by the register in serial form at the clock rate.

7.

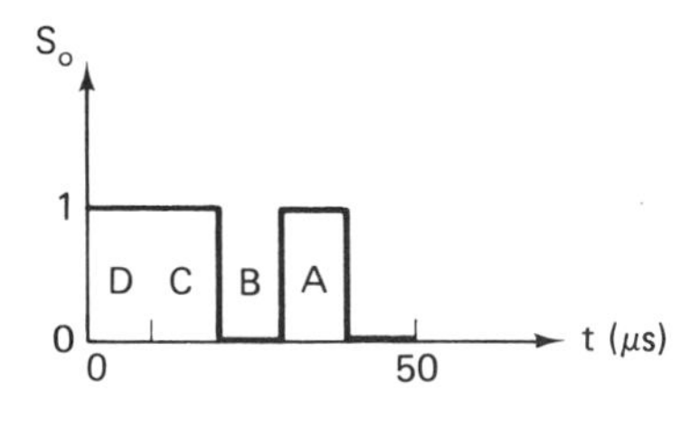

Section 17.4

11.

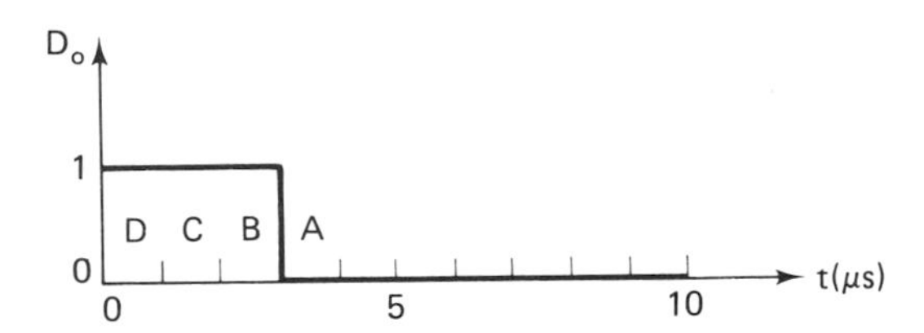

CHAPTER 18

Section 18.2

1. *N*-channel MOSFET, *P*-channel MOSFET; **3.** $V_o = 0$ V.

Section 18.3

5. $F = 0$; **7.** $V_F = 4.5$ V.

Section 18.4

9. $F = 1$.

Section 18.5

13. Output 7 = 0.

Section 18.6

15.

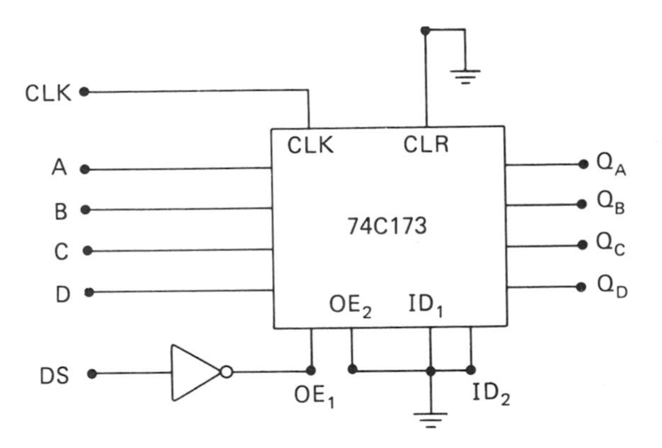

CHAPTER 19

Section 19.2

1. Clock, central processing unit, read only memory, random access memory, and input/output section; **5.** Parallel processing unit.

Section 19.3

9. ϕ_1, ϕ_2, ϕ_3, and ϕ_4.

Section 19.4

11. Address words = 12 bits, instruction words = 8 bits, data word = 4 bits; **13.** Read operations, write operations, or input/output operations.

Section 19.5

15. Nonvolatile; **17.** A/B_{11} and A/B_{12}.

Section 19.6

19. Volatile; **23.** Address location 217 in ROM 1.

Section 19.7

25. ϕ_4; **27.** GPI/O 0 is selected, and a *SET GROUP A* instruction is given to the device.

Section 19.9

31. ϕ_2; **35.** $1C_{16}$; **37.** ϕ_2 in the third cycle.

7400 NAND Gate and 7402 NOR Gate Operation

Appendix 1

7400 NAND GATE OPERATION

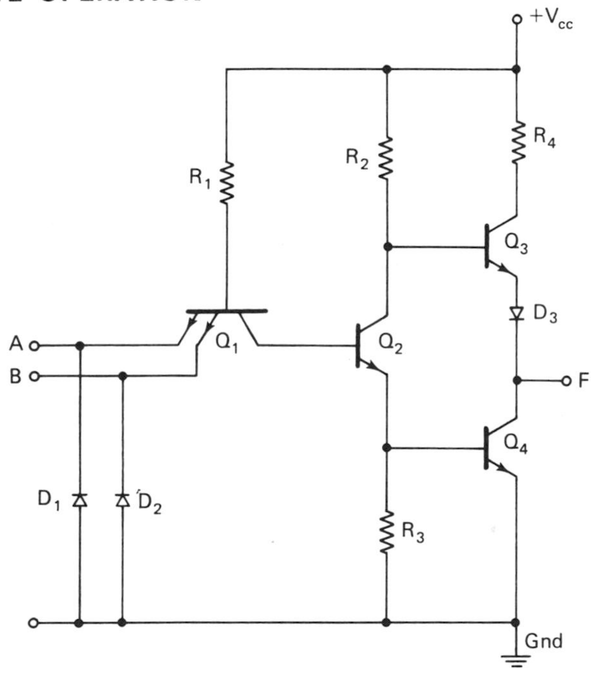

Input $A = B = 0$

When both inputs are logic 0, they are electrically at ground. In this case, the base-emitter junctions of transistor Q_1 turn ON. This makes the voltage at the base lead of Q_1 .7 V above ground.

For current to flow down the path through resistor R_1, the base to collector part of Q_1, and the base to emitter of Q_2, the voltage at the base of Q_1 must be above 1.4 V. Since it is just .7 V, no current flows from the base to the emitter of Q_2, and it is OFF.

With Q_2 electrically an open, current flows down through resistor R_2 and the base-emitter part of transistor Q_3. In turn, Q_3 goes ON, and the output F switches toward the $+V_{cc}$ supply voltage. When Q_2 is OFF, the base of Q_4 goes to ground through resistor R_3, and transistor Q_4 cuts OFF. This produces a 1 logic level at output F.

Input $A = 0, B = 1$

For this case, the operation of the NAND gate circuit is similar to that when both inputs are at the 0 logic level. However, just the base-emitter part of Q_1 corresponding to input A is conducting. This still holds the base of Q_1 at .7 V, and transistor Q_2 is again OFF. Therefore, Q_3 switches ON, and Q_4 goes OFF. The output stays at logic 1 with a voltage value close to $+V_{cc}$.

Input $A = 1, B = 0$

With this input condition, the operation of the circuit is identical to that for $A = 0$ and $B = 1$. But conduction at the B input holds the base of Q_1 at .7 V, Q_2 OFF, Q_3 ON, and Q_4 OFF. The output F remains at the 1 logic level.

Input $A = B = 1$

If both inputs are switched to logic 1 or $+V_{cc}$, the base-emitter parts of transistor Q_1 are turned OFF. Now current can flow down the path through R_1, the base to collector of Q_1, the base to emitter of Q_2, and R_3 to ground. In this way, transistor Q_2 switches ON.

Current flows down the path through resistor R_2, the collector to emitter of Q_2, and the base-emitter of Q_4. This current switches Q_4 ON. As Q_4 goes ON, output F switches to ground or logic 0. When the output is in the 0 logic state, the bias to transistor Q_3 is lost, and it turns OFF.

7402 NOR GATE OPERATION

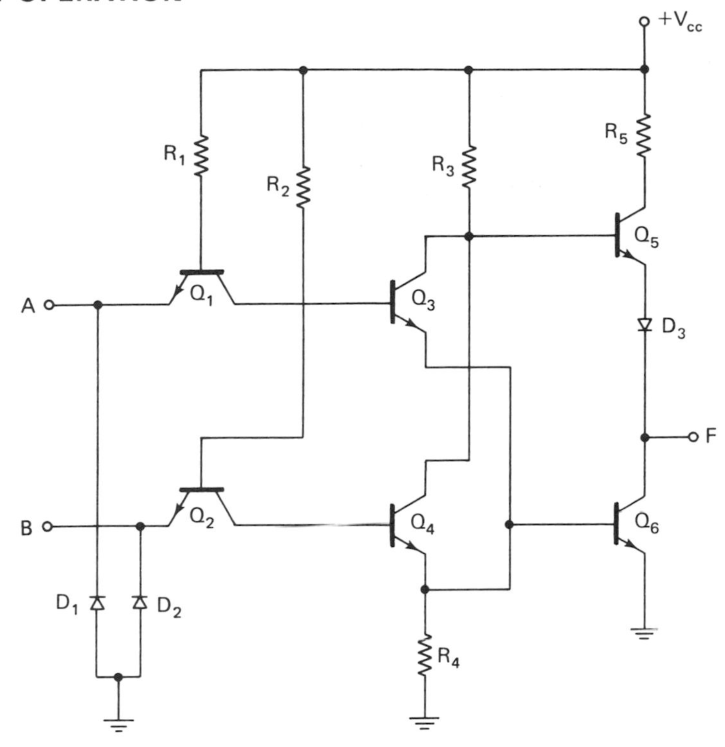

Input $A = B = 0$

Making both the A and B inputs logic 0 or 0 V, the base to emitter parts of transistors Q_1 and Q_2 conduct. For this reason, the bases of Q_1 and Q_2 are both held at .7 V.

With the bases of Q_1 and Q_2 at .7 V, none of the current flowing through resistors R_1 and R_2 takes the paths through transistor Q_3 or Q_4 to ground. So both Q_3 and Q_4 are OFF and act like open circuits.

When transistors Q_3 and Q_4 are OFF, current flows down through R_3 and the base-emitter of Q_5. This causes Q_5 to switch ON, and the output F goes to the 1 logic level. On the other hand, the base of Q_6 is returned to ground through resistor R_4, and Q_6 switches OFF.

Input $A = 0, B = 1$

If the B input is made logic 1 and A logic 0, transistor Q_1 stays conducting from base to emitter. However, Q_2 stops conducting, and current flows from base to collector instead of toward the emitter. This switches Q_4 ON, and it acts like a short circuit. Now current flows through R_3 to the base of Q_6, and it turns ON. As

Q_6 switches ON, the bias to Q_5 is lost, and it goes OFF. The new output F is logic 0 or 0 V.

Input $A = 1, B = 0$

For this input condition, the operation is similar to that described for $A = 0$ and $B = 1$. However, the base-emitter of Q_1 does not conduct; instead Q_2 turns ON. As Q_2 goes ON, Q_3 switches ON and Q_4 OFF. Again, Q_6 is biased ON through R_3, and Q_5 cuts OFF. This gives a logic 0 at output F.

Input $A = B = 1$

When both the A and B inputs are made logic 1 or $+V_{cc}$, the base-emitter parts of Q_1 and Q_2 do not conduct. In this case, current flows from the base to collector of Q_1 and Q_2 to the bases of Q_3 and Q_4, respectively, causing them to turn ON. In turn, Q_5 switches OFF and Q_6 ON. Therefore, the F output remains at logic 0 or 0 V.

PPS-4 Instruction Set* Appendix 2

	Mnemonics	I/D Bus OP Code Hex & Binary	Name	Symbolic Equation
CONTROL TRANSFER INSTRUCTIONS	T	80-BF 10	Transfer (1 cycle)	P(6:1) ← I(6:1)
	TM	DO-FF 1st word 11xx .... 2nd word from page 3	Transfer and Mark Indirect (2 cycles)	SB ← SA ← P P(12:7) ← 000011 P(6:1) ← I1(6:1) P(12:9) ← 0001 P(8:1) ← I2(8:1) I1 (6,5) ≠ 00
	TL	50-5F 1st word 0101 .... 2nd word	Transfer Long (2 cycles)	P(12:9) ← I1(4:1); P(8:1) ← I2(8:1)
	TML	01-03 * 1st word 0000 00xx 2nd word	Transfer and Mark Long (2 cycles)	SB ← SA ← P P(12:9) ← I1(4:1) P(8:1) ← I2(8:1) Note I1(2:1) ≠ 00
	SKC	15 0001 0101	Skip on Carry flip-flop (1 cycle)	Skip if C = 1
	SKZ	1E 0001 1110	Skip on Accumulator Zero (1 cycle)	Skip if A = 0
	SKBI	40-4F 0100	Skip if BL Equal to Immediate (1 cycle)	Skip if BL = I(4:1)
	SKF1	16 0001 0110	Skip if FF1 Equals 1 (1 cycle)	Skip if FF1 = 1
	SKF2	14 0001 0100	Skip if FF2 Equals 1 (1 cycle)	Skip if FF2 = 1
	RTN	05 0000 0101	Return (1 cycle)	P ← SA ↔ SB
	RTNSK	07 0000 0111	Return and Skip (1 cycle)	P ← SA ↔ SB P ← P+1
INPUT/OUTPUT INSTRUCTIONS	IOL	1C 1st word 0001 1100 2nd word	Input/Output Long (2 cycles)	$\bar{A}$ → Data Bus A ← $\overline{\text{Data Bus}}$ '2 → I/O Device
	DIA	27 0010 0111	Discrete Input Group A (1 cycle)	A ← DIA
	DIB	23 0010 0011	Discrete Input Group B (1 cycle)	A ← DIB
	DOA	1D 0001 1101	Discrete Output (1 cycle)	DOA ← A

The word "skip" or "ignore" as used in this instruction set means the instruction will be read from memory but not executed. Each skipped or ignored word will require one clock cycle time.

Instruction ADI, LD, EX, EXD, LDI, LB and LBL have a numeric value coded as part of the instruction in the immediate field. This numeric value must be in complementary form on the bus. All of these immediate fields which are inverted are shown in brackets.

3

*Courtesy of Rockwell International, Inc.

	Mnemonics	I/D Bus OP Code Hex & Binary	Name	Symbolic Equation
DATA TRANSFER INSTRUCTIONS (Cont)	LAX	12 0001 0010	Load Accumulator from X register (1 cycle)	A← X
	LXA	1B 0001 1011	Load X Register from Accumulator (1 cycle)	X← A
	LABL	11 0001 0001	Load Accumulator with BL (1 cycle)	A← BL
	LBMX	10 0001 0000	Load BM with X (1 cycle)	BM← X
	LBUA	04 0000 0100	Load BU with A (1 cycle)	BU← A← M
	XABL	19 0001 1001	Exchange Accumulator and BL (1 cycle)	A↔BL
	XBMX	18 0001 1000	Exchange BM and X (1 cycle)	X↔BM
	XAX	1A 0001 1010	Exchange Accumulator and X (1 cycle)	A↔X
	XS	06 0000 0110	Exchange SA and SB (1 cycle)	SA↔SB
	CYS	6F 0110 1111	Cycle SA register and accumulator (1 cycle)	A ← $\overline{SA(4:1)}$ SA(4:1) ← SA(8:5) SA(8:5) ← SA(12:9) SA(12:9) ← $\overline{A}$
	LB *	CO-CF 1st word 1100 .... 2nd word from page 3	Load B Indirect (2 cycles)	SB← SA← P P(12:5)← 0000 1100 P(4:1)← I1(4:1) BU← 0000 B(8:1)← [I2(8:1)] P← SA↔SB
	LBL *	00 1st word 0000 0000 2nd word	Load B Long (2 cycles	BU← 0000 B(8:1)← [I2(8:1)]
	INCB	17 0001 0111	Increment BL (1 cycle)	BL← BL+1 Skip on BL = 0000
	DECB	1F 0001 1111	Decrement BL (1 cycle)	BL← BL-1 Skip on BL = 1111
SPECIAL	SAG	13 0001 0011	Special Address Generation (1 cycle)	A/B Bus (12:5) ← 0000 0000 A/B Bus (4:1) ← BL(4:1) Contents of "B" remain unchanged

Only the first occurrence of an LB or LBL instruction in a consecutive string of LB or LBL will be executed. The program will ignore the remaining LB or LBL and execute the next valid instruction.

*xxxx – Indicates restrictions on bit patterns allowable in immediate field as specified in the symbolic equation description.

2

	Mnemonics	I/D Bus OP Code Hex & Binary	Name	Symbolic Equation
ARITHMETIC INSTRUCTIONS	AD	OB 0000 1011	Add (1 cycle)	C, A ← A+M
	ADC	OA 0000 1010	Add with carry-in (1 cycle)	C, A ← A+M+C
	ADSK	O9 0000 1001	Add and skip on carry-out (1 cycle)	C, A ← A+M Skip if C = 1
	ADCSK	O8 0000 1000	Add with carry-in and skip on carry-out (1 cycle)	C, A ← A+M+C Skip if C = 1
	ADI	60-6E 0110 xxxx Except 65	Add immediate and skip on carry-out (1 cycle)	A ← A+[I(4:1)] Skip if carry-out = one I(4:1) ≠ 0000, 1010
	DC	65 0110 0101	Decimal Correction (1 cycle)	A ← A+1010
LOGICAL INSTRUCTIONS	AND	OD 0000 1101	Logical AND (1 cycle)	A ← A∧M
	OR	OF 0000 1111	Logical OR (1 cycle)	A ← A∨M
	EOR	OC 0000 1100	Logical Exclusive-OR (1 cycle)	A ← A⊻M
	COMP	OE 0000 1110	Complement (1 cycle)	A ← $\bar{A}$
DATA TRANSFER INSTRUCTIONS	SC	20 0010 0000	Set Carry flip-flop (1 cycle)	C ← 1
	RC	24 0010 0100	Reset Carry flip-flop (1 cycle)	C ← 0
	SF1	22 0010 0010	Set FF1 (1 cycle)	FF1 ← 1
	RF1	26 0010 0110	Reset FF1 (1 cycle)	FF1 ← 0
	SF2	21 0010 0001	Set FF2 (1 cycle)	FF2 ← 1
	RF2	25 0010 0101	Reset FF2 (1 cycle)	FF2 ← 0
	LD	30-37 0011 0 . . .	Load Accumulator from Memory (1 cycle)	A ← M B(7:5) ← B(7:5)⊻ [I(3:1)]
	EX	38-3F 0011 1 . . .	Exchange Accumulator and Memory (1 cycle)	A ↔ M B(7:5) ← B(7:5)⊻ [I(3:1)]
	EXD	28-2F 0010 1 . . .	Exchange Accumulator and Memory and decrement BL (1 cycle)	A ↔ M B(7:5) ← B(7:5)⊻ [I(3:1)] BL ← BL-1 Skip on BL = 1111
	LDI	70-7F 0111	Load Accumulator Immediate (1 cycle)	A ← [I(4:1)]

Only the first occurrence of an LDI in a consecutive string of LDI's will be executed. The program will ignore the remaining LDI's and execute the next valid instruction.

1

Bibliography

BELL, DAVID A., *Solid State Pulse Circuits*. Reston, Va.: Reston Publishing Company, Inc., 1976.

DEBOO, GORDON J., and CLIFFORD N. BURROUS, *Integrated Circuits and Semiconductor Devices: Theory and Applications*, 2nd ed. New York: McGraw-Hill Book Company, 1977.

DEEM, BILL R., KENNETH MUCHOW, and ANTHONY ZEPPA, *Digital Computer Circuits and Concepts*. Reston, Va.: Reston Publishing Company, Inc., 1974.

FLORES, IVAN, *Computer Logic: The Functional Design of Digital Computers*. Englewood Cliffs, N.J.: Prentice-Hall, Inc., 1960.

GILLIE, Angelo C., *Binary Arithmetic and Boolean Algebra*. New York: McGraw-Hill Book Company, 1965.

KOHONEN, TEUVO, *Digital Circuits and Devices*. Englewood Cliffs, N.J.: Prentice-Hall, Inc., 1972.

MALMSTADT, H. V., and C. G. ENKE, *Digital Electronics for Scientists*. Menlo Park, Calif.: W. A. Benjamin, Inc., 1969.

MALVINO, ALBERT P., and DONALD LEACH, *Digital Principles and Applications*, 2nd ed. New York: McGraw-Hill Book Company, 1975.

MARCUS, MITCHELL P., *Switching Circuits for Engineers*, 3rd ed. Englewood Cliffs, N.J.: Prentice-Hall, Inc., 1960.

NATIONAL SEMICONDUCTOR CORPORATION, *CMOS Integrated Circuits*. Santa Clara, Calif.: National Semiconductor Corporation, 1975.

RICHARDS, R. K., *Digital Design*. New York: John Wiley & Sons, Inc., 1971.

ROCKWELL INTERNATIONAL CORPORATION, *PPS-4 Microcomputer Basic Devices Product Description* (Document No. 29003, Rev. 2). Anaheim, Calif.: Rockwell International Corporation, Microelectronics Devices Division, May 1976.

TEXAS INSTRUMENTS, INC., *Designing with TTL Integrated Circuits* (Robert L. Morris and John R. Miller, eds.). New York: McGraw-Hill Book Company, 1971.

TEXAS INSTRUMENTS, INC. (The Engineering Staff, Components Group), *The TTL Data Book for Design Engineers*. Dallas: Texas Instruments, Inc., 1973.

WICKES, WILLIAM E., *Logic Design with Integrated Circuits*. New York: John Wiley & Sons, Inc., 1968.

Index